T0280291

Graph Theory

Textbooks in Mathematics
Series editors:
Al Boggess, Kenneth H. Rosen

Real Analysis
With Proof Strategies
Daniel W. Cunningham

Train Your Brain
Challenging Yet Elementary Mathematics
Bogumil Kaminski, Pawel Pralat

Contemporary Abstract Algebra, Tenth Edition
Joseph A. Gallian

Geometry and Its Applications
Walter J. Meyer

Linear Algebra
What You Need to Know
Hugo J. Woerdeman

Introduction to Real Analysis, 3rd Edition
Manfred Stoll

Discovering Dynamical Systems Through Experiment and Inquiry
Thomas LoFaro, Jeff Ford

Functional Linear Algebra
Hannah Robbins

Introduction to Financial Mathematics
With Computer Applications
Donald R. Chambers, Qin Lu

Linear Algebra
An Inquiry-based Approach
Jeff Suzuki

Mathematical Modeling in the Age of the Pandemic
William P. Fox

Games, Gambling, and Probability
An Introduction to Mathematics
David G. Taylor

Financial Mathematics
A Comprehensive Treatment in Discrete Time
David G. Taylor

Graph Theory
An Introduction to Proofs, Algorithms, and Applications
Karin R. Saoub

https://www.routledge.com/Textbooks-in-Mathematics/book-series/CANDHTEXBOOMTH

Graph Theory
An Introduction to Proofs, Algorithms, and Applications

Karin R. Saoub

CRC Press
Taylor & Francis Group
Boca Raton London New York

CRC Press is an imprint of the
Taylor & Francis Group, an **informa** business

A CHAPMAN & HALL BOOK

First edition published 2021
by CRC Press
6000 Broken Sound Parkway NW, Suite 300, Boca Raton, FL 33487-2742

and by CRC Press
2 Park Square, Milton Park, Abingdon, Oxon, OX14 4RN

© 2021 Taylor & Francis Group, LLC

CRC Press is an imprint of Taylor & Francis Group, LLC

The right of Karin R. Saoub to be identified as author of this work has been asserted by her in accordance with sections 77 and 78 of the Copyright, Designs and Patents Act 1988.

Reasonable efforts have been made to publish reliable data and information, but the author and publisher cannot assume responsibility for the validity of all materials or the consequences of their use. The authors and publishers have attempted to trace the copyright holders of all material reproduced in this publication and apologize to copyright holders if permission to publish in this form has not been obtained. If any copyright material has not been acknowledged please write and let us know so we may rectify in any future reprint.

Except as permitted under U.S. Copyright Law, no part of this book may be reprinted, reproduced, transmitted, or utilized in any form by any electronic, mechanical, or other means, now known or hereafter invented, including photocopying, microfilming, and recording, or in any information storage or retrieval system, without written permission from the publishers.

For permission to photocopy or use material electronically from this work, access www.copyright.com or contact the Copyright Clearance Center, Inc. (CCC), 222 Rosewood Drive, Danvers, MA 01923, 978-750-8400. For works that are not available on CCC please contact mpkbookspermissions@tandf.co.uk

Trademark notice: Product or corporate names may be trademarks or registered trademarks and are used only for identification and explanation without intent to infringe.
ISBN: 978-1-138-36140-9 (hbk)

Library of Congress Cataloging-in-Publication Data

Names: Saoub, Karin R., author.
Title: Graph theory : an introduction to proofs, algorithms, and
applications / Karin R. Saoub.
Description: Boca Raton : CRC Press, 2021. | Series: Textbooks in
mathematics | Includes bibliographical references and index. | Summary:
"Graph theory is the study of interactions, conflicts, and connections.
The relationship between collections of discrete objects can inform us
about the overall network in which they reside, and graph theory can
provide an avenue for analysis. This text, for the first undergraduate
course, will explore major topics in graph theory from both a
theoretical and applied viewpoint. Topics will progress from
understanding basic terminology, to addressing computational questions,
and finally ending with broad theoretical results. Examples and
exercises will guide the reader through this progression, with
particular care in strengthening proof techniques and written
mathematical explanations. Current applications and exploratory
exercises are provided to further the reader's mathematical reasoning
and understanding of the relevance of graph theory to the modern
world"-- Provided by publisher.
Identifiers: LCCN 2020053884 (print) | LCCN 2020053885 (ebook) | ISBN
9781138361409 (hardback) | ISBN 9780367743758 (paperback) | ISBN
9781138361416 (ebook)
Subjects: LCSH: Graph theory.
Classification: LCC QA166 .S227 2021 (print) | LCC QA166 (ebook) | DDC
511/.5--dc23
LC record available at https://lccn.loc.gov/2020053884
LC ebook record available at https://lccn.loc.gov/2020053885

ISBN: 978-0-367-74375-8 (pbk)
ISBN: 978-1-138-36141-6 (ebk)

Typeset in Computer Modern font
by KnowledgeWorks Global Ltd

For my children
who inspire me to never stop learning
and for my students
who inspire me to continue writing

Contents

Preface

At its heart, graph theory is the mathematical study of interactions, conflicts, and connections. The relationship between collections of discrete objects can inform us about the overall network in which they reside, and graph theory can provide one avenue for analysis. In our ever more connected world, understanding the information a connection, or the lack thereof, can provide is extremely powerful.

This text will explore major topics in graph theory from both theoretical and applied viewpoints. Topics will progress from understanding basic terminology, to addressing computational questions, and finally end with broad theoretical results. Examples and exercises will guide the reader through this progression, with particular care in strengthening proof techniques and written mathematical explanations. Current applications and exploratory exercises will be provided where appropriate to further the reader's mathematical reasoning and understanding of the relevance of graph theory to the modern world.

The first chapter introduces basic graph theory terminology and mathematical modeling using graphs. Tournaments are used to solidify understanding of terminology and provide an application accessible to the average undergraduate student, followed by some standard graph theory methodology. Graph isomorphism is discussed to provide a theoretical counterpoint and practice in graph drawing. The chapter includes a review of proof techniques featured throughout the book.

The second chapter introduces three major route problems: eulerian circuits, hamiltonian cycles, and shortest paths. Each topic is introduced through its historical origin, followed by a discussion of more modern applications and theoretical implications. These topics allow the reader to delve into processes on a graph, and provide a few areas for practice with graph theory proofs.

The third chapter focuses entirely on trees—terminology, applications, and theory. Algorithms for finding a minimum spanning tree are discussed, as well as counting the number of different spanning trees. Trees provide ample areas for improving skills in induction, contradiction proofs, and counting techniques.

The fourth chapter begins with the more theoretical topic of connectivity, a discussion of Menger's Theorem, and ends with flow and capacity and additional applications. Some modern applications are also discussed including centrality measures and their use for network analysis.

The subsequent three chapters each focus around a major graph concept: matching, coloring, and planarity. The standard theoretical aspect of these topics are included, but each chapter brings in a modern application or approach. These include the Stable Marriage Problem, on-line coloring and list coloring, and edge-crossing and thickness.

The end of the book includes appendices that cover some prerequisite material on set theory, functions, and matrix multiplication, as well as a discussion on algorithm run-time and pseudocode for some of the algorithms appearing throughout the book. There are also Hints and Solutions to selected exercises provided at the back of the book.

Advice for Students

Reading a mathematics textbook takes skill and more effort than reading your favorite novel at the beach. Professors often complain that their students are not getting enough out of the readings they assign, but fail to realize that most students have not been taught how to read mathematics.

My advice can be boiled down to this one thing: write while you read. Have paper and pencil next to your book anytime you read mathematics. You should expect to work through examples, draw graphs, and play around with the concepts. We learn by doing, not passively reading or watching someone perform mathematics.

This book contains examples often posed in the form of a question. You should attempt to find the solution before reading the one provided. In addition, some definitions and concepts can get technical (as happens in mathematics) and the best way to truly understand these is through working examples. At times, details of an example, especially if it is the second or third of a type, will be left for the reader or will appear in the Exercises.

While understanding *how* to apply a concept is an important part of mathematics, we cannot ignore the benefits of working through the *why*. This book will give you opportunities to test both aspects of graph theory, with theoretical results intended to strengthen your proof-writing skills. Some smaller results will be left as exercises, as well as some small pieces of large concepts. Do not shy away from the challenge of the theory—we gain better insight into the richness of mathematics when we push ourselves to struggle with its complexity.

Advice for Instructors

Each chapter is intended to give a good overview of a major graph theory topic. Within each chapter, these topics are explored from a computational and theoretical aspect, as well as through various applications. A one-semester course should be able to cover the basics of each chapter, but should at minimum cover the majority of Chapters 1–3 and some portion of the remaining four chapters. For a truly introductory course in graph theory, some proofs and exercises should be lightly covered or omitted. Conversely, for more advanced students, the theoretical aspects of graph theory should be emphasized. With either group, some computational results and exercises should be prioritized to help students better understand the concepts; the emphasis on applications is left to the discretion of the instructor.

The chapters build upon each other, both in terms of terminology and connections between topics; however, some sections can easily be omitted without impacting later chapters, as summarized below.

Essential Sections	
1.1	Introduction to graphs and tournaments
1.2	Provides many basic terms used throughout text
1.5	Introductory Proofs (Handshaking Lemma)
2.1	Eulerian circuits and route terminology, cover through 2.1.3
2.2	Hamiltonian cycles, cover up to 2.2.1
3.1	Tree definition and spanning trees, cover up to 3.1.1
3.2	Tree properties
4.1	Connectivity definition and major results
4.2	Menger's Theorem
5.1	Matching basics, cover through 5.1.1
5.2	General graph matching should be discussed briefly
6.1	Four Color Theorem and coloring introduction
6.2	Basic vertex coloring results and techniques
7.1	Planar Graphs, cover up to 7.1.3

Additional Sections for Theory-Focused Course	
1.3	Graph isomorphism
1.4	Matrix representation
1.7	Tournaments
2.2.2	Hamiltonian tournaments
2.3.3	Distance, Diameter, and Radius
4.3	2-connected graphs, ear decomposition
5.1.2	Additional proofs of Hall's Theorem
5.4	Factors
6.3	Edge-coloring and 1-factors, Ramsey numbers
6.4	Additional proofs, especially in 6.4.1, 6.4.2, and 6.4.4
7.1.3	Proof of Kuratowski's Theorem
7.2	Planar graph coloring
7.3	Edge crossing

	Additional Sections for Application-Focused Course
1.4	Matrix representation for applied problems
1.6	Degree sequence
1.7	Tournaments
2.1.4	Eulerian circuit algorithms
2.1.5	Eulerian circuit applications
2.2.1	Traveling Salesman Problem
2.2.2	Hamiltonian circuits in tournaments
2.3.1	Dijsktra's Algorithm
2.3.2	Counting Walks using Matrices
3.3	Rooted Trees, Search Trees
3.4	Additional Tree Applications
3.4.1	Traveling Salesman Problem using trees
4.4	Network Flow
4.5	Centrality Measures
5.3	Stable Matching
6.4	Coloring variations (includes additional applications)

Exercises appear in the last section of each chapter. The first group of exercises are more computational and process based; these provide good practice with terminology and algorithms. The second group of problems are proof based and allow for strengthening of proof writing. Some of these can be tricky and have hints in the back of the book. The last few exercises of each chapter are more exploratory or advanced. Within each grouping, the problems are organized in the same order as the sections of the chapter.

Thanks

I owe a huge debt to the many people who made this book possible.

To my colleagues Adam Childers, Chris Lee, Roland Minton, Maggie Rahmoeller, Hannah Robbins, and David Taylor for their unwavering support. In particular, David Taylor who provided a great sounding board for my ideas, gave advice on formatting, and fixed the coding that I could not.

To my wonderful Roanoke College students, especially my Spring 2020 Math 268 students who tested this book during a historic semester, provided wonderful feedback, and found some ridiculous typos that escaped my numerous proof-reading attempts. A special thanks to Lucas Figlietti who honed his LaTeX and computer science skills with typing up the pseudocode appearing in the Appendix.

To my sister-in-law, Leena Saoub Saunders, for her expertise in designing the cover of this book.

To my parents, David and Pamela Steece, who always supported my dreams and taught me the benefit of hard work.

To my husband Samer, and children, Layla and Rami, who bring joy to my life and endured my excitement and exasperation while writing this book.

To my CRC Press team, especially senior editor Bob Ross, project editor Michele Dimont, and project manager Ashraf Reza, for their guidance through the long process of writing a book.

Finally, a very special thanks goes to Ann Trenk, my graph theory professor at Wellesley College, and Hal Kierstead, my graduate advisor at Arizona State University. Through their excellent teaching and mentorship, I discovered my life's passion and became a much better mathematician and writer.

Dr. Karin R. Saoub
MCSP Department
Roanoke College
Salem, Virginia 24153
saoub@roanoke.edu

1

Graph Models, Terminology, and Proofs

This chapter will introduce you to some basic Graph Theory terminology and provide some motivation for the study of graphs. We begin by describing a specific type of graph called a tournament, and follow with a few sections outlining important terms and operations on graphs. This chapter also provides a basic review of proof techniques and concludes by revisiting tournaments.

1.1 Tournaments

Consider the following scenario:

> The Roanoke Soccer League is planning their end-of-season tournament. Each of the five teams (Aardvarks, Bears, Cougars, Ducks, and Eagles) plays every other team exactly once and no ties are allowed. The tournament director must determine how many games are needed, how to schedule the games, and how to determine a winner once the tournament is completed.

The soccer tournament described above is often referred to as a round-robin tournament. While we can describe the tournament in words, or list the game outcomes in a table, it is often useful to provide a visual representation. One method, and the one we will continue to use throughout this book, is to model the information as a *graph*.

We will formally describe a graph next section, but for now think of a graph as a collection of dots (which we call *vertices*) on the page with lines (called *edges*) connecting the dots to indicate some relationship between them. In terms of the Roanoke Soccer League, we could represent each team as a *vertex* and put an edge between a pair of vertices if they have played each other. The following graphs G_1 and G_2 depict a possible way to run the first few games of the tournament and G_3 is the graph when all games of the tournament have been played (these are called complete graphs and will be discussed later).

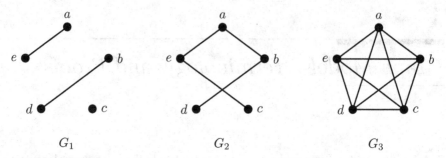

Using the graphs above, we can help the tournament director answer at least one of the questions posed. The number of games needed is the same as the number of edges in the graph G_3 and without any complicated mathematics, we can easily count these and determine 10 total games are needed.

What about the other questions for the tournament director? We need to understand not just which teams played each other, but also the outcomes of these games. One way to do this is to add an arrow to each edge indicating a direction, what we will call a *directed edge* or *arc*. Once directions have been added to each of the edges, we now refer to the graph as a *digraph*, short for directed graph. The digraph shown below indicates that the Aardvarks won all their games, the Bears beat the Cougars and Ducks, the Cougars beat the Ducks and Eagles, the Ducks lost all their games, and the Eagles won their games against the Bears and Ducks.

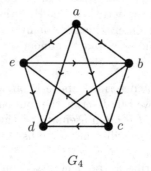

G_4

For the specific situation shown above, the winner of the tournament would be the Aardvarks since they are the only team to win all four of their games. The number of wins for a team is coded into the digraph by looking at the number of arcs leaving a vertex (what we will call out-degree later). But how could we determine a winner if no team won all their games? And how do we determine a schedule for the games? These questions will be addressed later (see Sections 1.7 and 6.3), but for now we will move onto formalizing some definitions and concepts that will be needed throughout the remainder of the book.

1.2 Introduction to Graph Models and Terminology

An integral component of mathematics is precise (and appropriate) definitions. Throughout this book, we will use an example to motivate and gain intuition about concepts and then provide the precise definitions. To that end, we give the definition of a graph below. Note that many aspects of graph theory rely on basic set theory concepts (mainly the subset relationship); see Appendix A if you need a review of set theory.

Definition 1.1 A **graph** G consists of two sets: $V(G)$, called the vertex set, and $E(G)$, called the edge set. An **edge**, denoted xy, is an unordered pair of vertices. We will often use G or $G = (V, E)$ as short-hand.

Example 1.1 Let G_4 be a graph where $V(G_4) = \{a, b, c, d, e\}$ and $E(G_4) = \{ab, cd, cd, bb, ad, bc\}$. Although G_4 is defined by these two sets, we generally use a visualization of the graph where a dot represents a vertex and an edge is a line connecting the two dots (vertices). A drawing of G_4 is given below.

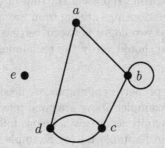

Note that two lines were drawn between vertices c and d as the edge cd is listed twice in the edge set. In addition, a circle was drawn at b to indicate an edge (bb) that starts and ends at the same vertex.

We will often need to describe the size of a graph. This term is ambiguous, as we may be discussing the number of vertices or edges. We will use the following notation to make this clear.

Definition 1.2 The number of vertices in a graph G is denoted $|V(G)|$, or more simply $|G|$. The number of edges is denoted $|E(G)|$ or $\|G\|$.

Using this notation we see that graph G_4 from Example 1.1 above satisfies $|G_4| = 5$ and $\|G_4\| = 6$.

It should be noted that the drawing of a graph can take many different forms while still representing the same graph. The only requirement is to faithfully record the information from the vertex set and edge set. We often draw graphs with the vertices in a circular pattern (as shown in Example 1.1), though in some instances other configurations better display the desired information. The best configuration is the one that reduces complexity or best illustrates the relationships arising from the vertex set and edge set.

Example 1.2 Consider the graph G_4 from Example 1.1. Below are two different drawings of G_4.

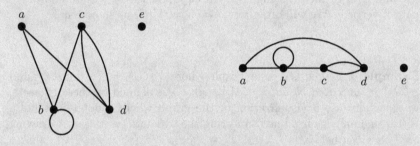

To verify that these drawings represent the same graph from Example 1.1, we should check the relationships arising from the vertex set and edge set. For example, there are two edges between vertices c and d, a loop at b, and no edges at e. You should verify the remaining edges.

To discuss and prove properties of graphs, we need the proper terminology. The graph given in the examples above are good references for this initial terminology. Some initial definitions are given below, followed by the appropriate references to the graph in Example 1.1 (or Example 1.2).

Definition 1.3 Let G be a graph.

- If xy is an edge, then x and y are the **endpoints** for that edge. We say x is **incident to** edge e if x is an endpoint of e.

- If two vertices are incident to the same edge, we say the vertices are **adjacent**, denoted $x \sim y$. Similarly, if two edges share an endpoint, we say they are adjacent. If two vertices are adjacent, we say they are **neighbors** and the set of all neighbors of a vertex x is denoted $N(x)$.

    – ab and ad are adjacent edges in G_4 since they share an endpoint, namely vertex a

- $a \sim b$, that is a and b are adjacent vertices as ab is an edge of G_4
  - $N(d) = \{a, c\}$ and $N(b) = \{a, b, c\}$

- If two vertices (or edges) are not adjacent then we call them *independent*.

- If a vertex is not incident to any edge, we call it an *isolated vertex*.

  - e is an isolated vertex of G_4

- If both endpoints of an edge are the same vertex, then we say the edge is a *loop*.

  - bb is a loop in G_4

- If there is more than one edge with the same endpoints, we call these *multi-edges*.

  - cd is a multi-edge of G_4

- If a graph has no multi-edges or loops, we call it *simple*.

- The *degree* of a vertex v, denoted $\deg(v)$, is the number of edges incident to v, with a loop adding two to the degree. If the degree is even, the vertex is called *even*; if the degree is odd, then the vertex is *odd*.

  - $\deg(a) = 2$, $\deg(b) = 4$, $\deg(c) = 3$, $\deg(d) = 3$, $\deg(e) = 0$

- If all vertices in a graph G have the same degree k, then G is called a *k-regular* graph. When $k = 3$, we call the graph **cubic**.

When examining graphs, especially if they are particularly large, we may want to discuss a smaller portion of the graph, called a *subgraph*. For example, graphs G_1 and G_2 shown on page 2 display when a portion of the total games have been played in the soccer tournament, and these are both subgraphs of graph G_3.

Definition 1.4 A *subgraph* H of a graph G is a graph where H contains some of the edges and vertices of G; that is, $V(H) \subseteq V(G)$ and $E(H) \subseteq E(G)$.

Note that if a subgraph H contains the edge ab then it necessarily contains both of its endpoints (a and b).

Example 1.3 Consider the graph G below. Find two subgraphs of G, both of which have vertex set $V' = \{a, b, c, f, g, i\}$.

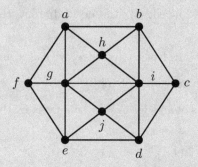

Solution: Two possible solutions are shown below. Note that the graph H_1 on the left contains every edge from G amongst the vertices in V', whereas the graph H_2 on the right does not since some of the available edges are missing (namely, $ab, af, ci,$ and gi).

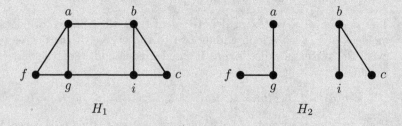

The graph shown on the left above is a special type, called an induced subgraph, since all the edges are present between the chosen vertices. Another special type of subgraph, called a spanning subgraph, includes all the vertices of the original graph.

Definition 1.5 Given a graph $G = (V, E)$, an ***induced subgraph*** is a subgraph $G[V']$ where $V' \subseteq V$ and every available edge from G between the vertices in V' is included.

We say H is a ***spanning subgraph*** if it contains all the vertices but not necessarily all the edges of G; that is, $V(H) = V(G)$ and $E(H) \subseteq E(G)$.

Example 1.4 Find a spanning subgraph of the graph G from Example

1.3 above.

Solution: Two possible solutions are shown below. Note that both graphs contain all the vertices from G, but only in the graph H_4 could we move between any two vertices in the graph (which we will later call connected). Spanning subgraphs similar to H_4 will be studied in Chapter 3.

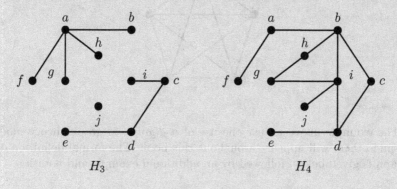

In a general graph xy and yx are treated equally, though it is customary to write them in alphabetical order. An edge is intended to convey some sort of relationship between two discrete objects that are modeled as vertices. Using the soccer tournament example from above, the vertices are the teams and an edge displays that two teams have played one another. Additional information is sometimes required beyond a simple pairing. For example we may want to indicate a direction (such as who won a game, flow along pipes, or street direction) or some quantity associated with the edge (such as cost, distance, or probability). Each of these concepts give way to modifications on a basic graph, and can be seen in the following examples.

1.2.1 Digraphs

Many theorems and applications of graph theory relate to symmetric relationships between objects, as we view edge ab the same as edge ba. However, scenarios exist where relationships between discrete objects need not be symmetric, and in some cases asymmetry would better display the requisite information.

Example 1.5 Consider the soccer tournament above, where the Aardvarks won all their games, the Bears, Cougars, and Eagles each won two games, and the Ducks did not win any game. To model which team won a game we can add a direction to any edge, which we now call a *directed* *edge* or *arc*, where if $a \rightarrow b$ then team A beat team B. We refer to this structure as a *digraph*, short for directed graph, when the edges now have a direction associated to them.

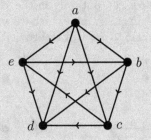

The example above is just one use of a *digraph* in graph theory modeling, and others will appear throughout this book. The formal definition of a digraph is given below, followed by an additional example and notation.

Definition 1.6 A ***directed graph***, or ***digraph***, is a graph $G = (V, A)$ that consists of a vertex set $V(G)$ and an *arc set* $A(G)$. An ***arc*** is an ordered pair of vertices.

Digraphs have many similar properties to (undirected) graphs. Looking at the digraph above, we can see that the number of wins is modeled as the number of arcs coming from a team's vertex, and the number of losses is the number of arcs entering the vertex.

Example 1.6 Let G_5 be a digraph where $V(G_5) = \{a, b, c, d\}$ and $A(G_5) = \{ab, ba, cc, dc, db, da\}$. A drawing of G_5 is given below.

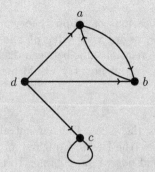

Analogous definitions to those in Definition 1.3 exist for digraphs. A few of these are listed below along with the appropriate references to G_5 from Example 1.6. Other directed versions of previously defined terminology should be obvious based on context and will appear later in the text as appropriate.

Definition 1.7 Let $G = (V, A)$ be a digraph.

- Given an arc xy, the **head** is the starting vertex x and the **tail** is the ending vertex y.

    - a is the head of arc ab and the tail of arcs da and ba from G_5

- Given a vertex x, the **in-degree** of x is the number of arcs for which x is a tail, denoted $\deg^-(x)$. The **out-degree** of x is the number of arcs for which x is the head, denoted $\deg^+(x)$.

    - $\deg^-(a) = 2$, $\deg^-(b) = 2$, $\deg^-(c) = 2$, $\deg^-(d) = 0$
    - $\deg^+(a) = 1$, $\deg^+(b) = 1$, $\deg^+(c) = 1$, $\deg^+(d) = 3$

- The **underlying graph** for a digraph is the graph $G' = (V, E)$ which is formed by removing the direction from each arc to form an edge.

Knowing the degrees in a graph or digraph can tell you a lot of information, but need not uniquely determine the underlying graph structure. For example, both digraphs below have the same in-degrees of $3, 2, 1, 0$ but one contains a loop whereas the other does not.

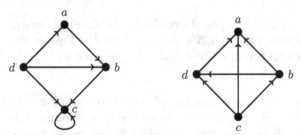

Since tournaments have a more regular structure (namely exactly one arc between any two vertices), their degree sequence is more restrictive. We will investigate these further in Section 1.7, where we revisit some properties of tournaments.

1.2.2   Weighted Graphs

As seen above, digraphs are used to model asymmetric relationships between discrete objects. We now consider a different edge relationship, where instead of direction we are concerned with quantity. These graphs are called *weighted graphs*.

Example 1.7 Sam wants to visit 4 national parks over the summer. To save money, he needs to minimize his driving distance. The graph below has weights along each edge indicating the driving distance between his home (in Boise, Idaho) and the four national parks he will visit.

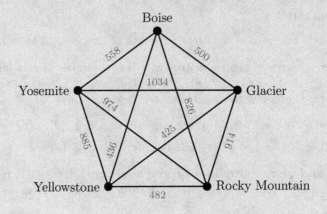

A solution to his question will be revisited in Section 2.2 on hamiltonian cycles; for now, we will end our discussion with the formal definition of a weighted graph and another example modeled by a weighted graph.

Definition 1.8 A *weighted graph* $G = (V, E, w)$ is a graph where each of the edges has a real number associated with it. This number is referred to as the *weight* and denoted $w(xy)$ for the edge xy.

Note that a weighted graph can also refer to a graph in which each of the vertices is assigned a weight, and denoted $w(v)$ for a vertex v. In the next few chapters, we will focus on graphs in which the edges are weighted; the vertex version will be addressed in Chapter 6. Also, the weight associated with an edge can represent more than just distance. For example, we may be interested in time, cost or some other measure related to the connection between two discrete objects. Choose the appropriate measure based upon the scenario in question.

Example 1.8 Adam comes to you with a new game. He flips a coin and you roll a die. If he gets heads and you roll an even number, you win $2; if he gets heads and you roll an odd number, you pay him $3. If he gets tails and you roll either 1 or 4, you win $5; if he gets tails and any of 2, 3, 5, or 6 is rolled, you pay him $2. What is the probability you win $5? What is the probability you win any amount of money?

Solution: A probability tree is a graph with the vertices representing possible outcomes of each part of the experiment (here a coin and dice game) and the edges are labeled with the probability that the outcome occurred. To find the probability of any final outcome, multiply along the path from the initial vertex to the ending result. The tree below has the edges labeled and the final probabilities calculated for Adam's game.

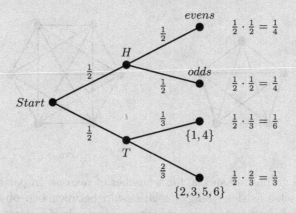

Using the probability tree above, the probability you win $5 (which requires tails and a 1 or 4) is $\frac{1}{6}$ and the probability you win any money is $\frac{1}{6} + \frac{1}{4} = \frac{5}{12}$. Do not play this game with Adam! He is more likely to win than you! Further discussions of trees will be seen in Chapter 3.

1.2.3   Complete Graphs

The weighted graph in Example 1.7 and G_3 from page 2 have the same underlying structure. If we remove the weights of edges and the vertex names, we would not be able to distinguish between the two graphs since in both graphs every pair of distinct vertices is joined by an edge. These graphs are called *complete graphs*.

Definition 1.9 A simple graph G is **complete** if every pair of distinct vertices is adjacent. The complete graph on n vertices is denoted K_n.

The first six complete graphs are shown on the next page.

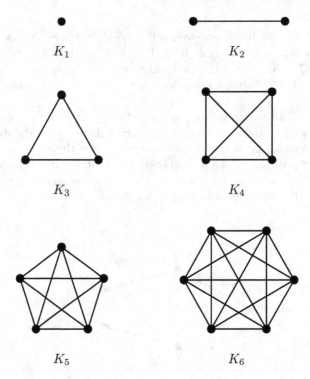

K_1 K_2

K_3 K_4

K_5 K_6

Complete graphs are special for a number of reasons. In particular, if you think of an edge as describing a relationship between two objects, then a complete graph represents a scenario where every pair of vertices satisfies this relationship. Other useful properties of complete graphs are given below.

Properties of K_n

(1) Each vertex in K_n has degree $n-1$.

(2) K_n has $\dfrac{n(n-1)}{2}$ edges.

(3) K_n contains the most edges out of all simple graphs on n vertices.

Complete graphs will periodically appear throughout the book. In many cases, we will be looking for the largest complete graph that appears as a subgraph. This is called the *clique-size* of a graph.

Definition 1.10 The ***clique-size*** of a graph, $\omega(G)$, is the largest integer n such that K_n is a subgraph of G but K_{n+1} is not.

Example 1.9 Find $\omega(G)$ for each of the graphs shown below.

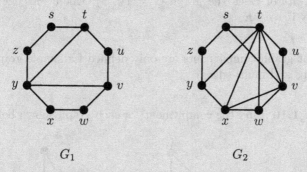

G_1 G_2

Solution: First note that G_1 does not contain any triangles (K_3) but does have an edge and so contains K_2. Thus $\omega(G_1) = 2$. Next, in G_2 the vertices t, v, w, x are all adjacent, as shown below, but we cannot find a collection of 5 vertices that are all adjacent (not enough vertices have degree at least 4). Thus $\omega(G_2) = 4$.

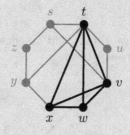

    Knowing the clique-size of a graph is often a tool for determining other useful properties about a graph, such as the chromatic number (see Chapter 6) or if it is planar (see Chapter 7).

1.2.4 Graph Complements

Consider a graph representing friendships. Given a collection of people, we could form a graph where an edge exists between two vertices if those people are friends. But what, if instead, we want to know who are not friends with each other? Perhaps a teacher wants to avoid friends talking during class and so will not seat them at the same table. This new graph would include all the edges missing from the original graph created, and is called the *graph complement*.

Definition 1.11 Given a simple graph $G = (V, E)$, define the *complement* of G as the graph $\overline{G} = (V, \overline{E})$, where an edge $xy \in \overline{E}$ if and only if $xy \notin E$.

Note that graph complements are only defined for simple graphs (graphs without loops and multi-edges).

Example 1.10 Find the complements of each graph shown below.

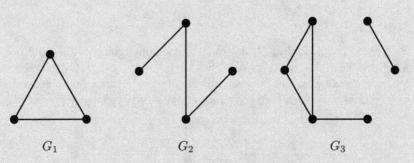

G_1 G_2 G_3

Solution: For each graph we simply add an edge where there wasn't one before and remove the current edges.

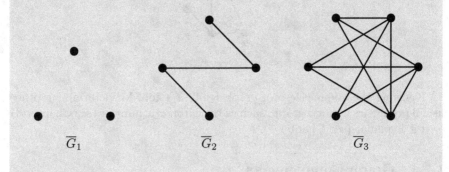

$\overline{G}_1$ $\overline{G}_2$ $\overline{G}_3$

Notice that if we have a graph G on n vertices and add every edge in $\overline{G}$ to the edges of G, then the resulting graph is simply K_n.

1.2.5   Bipartite Graphs

As we have already seen, problems that can be modeled by a graph need to consist of distinct objects (such as people or places) and a relationship between them. The proper model will allow the graph structure, or properties of the graph, to answer the question being asked. If we want to display the relationship between different types of objects, we would use a *bipartite graph*.

Definition 1.12 A graph G is *bipartite* if the vertices can be partitioned into two sets X and Y so that every edge has one endpoint in X and the other in Y.

Example 1.11 Three student organizations (Student Government, Math Club, and the Equestrian Club) are holding meetings on Thursday afternoon. The only available rooms are 105, 201, 271, and 372. Based on membership and room size, the Student Government can only use 201 or 372, Equestrian Club can use 105 or 372, and Math Club can use any of the four rooms. Draw a graph that depicts these restrictions.

Solution: Each organization and room is represented by a vertex, and an edge denotes when an organization is able to use a room.

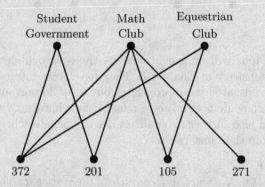

Note that edges do not occur between two organizations or between two rooms, as these would be nonsensical in the context of the problem. The graph above is a bipartite graph.

In the example above, there are some edges that could be added to the graph while still keeping the graph bipartite. Just as we defined a complete graph as the simple graph with the most edges, we similarly define a complete bipartite graph.

Definition 1.13 $K_{m,n}$ is the *complete bipartite graph* where $|X| = m$ and $|Y| = n$ and every vertex in X is adjacent to every vertex in Y.

Below are a few complete bipartite graphs. It is customary to write m and n in increasing order (so $K_{2,3}$ versus $K_{3,2}$), but it is not required.

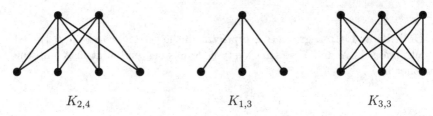

$$K_{2,4} \qquad\qquad K_{1,3} \qquad\qquad K_{3,3}$$

When $m = 1$, we call $K_{1,n}$ a *star* since we could draw these with a singular vertex in the center and the remaining vertices surrounding it, as seen below with $K_{1,5}$ and $K_{1,8}$.

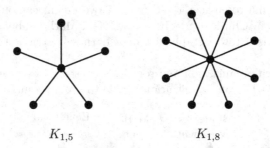

$$K_{1,5} \qquad\qquad K_{1,8}$$

Bipartite graphs will appear at various times throughout this text, but will be used extensively in Chapter 5. They have some interesting properties that will be investigated as appropriate, but most importantly we want to remember that they are used to show relationships between distinct groups of objects. We can also further generalize bipartite graphs where we break the vertices into more than just two sets.

Definition 1.14 A graph G is **k-partite** if the vertices can be partitioned into k sets $X_1, X_2, \ldots X_k$ so that every edge has one endpoint in X_i and the other in X_j where $i \neq j$.

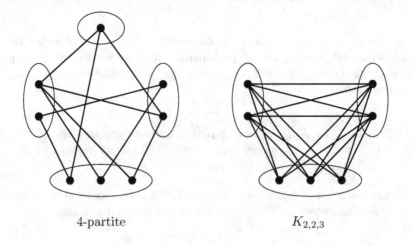

4-partite $K_{2,2,3}$

When $k = 3$, we call the graph tripartite rather than a 3-partite. In addition, we can add the adjective complete to any k-partite graph, and we simply want to include all possible edges so that the graph remains k-partite. Above are drawings of a 4-partite graph and the complete tripartite graph $K_{2,2,3}$.

1.2.6   Graph Combinations

As graphs are built from sets of vertices and edges, some operations on sets have natural translations onto graphs (for a review of set theory, see Appendix A). We will focus on a few that will appear at times throughout this book.

Definition 1.15 Given two graphs G and G the ***union*** $G \cup H$ is the graph with vertex-set $V(G) \cup V(H)$ and edge-set $E(G) \cup E(H)$.

If the vertex-sets are disjoint (that is $V(G) \cap V(H) = \emptyset$) then we call the disjoint union the ***sum***, denoted $G + H$.

Note that $G + H$ is just a special type of union, and so unless we want to explicitly use or note that the vertex sets are disjoint, it is customary to use the union notation.

Example 1.12 Find the sum $K_3 + H_1$ and the union $H_1 \cup H_4$ using the graphs from Examples 1.3 and 1.4.

Solution: First note that, since we are finding the sum $K_3 + H_1$, we are assuming the vertex sets are disjoint. Thus the resulting graph is simply the graph below.

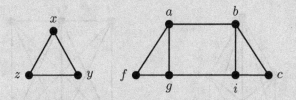

Next, since H_1 and H_4 are subgraphs of the same graph and have some edges in common, their union will consist of all the edges in at least one of H_1 and H_4, where we do not draw (or list) an edge twice if it appears in both graphs, as shown in the following graph.

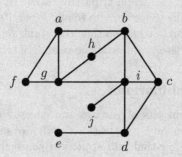

Definition 1.16 The *join* of two graphs G and H, denoted $G \vee H$, is the sum $G + H$ together with all edges of the form xy where $x \in V(G)$ and $y \in V(H)$.

Example 1.13 Find the join of K_3 and the graph G below consisting of three vertices and two edges, as well as the join $G \vee G$.

Solution: The join $K_3 \vee G$ is shown below on the left. Note that every vertex from K_3 is adjacent to all those from G, but this is not K_6 since the edge ac is missing. The join $G \vee G$ is on the right below.

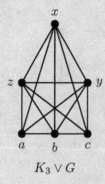
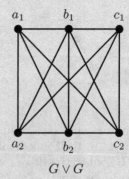

$$K_3 \vee G \qquad\qquad G \vee G$$

Each of the definitions above can be generalized to more than 2 graphs, just as we can describe a union of more than two sets.

1.3 Isomorphisms

In Example 1.2 we showed two different modes for drawing the graph in Example 1.1. At the time, we focused on the fact that we were dealing with the same set of vertices and verified the edge set was maintained in the new drawings. However, two graphs with distinct vertex sets can still produce the same edge relationships (see the discussion of complete graphs on page 11); more technically these graphs are called *isomorphic* if every vertex from G_1 can be paired with a unique vertex from G_2 so that corresponding edges from G_1 are maintained in G_2.

Definition 1.17 Two graphs G_1 and G_2 are ***isomorphic***, denoted $G_1 \cong G_2$, if there exists a bijection $f : V(G_1) \to V(G_2)$ so that $xy \in E(G_1)$ if and only if $f(x)f(y) \in E(G_2)$.

Throughout this section we will only consider simple graphs (those without multi-edges or loops). Similar definitions and results exist for multi-graphs and digraphs. The definition of isomorphic uses a special function, called a bijection, between the vertices of G_1 and G_2; for a review of functions see Appendix B.

Later we will list some of the common properties that must be maintained with isomorphic graphs, called graph invariants. But to begin, it should be easy to name a few things that are quick to check:

- number of vertices

- number of edges

- vertex degrees

By no means is this list comprehensive, but it allows for a quick check before working on more complex ideas. Note that defining the bijection is essentially just providing the vertex pairings, so we will list them explicitly and then check that the edge relationships are maintained.

Example 1.14 Determine if the following pair of graphs are isomorphic. If so, give the vertex pairings; if not, explain what property is different among the graphs.

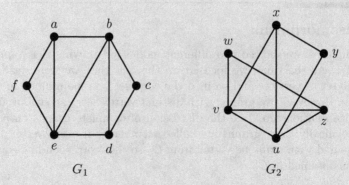

$$G_1 \qquad\qquad\qquad G_2$$

Solution: First note that both graphs have six vertices and nine edges, with two vertices each of degrees 4, 3, and 2. Since corresponding vertices must have the same degree, we know b must map to either u or v. We start by trying to map b to v. By looking at vertex adjacencies and degree, we must have e map to u, c map to w, and a map to x. This leaves f and d, which must be mapped to y and z, respectively. The chart below show the vertex pairings and checks for corresponding edges.

$V(G_1) \longleftrightarrow V(G_2)$	Edges	
$a \longleftrightarrow x$	$ab \longleftrightarrow xv$	✓
$b \longleftrightarrow v$	$ae \longleftrightarrow xu$	✓
$c \longleftrightarrow w$	$af \longleftrightarrow xy$	✓
$d \longleftrightarrow z$	$bc \longleftrightarrow vw$	✓
$e \longleftrightarrow u$	$bd \longleftrightarrow vz$	✓
$f \longleftrightarrow y$	$be \longleftrightarrow vu$	✓
	$cd \longleftrightarrow wz$	✓
	$de \longleftrightarrow zu$	✓
	$ef \longleftrightarrow uy$	✓

Since all edge relationships are maintained, we know G_1 and G_2 are isomorphic.

Example 1.15 Determine if the following pair of graphs are isomorphic. If so, give the vertex pairings; if not, explain what property is different among the graphs.

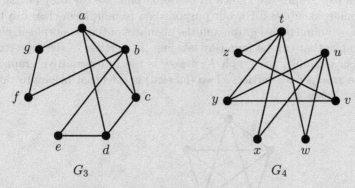

G_3 G_4

Solution: First note that both graphs have seven vertices and ten edges, with two vertices each of degrees 4 and 3, and three vertices of degree 2. As in the previous example, we know corresponding vertices must have the same degree, and so the vertices of degree 4 in G_3, a and b, must map to the vertices of degree 4 in G_4, namely t and u. However, in G_3 the degree 4 vertices (a and b) are adjacent, whereas in G_4 there is no edge between the degree 4 vertices (t and u). Thus G_3 and G_4 are not isomorphic.

The previous example illustrates that no one property guarantees two graphs are isomorphic. In fact, simply having the same number of vertices of each degree is not enough. The theorem below lists the more useful graph invariants.

Theorem 1.18 Assume G_1 and G_2 are isomorphic graphs. Then G_1 and G_2 must satisfy any of the properties listed below; that is, if G_1

- is connected

- has n vertices

- has m edges

- has m vertices of degree k

- has a cycle of length k (see Section 2.1.2)

- has an eulerian circuit (see Section 2.1.3)

- has a hamiltonian cycle (see Section 2.2)

then so too must G_2 (where n, m, and k are non-negative integers).

When two graphs are known to be isomorphic, we say they belong to the same isomorphism class. For our purposes, an isomorphism class can be represented by an unlabeled graph and the members of the isomorphism class are those graphs that can be found by labeling the vertices of the representative graph. For example, the graph K_5 below is the representative graph for all complete graphs on 5 vertices. Two (labeled) graphs that belong to this class are shown below it.

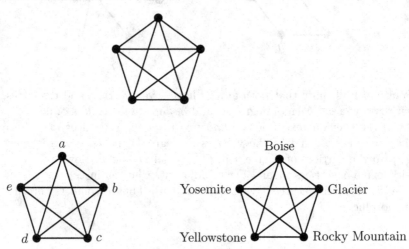

Throughout the majority of this text, we will gloss over graph isomorphism. This is in part because when modeling a problem, we only care about whether the graph in question adequately models the given information. Moreover, properties about graphs are truly properties about every graph within an isomorphism class, but we will not make that distinction when proving broad graph theoretical results. We will not make much, if any, distinction between multiple ways to draw the same set of information until we reach Chapter 7. In general, graph isomorphism is less applicable to real world scenarios; however, they will be briefly mentioned in Sections 1.6, 5.4, and 6.2.

1.4   Matrix Representation

The graphs we have encountered in this book so far are fairly small and can be described easily in terms of the vertex and edge sets. However, very large graphs (such as those modeling the spread of an infectious disease, the connections within a terrorist organization, or the results from a season of NCAA Division 1 football) would be unwieldy without additional resources. One way to tackle large graphs is to represent them in such a way that a computer program can perform the required analysis. One method, which we will use at various times throughout this book, is to form the *adjacency matrix* $A(G)$ of the graph G.

Definition 1.19 The **adjacency matrix** $A(G)$ of the graph G is the $n \times n$ matrix where vertex v_i is represented by row i and column i and the entry a_{ij} denotes the number of edges between v_i and v_j.

Example 1.16 Find the adjacency matrix for the graph G_4 from Example 1.1.

Solution:

$$
\begin{array}{c c c c c c}
 & a & b & c & d & e \\
a & \begin{bmatrix} 0 \\ 1 \\ 0 \\ 1 \\ 0 \end{bmatrix} & \begin{matrix} 1 \\ 1 \\ 1 \\ 0 \\ 0 \end{matrix} & \begin{matrix} 0 \\ 1 \\ 0 \\ 2 \\ 0 \end{matrix} & \begin{matrix} 1 \\ 0 \\ 2 \\ 0 \\ 0 \end{matrix} & \begin{matrix} 0 \\ 0 \\ 0 \\ 0 \\ 0 \end{bmatrix}
\end{array}
$$

Note that the entry $(2,2)$ represents the loop at b and the entries $(3,4)$ and $(4,3)$ show that there are two edges between c and d. The column for e has all 0's since e is an isolated vertex.

A few interesting properties of the adjacency matrix can be seen. First, the matrix is symmetric along the main diagonal since if there is an edge $v_i v_j$ then it will be accounted for in both the entry (i,j) and (j,i) in the matrix. Second, the main diagonal represents all loops in the graph. Finally, the degree of a vertex can be easily calculated from the adjacency matrix by adding the entries along the row (or column) representing the vertex but double any item along the diagonal. In the matrix above, we would get $\deg(a) = 2$ and $\deg(b) = 4$, which matches the graph representation from Example 1.1.

While we will often start with the graph and form its adjacency matrix, we can work in reverse as well. The example below demonstrates how to draw the graph given a matrix.

Example 1.17 Draw the graph whose adjacency matrix is shown below.

$$
\begin{bmatrix}
0 & 1 & 0 & 0 & 1 & 1 \\
1 & 0 & 1 & 1 & 0 & 0 \\
0 & 1 & 1 & 0 & 0 & 0 \\
0 & 1 & 0 & 0 & 3 & 0 \\
1 & 0 & 0 & 3 & 0 & 1 \\
1 & 0 & 0 & 0 & 1 & 0
\end{bmatrix}
$$

Solution: Since the matrix has 6 rows and columns, we know that the graph must have 6 vertices. We will label them as $v_1, v_2, \ldots, v_6$.

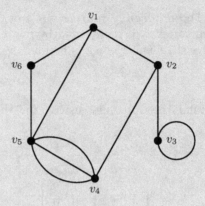

Certain matrix operations will prove to have meaning for the graph being represented; these will appear as appropriate throughout the book. For example, the next chapter will show how matrices can be used to find routes of a certain length (see Section 2.3). For a review of basic matrix rules and operations, see Appendix C.

1.5   Proof Techniques

Graph Theory as a mathematical discipline straddles the distinction between applied and theoretical mathematics. As we've already seen in this chapter, graphs can be used to model various scenarios, especially some complicated modern applications. Alternatively, isomorphisms comprise a very theoretical aspect of graph theory. Within each of these areas, however, we use proofs to deepen and demonstrate our understanding of graphs and their properties.

While this book relies on a basic understanding of logic, proof structure, and proof techniques, it is by no means expected that the reader is a proficient writer of proofs. This section is meant to review the basics of mathematical proof, and introduce some early graph results that can be proven with little intuition about graphs and their structure. For a more complete introduction to logic and proofs, see *Discrete Mathematics* by Susanna Epp [31].

1.5.1 Direct Proofs

Most mathematical statements have an underlying conditional form; that is, they can be written as "If . . ., then . . .." For example, we may say "The sum of two odd integers is even" but we are, in fact, making the conditional statement "If x and y are odd integers, then $x + y$ is even." Writing a statement in the standard if–then form allows the logical structure to stand out and provides guidance into the format of the argument.

In logical symbols, conditional statements are given as $p \to q$. A direct proof begins by assuming the premise of the conditional (p) and uses logic, definitions, and previously proven theorems to show the conclusion (q) is true. The example below uses the definition of odd, even, and the assumption that the sum of two integers is still an integer.

Proposition 1.20 The sum of two odd integers is even.

Proof: Assume x and y are odd integers. Then there exist integers n and m such that $x = 2n+1$ and $y = 2m+1$. Thus $x+y = (2n+1)+(2m+1) = 2(n + m + 1) = 2k$, where k is the integer given by $n + m + 1$. Therefore $x + y$ is even.

A proper mathematical proof should be self-contained (all variables are defined), concise (no extra information is included), and complete (the proper conclusion is reached). The theorem below can be considered one of the first results in graph theory (and in some publications is referred to as "The First Graph Theorem"), as it was published in 1736 by Leonhard Euler. Euler was a prolific mathematician, often called the "Father of Graph Theory" as he was the first to formalize what we now call a graph. His work will be discussed throughout this book, but especially in Chapter 2.

Theorem 1.21 (Handshaking Lemma) Let $G = (V, E)$ be a graph and $|E|$ denote the number of edges in G. Then the sum of the degrees of the vertices equals twice the number of edges; that is if $V = \{v_1, v_2, \ldots, v_n\}$, then

$$\sum_{i=1}^{n} \deg(v_i) = \deg(v_1) + \deg(v_2) + \cdots + \deg(v_n) = 2|E|.$$

Proof: Let $G = (V, E)$ be a graph with $V = \{v_1, v_2, \ldots, v_n\}$. Any edge $e = v_i v_j$ of G will be counted once in the total $|E|$. Since each edge is defined by its two endpoints, this edge will add one to the count of both

$\deg(v_i)$ and $\deg(v_j)$. Thus every edge of G will add two to the count of the sum of the degrees. Thus $\deg(v_1) + \deg(v_2) + \cdots + \deg(v_n) = 2|E|$.

While this result does not appear to be ground breaking, it does provide information about graphs that will be useful for many results later. In particular, the following is a direct consequence of the Handshaking Lemma; the proof appears in Exercise 1.15.

Corollary 1.22 Every graph has an even number of vertices of odd degree.

1.5.2   Indirect Proofs

Direct proofs can be considered the preferable method of proof as their structure models the statement they are proving. However, some statements are either impossible or much more difficult to prove in this way and a different technique is needed. Classic examples of this include proving there are infinitely many primes or that $\sqrt{2}$ is irrational. While these are great examples, better examples exist in graph theory for the usefulness of indirect proofs.

There are two main types of indirect proofs: contradiction and contraposition. For a Proof by Contradiction, we assume the negation of the statement is true. Through logic, definitions, and previous results, we show a contradiction must be occurring, thus proving the original statement must be true. An example from elementary number theory is shown below.

Proposition 1.23 For any integer n, if n^2 is odd then n is odd.

Proof: Suppose for a contradiction that n^2 is odd but n is even. Then $n = 2k$ for some integer k and $n^2 = (2k)^2 = 4k^2 = 2j$ where j is the integer $2k^2$. Thus n^2 is both even and odd, a contradiction. Therefore if n^2 is odd then n is also odd.

For a Proof by Contraposition, we use a direct proof on the contrapositive ($\sim q \rightarrow \sim p$) of the original conditional statement ($p \rightarrow q$). Since the contrapositive is logically equivalent to the original statement, this shows the intended result to be true. The statement above can also be proven using the contrapositive, as shown below.

Proof: Suppose n is not odd. Then n is even and $n = 2k$ for some integer k. Then $n^2 = (2k)^2 = 4k^2 = 2j$ where j is the integer $2k^2$, and so n^2 is even. Thus if n^2 is odd, it must be that n is also odd.

Note that both indirect proof techniques can be used on (appropriate) conditional statements, but some statements can only use a contradiction argument. In graph theory, it is often useful to assume some property of a graph does not hold and then use that assumption to find a contradiction to another known graph property.

Proposition 1.24 For every simple graph G on at least 2 vertices, there exist two vertices of the same degree.

Proof: Suppose for a contradiction that G is a simple graph on n vertices, with $n \geq 2$, in which no two vertices have the same degree. Since there are no loops and each vertex can have at most one edge to any other vertex, we know the maximum degree for any vertex is $n - 1$ and the minimum degree is 0. Since there are exactly n integers from 0 to $n - 1$, we know there must be exactly one vertex for each degree between 0 and $n-1$. But the vertex of degree $n - 1$ must then be adjacent to every other vertex of G, which contradicts the fact that a vertex has degree 0. Thus G must have at least two vertices of the same degree.

1.5.3 Mathematical Induction

The last proof technique we review is quite useful when studying discrete objects, especially objects that can easily be transformed into ones of smaller size. Mathematical induction relies on a two step process. In the first step (sometimes referred to as the base case or basis step) we show the statement to be proved holds for a specific value or size. In the second step (called the induction step) we assume that the statement holds for some unknown value and then show the statement also holds for the next value.

The power of induction is that we are proving a statement that holds for an infinite number of objects but only need to prove two very specific items. As we will see throughout this book, graphs naturally lend themselves to induction proofs by our ability to take a graph of a specific size (usually in terms of the number of vertices or edges) and make it smaller by removing either an edge or a vertex. This technique is shown in the result below about the number of edges in a complete graph.

Proposition 1.25 The complete graph K_n has $\dfrac{n(n - 1)}{2}$ edges.

Proof: Argue by induction on n. If $n = 1$ then K_1 is just a single vertex and has $0 = \frac{1(0)}{2}$ edges.

Suppose for some $n \geq 1$ that K_n has $\dfrac{n(n-1)}{2}$ edges. We can form K_{n+1} by adding a new vertex v to K_n and adjoining v to all the vertices from K_n. Thus K_{n+1} has n more edges than K_n and so by the induction hypothesis has

$$n + \frac{n(n-1)}{2} = \frac{2n + n(n-1)}{2} = \frac{n(2 + n - 1)}{2} = \frac{n(n+1)}{2}$$

edges.

Thus by induction we know K_n has $\dfrac{n(n-1)}{2}$ edges for all $n \geq 1$.

One final note about results and proofs. Throughout this book various items will be labeled as Theorem, Proposition, Lemma, or Corollary. All four of these are just labels for a result that will be proven. The name of the result has more to do with its relative place within Graph Theory. In particular, a lemma is a result whose main purpose is to prove something (relatively) minor that will be useful in the proof of a theorem, usually appearing very closely after the lemma. A corollary is a result that is again (relatively) minor but is now almost an immediate result from a theorem, and often listed immediately following the theorem from which it is based. Proposition and Theorem are often used interchangeably, but we view theorems to be more substantial, either in terms of the content or the possibility for application.

1.6   Degree Sequence

As noted above, the degree of a vertex can be seen in the adjacency matrix for a graph by summing the entries along its corresponding row or column. Here we switch gears a bit to look at what conditions must be placed on a sequence of integers so that they could be the degrees of a simple graph.

Definition 1.26 The *degree sequence* of a graph is a listing of the degrees of the vertices. It is customary to write these in decreasing order. If a sequence is a degree sequence of a simple graph then we call it *graphical*.

Example 1.18 Explain why neither $4, 4, 2, 1, 0$ nor $4, 4, 3, 1, 0$ can be graphical.

Solution: The first sequence sums to 11, but we know the sum of the degrees of a graph must be even by the Handshaking Lemma. Thus it cannot be a degree sequence.

The second sequence sums to 12, so it is at least even. However, in a simple graph with 5 vertices, a vertex with degree 4 must be adjacent to all the other vertices, which would mean no vertex could have degree 0. Thus the second sequence cannot be a degree sequence.

The example above is one instance of the general question called the *graph realization problem*. It is easy to confirm certain sequences are not degree sequences using graph properties (such as the Handshaking Lemma or isolated vertices), but we would like to have a procedure that will always answer this question. The theorem below, attributed to the Czech mathematician Václav Havel and the Iranian-American mathematician Seifollah Louis Hakimi, provides an iterative approach to answering the graph realization problem.

Theorem 1.27 (Havel-Hakimi Theorem) An increasing sequence S : $s_1, s_2, \ldots, s_n$ (for $n \geq 2$) of nonnegative integers is graphical if and only if the sequence

$$S' : s_2 - 1, s_3 - 1, \ldots, s_{s_1} - 1, s_{s_{n+1}}, \ldots, s_n$$

is graphical.

The intention behind this procedure is to peel off the first vertex, removing one from the degrees of its neighbors. We continue this process until we arrive at a sequence that either can easily be turned into a graph or contains negative integers and therefore cannot be a degree sequence. Since the theorem is a biconditional statement, we know that any sequence that appeared throughout this iterative process has the same graphical answer as the last sequence considered. Note, we may need to reorder the vertices after the peeling procedure to ensure the sequence remains decreasing.

Example 1.19 Determine if either of $S : 4, 4, 2, 1, 1, 0$ or $T : 4, 3, 3, 2, 1, 1$ is graphical.

Solution: Applying the Havel-Hakimi Theorem to S, we note the first term of the sequence is 4, and so we eliminate the first term and subtract 1 from

the next 4 terms and leaving the last one alone. This gives $S_1 : 3, 1, 0, 0, 0$.
In this new sequence the first term is 3, so we eliminate it and subtract
1 from the next 3 terms of S_1, producing $S_2 : 0, -1, -1, 0$. This last
sequence cannot be graphical since degrees cannot be negative. Thus S_1
and S cannot be graphical either.

Using the same procedure on T, after the first iteration we get
$2, 2, 1, 0, 1$. We reorder this to make it decreasing as $T_1 : 2, 2, 1, 1, 0$. After the second iteration we have $1, 0, 1, 0$, which is again reordered to
$T_2 : 1, 1, 0, 0$. At this point we can stop since it is not too difficult to see
this sequence is graphical: the two vertices of degree 1 are adjacent and
the other two vertices are isolated (see below). This means that T_2, T_1,
and the original sequence T are all graphical.

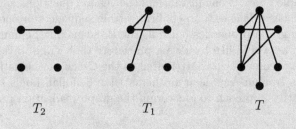

T_2 T_1 T

While it may be tempting to work this procedure backwards to determine
what the original graph looks like, we would be giving this theorem too much
weight. Consider the two graphs, G_5 and G_6 shown below.

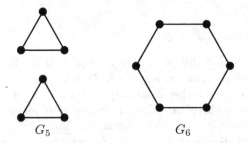

G_5 G_6

Graph G_5 contains two pieces, each of which is the graph K_3 (these pieces
are called components, which we formally define in the next chapter). It's
degree sequence is $2, 2, 2, 2, 2, 2$. Note G_6 has the same degree sequence but
is just one big circle (which we will define as a cycle in the next chapter).
It should be fairly obvious that these two graphs are not isomorphic. Thus
degree sequences cannot uniquely determine a graph.

1.7 Tournaments Revisited

Now that we have a bit more familiarity with graphs in general, we return once again to tournaments. Note that these will reappear in Sections 2.2 and 6.3 where we look at two other operations on tournaments. For now, we will investigate further the degrees of a tournament and their matrix representation.

1.7.1 Score Sequence

Let us return to the Roanoke Soccer League. The table below summarizes the wins for each team.

Team	Teams they Beat
Aardvarks	Bears, Cougars, Ducks, Eagles
Bears	Cougars, Ducks
Cougars	Ducks, Eagles
Ducks	
Eagles	Bears, Ducks

Although we gave the complete list of which teams won their games, we could have left out any one of the rows and still obtain the same graph due to the fact that each team must play every other team and exactly one team in a pair can win. Moreover, the number of wins for a team can be seen in the digraph by simply computing the out-degree of a vertex. Recall that the out-degree of a vertex, denoted $\deg^+(x)$, is the number of arcs pointing out of x. Thus in the example above, $\deg^+(a) = 4$ and $\deg^+(d) = 0$.

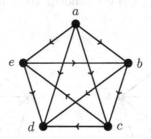

Suppose we are less concerned with the specific outcomes of a round-robin tournament but rather with the relationships from all possible outcomes. In this case, we do not care if a specific team won four games but how many ways there are for any team to win four games. For example, if outcomes above stayed the same except the Eagles beat the Cougars and the Ducks beat the

Cougars and the Eagles, then a simple relabeling of the graph (shown below) gives the same structure as above.

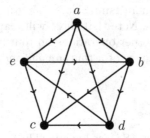

These two graphs are in fact isomorphic. When comparing the wins for all teams, we see that both graphs had one team that lost all their games, three teams that won twice, and one team that won four games; in short we had wins of $0, 2, 2, 2, 4$. This listing of the wins for a tournament is called a *score sequence*.

Definition 1.28 The *score sequence* of a tournament is a listing of the out-degrees of the vertices. It is customary to write these in increasing order.

The score sequence is analogous of the degree sequence of a graph and can provide a lot of information about a tournament, but does not give all possible information. For example, consider again the Roanoke Soccer League teams. Suppose in one year we had wins as shown on the left and the next year only one game was different, where the Ducks beat the Eagles as opposed to the original outcome of the Eagles beating the Ducks. Both scenarios have the same score sequence (1,2,2,2,3) even though the graph itself has changed by only one flip of an arc (see below).

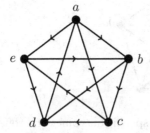 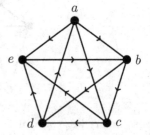

However, these are quite different outcomes in terms of how you might view a ranking of the teams. In the original scenario (shown above on the left) the three 2-win teams each beat one of the other 2-win teams and the

1-win team. These three teams are in essence interchangeable. However, in the new scenario (shown in the graph above on the right), the three 2-win teams have very different win structures and the strength of an opposing team might play into final rankings. For example, the Ducks beat the Aardvarks (3-wins) and the Eagles (1-win), whereas the Cougars beat the Ducks (2-wins) and the Eagles (1-win). Does this mean the Ducks are a better team than the Cougars since they beat a 3-win team over a 2-win team? As you can see, ranking teams in a round-robin style tournament is non-trivial!

Returning to the notion of a score sequence, it is quite easy to produce one when the tournament is given to you (simply calculate the out-degrees of the vertices). What is more challenging is producing a tournament given a score sequence. Moreover, how do you know a given sequence could even represent the out-degrees of a tournament?

We begin with some simple necessary conditions for the score sequence of T_n. First note that the maximum out-degree of any vertex is $n - 1$ since it can have at most one arc to each of the other vertices in K_n. Moreover, at most one vertex can have out-degree $n - 1$ since each pair of vertices has an arc between them. Similarly, the minimum out-degree of any vertex is 0 and at most one vertex can have out-degree 0. Finally, recall that the number of edges in a complete graph K_n is $\frac{n(n-1)}{2}$. In a graph (not digraph) the total degree is always twice the number of edges since each edge adds 2 to the degree count, one for each endpoint. However, in a digraph the sum of the out-degrees equals the total number of arcs (see Exercise 1.24). Thus in a tournament T_n the sum of the out-degrees equals the number of arcs, which is $\frac{n(n-1)}{2}$. These properties are summarized below.

Properties of Score Sequences

The score sequence of any tournament T_n must satisfy the following:

- $s_1, s_2, \ldots, s_n$ is a sequence of integers satisfying $0 \leq s_k \leq n - 1$ for all $k = 1, 2, \ldots, n$.

- at most one s_k equals 0

- at most one s_k equals $n - 1$

- $s_1 + s_2 + \cdots + s_n = \frac{n(n-1)}{2}$

Though these properties are necessary, they are not sufficient; that is, a sequence can satisfy all of the properties listed above yet still not represent the score sequence of a tournament. Below we will discuss two conditions that are both necessary and sufficient and use them to determine if a given sequence is a score sequence of a tournament. The first is closely related to the properties listed above and is easier in its application to a given sequence; the second is more complicated and is based the Havel-Hakimi Theorem from Section 1.6, but also provides a method for drawing a tournament with the given score sequence.

Theorem 1.29 An increasing sequence $S : s_1, s_2, \ldots, s_n$ (for $n \geq 2$) of nonnegative integers is a score sequence if and only if

$$s_1 + s_2 + \cdots + s_k \geq \frac{k(k-1)}{2}$$

for each k between 1 and n with equality holding at $k = n$.

One additional benefit of this result is that if at any point the inequalities fail to hold, then we do not need to check the remaining inequalities and simply state the sequence is not a score sequence of a tournament.

Example 1.20 Determine if the sequence $1, 2, 2, 3, 3, 4$ is the score sequence of a tournament.

Solution: This sequence has length 6, so we will check the inequality above for $k = 1, 2, \ldots 5$, with equality for $k = 6$.

k	$s_1 + \cdots + s_k$	$\frac{k(k-1)}{2}$
1	1	0
2	$1 + 2 = 3$	1
3	$1 + 2 + 2 = 5$	3
4	$1 + 2 + 2 + 3 = 8$	6
5	$1 + 2 + 2 + 3 + 3 = 11$	10
6	$1 + 2 + 2 + 3 + 3 + 4 = 15$	15

Since the inequality $s_1 + \cdots + s_k \geq \frac{k(k-1)}{2}$ holds for each row in the table above, we know that $1, 2, 2, 3, 3, 4$ is a score sequence for T_6.

The next result works by modifying a sequence by removing the last item, which corresponds to deleting one vertex of a tournament and examining the resulting smaller tournament. In theory, this process would continue until either a sequence violates the properties listed above or until a single value remains.

Theorem 1.30 An increasing sequence $S : s_1, s_2, \ldots, s_n$ (for $n \geq 2$) of nonnegative integers is a score sequence of a tournament if and only if the sequence $S_1 : s_1, s_2, \ldots, s_{s_n}, s_{s_n+1} - 1, \ldots, s_{n-1} - 1$ is a score sequence.

The theorem above is almost identical for the one in Section 1.6 on degree sequences of a graph (see Theorem 1.27). The new sequence S_1 is created by deleting s_n, rewriting the first s_n terms of S, and then subtracting 1 from any remaining terms. We repeat the process thereby creating shorter sequences. In practice, so long as none of the Score Sequence Properties (see page 33) have been violated, we stop when the sequence reaches length three. There are only two possible tournaments on three vertices, as shown below, and so it is quick to verify if a sequence is a possible score sequence of T_3.

$1, 1, 1$

$0, 1, 2$

Example 1.21 Determine if the sequence $1, 2, 3, 3, 3, 3$ is the score sequence of a tournament.

Solution: Let S be the sequence $1, 2, 3, 3, 3, 3$. Then $n = 6$ and $s_n = s_6 = 3$. Thus we form a new sequence S_1 by removing the last term, rewriting the first $s_n = 3$ terms and then subtracting 1 from each of the remaining terms (s_4 and s_5). This produces the sequence S_1 below. Note that we need this sequence in increasing order to continue, so we rewrite this as S_1' below:

$$S_1 : 1, 2, 3, 2, 2 \qquad S_1' : 1, 2, 2, 2, 3$$

We perform this procedure again on S_1', where now $n = 5$ and $s_5 = 3$. So we get a new sequence S_2 by removing the last term, rewriting the first $s_n = 3$ terms and then subtracting 1 from each of the remaining terms (s_4). This produces the sequence S_2 and its increasing form S_2' below:

$$S_2 : 1, 2, 2, 1 \qquad S_2' : 1, 1, 2, 2$$

Finally, we perform this procedure one last time on S_2', where $n = 4$ and $s_4 = 2$. We get the sequence

$$S_3 : 1, 1, 1$$

which is one of the two score sequences for T_3. Thus by the theorem above we know that S is the score sequence of a tournament on 6 vertices.

Although this second method is more complex, it has the added benefit of providing a blueprint for how to draw a tournament with a given score sequence. We will work backwards, beginning by creating the T_3 tournament found, and using the previous score sequences to determine how new vertices and their arcs are added.

Example 1.22 Using the results from Example 1.21, draw a tournament with score sequence $1, 2, 3, 3, 3, 3$.

Solution: We begin by forming the T_3 tournament with score sequence $1, 1, 1$. To aid in understanding how new vertices are added, we will identify vertices in the digraph with their value in the score sequence at each step. Here we are starting with the final sequence from Example 1.21, namely S_3 which is $1, 1, 1$.

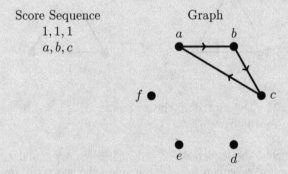

Next we consider the sequence $1, 1, 2, 2$ identified above as S_2'. The addition of d and its arcs is shown below. Note that since the out-degrees of a and b did not change, we know there must be arcs $d \to a$ and $d \to b$. Since the out-degree of c increased by 1, we know to add the arc $c \to d$.

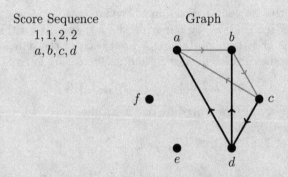

Next we consider the sequence $1, 2, 2, 2, 3$ identified above as S_1'. The addition of e and its arcs is shown below. Note that since the out-degrees of a, c, and d did not change, we know there must be arcs $e \to a, e \to c$, and $e \to d$. Since the out-degree of b increased by 1, we know to add the arc $b \to e$.

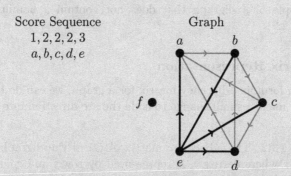

Score Sequence
$1, 2, 2, 2, 3$
a, b, c, d, e

Graph

Finally we consider the original sequence $S : 1, 2, 3, 3, 3, 3$. The addition of f and its arcs is shown below. Note that since the out-degrees of a, b, and e did not change, we know there must be arcs $f \to a, f \to b$, and $f \to e$. Since the out-degree of c and d increased by 1, we know to add the arcs $c \to f$ and $d \to f$.

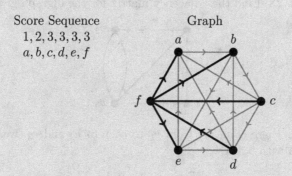

Score Sequence
$1, 2, 3, 3, 3, 3$
a, b, c, d, e, f

Graph

The example above shows that once we know a sequence is in fact a score sequence of a tournament, then we can find a way to produce such a tournament. However, as noted above, more than one tournament (in the sense that they are not isomorphic) can exist with the same score sequence, so this process may not find every possible tournament with a given score sequence. There is one exception though, namely the tournament whose score sequence consists of distinct entries. These are called *transitive tournaments*.

Definition 1.31 A tournament T_n is **transitive** if it does not contain any directed cycles. It has score sequence $0, 1, 2, \ldots, n-1$.

Transitive tournaments are unique, up to vertex relabeling, and provide an example of a digraph that does not contain a hamiltonian cycle (see Section 2.2.2).

1.7.2 Matrix Representation

Just as we can form an adjacency matrix for a graph, we can do the same for a digraph. We make a small change to code the arc direction into the matrix.

Definition 1.32 The adjacency matrix $A(G)$ of the digraph G is the $n \times n$ matrix where vertex v_i is represented by row i and column i and the entry a_{ij} denotes the number of arcs from v_i to v_j.

One major consequence of this change is that the adjacency matrix of a digraph need not be symmetric.

Example 1.23 Find the adjacency matrix for the digraph given below.

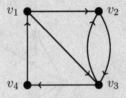

Solution: As there is an arc from v_1 to each of v_2 and v_3, we fill in the first column with $0, 1, 1, 0$.

$$\begin{bmatrix} 0 & 1 & 1 & 0 \\ & & & \\ & & & \\ & & & \end{bmatrix}$$

Next, we see there is only one arc from v_2, the one to v_3.

$$\begin{bmatrix} 0 & 1 & 1 & 0 \\ 0 & 0 & 1 & 0 \\ & & & \\ & & & \end{bmatrix}$$

Now, v_3 has two arcs out of it, one to v_2 and another to v_4.

$$\begin{bmatrix} 0 & 1 & 1 & 0 \\ 0 & 0 & 1 & 0 \\ 0 & 1 & 1 & 0 \end{bmatrix}$$

Finally, v_4 had one arc out to v_1, and our adjacency matrix is complete.

$$\begin{bmatrix} 0 & 1 & 1 & 0 \\ 0 & 0 & 1 & 0 \\ 0 & 1 & 1 & 0 \\ 1 & 0 & 0 & 0 \end{bmatrix}$$

As before, we can draw the digraph based on the matrix given, taking care to consider the direction of the arc.

Example 1.24 Draw the digraph with the adjacency matrix given below.

$$\begin{bmatrix} 0 & 1 & 0 & 0 \\ 0 & 0 & 1 & 0 \\ 0 & 0 & 1 & 0 \\ 1 & 1 & 1 & 0 \end{bmatrix}$$

Solution: The corresponding digraph is given below. Note that there is a loop at v_3, as can be seen by the entry a_{33} in the matrix.

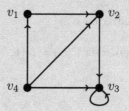

When we turn our attention back to tournaments, we see some special properties occurring. Consider the tournament on 5 vertices and its corresponding adjacency matrix given below.

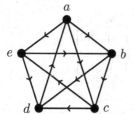

$$\begin{bmatrix} 0 & 1 & 1 & 0 & 1 \\ 0 & 0 & 1 & 1 & 0 \\ 0 & 0 & 0 & 1 & 1 \\ 1 & 0 & 0 & 0 & 0 \\ 0 & 1 & 0 & 1 & 0 \end{bmatrix}$$

First note that the main diagonal will always contain zeros since there are no loops in a tournament. Since the underlying graph of a tournament on n vertices is the complete graph K_n, we know each entry must be a 1 or a 0. These matrices are often called binary matrices or logical matrices in linear algebra. In addition, if we sum along any row we get the out-degree of a vertex and if we sum along any column we get the in-degree, and we know the column sum and row sum must add to $n-1$. Moreover, if $a_{ij} = 1$ then $a_{ji} = 0$.

1.8   Exercises

1.1 Let G be a graph with vertex set $V(G) = \{a, b, c, d, e, f\}$ and edge set $E(G) = \{ab, ae, bc, cc, de, ed\}$.
   (a) Draw G.
   (b) Is G simple?
   (c) List the degrees of every vertex.
   (d) Find all edges incident to b.
   (e) List all the neighbors of a.
   (f) Give the adjacency matrix for G.

1.2 Let G be a graph with vertex set $V(G) = \{a, b, c, d\}$ and edge set $E(G) = \{ab, ad\}$.
   (a) Draw G.
   (b) Is G simple?
   (c) List the degrees of every vertex.
   (d) Give the adjacency matrix for G.
   (e) Draw $\overline{G}$.
   (f) Give the adjacency matrix for $\overline{G}$.

1.3 Let G be a graph with vertex set $V(G) = \{a, b, c, d, e, f\}$ and edge set $E(G) = \{ad, ae, bd, bf, cd, ce, cf\}$.
   (a) Draw G.
   (b) Is G simple?
   (c) Is G bipartite?
   (d) List the degrees of every vertex.
   (e) Give the adjacency matrix for G.

1.4 Draw the graph for each of the adjacency matrices given below.

   (a)
$$\begin{bmatrix} 0 & 2 & 0 & 1 \\ 2 & 0 & 1 & 0 \\ 0 & 1 & 1 & 1 \\ 1 & 0 & 1 & 0 \end{bmatrix}$$

   (b)
$$\begin{bmatrix} 0 & 1 & 2 & 1 \\ 1 & 2 & 1 & 0 \\ 2 & 1 & 0 & 0 \\ 1 & 0 & 0 & 0 \end{bmatrix}$$

(c)

$$\begin{bmatrix} 2 & 0 & 0 & 1 & 0 \\ 0 & 0 & 3 & 0 & 1 \\ 0 & 3 & 0 & 1 & 0 \\ 1 & 0 & 1 & 0 & 0 \\ 0 & 1 & 0 & 0 & 0 \end{bmatrix}$$

(d)

$$\begin{bmatrix} 0 & 1 & 0 & 0 & 1 \\ 1 & 0 & 1 & 0 & 0 \\ 0 & 1 & 0 & 1 & 0 \\ 0 & 0 & 1 & 0 & 1 \\ 1 & 0 & 0 & 1 & 0 \end{bmatrix}$$

1.5 Draw the digraph for each of the adjacency matrices given below.

(a)

$$\begin{bmatrix} 0 & 1 & 1 & 1 \\ 0 & 0 & 0 & 0 \\ 0 & 0 & 1 & 0 \\ 0 & 0 & 1 & 0 \end{bmatrix}$$

(b)

$$\begin{bmatrix} 0 & 1 & 0 & 0 \\ 0 & 0 & 1 & 0 \\ 0 & 0 & 0 & 1 \\ 1 & 0 & 0 & 0 \end{bmatrix}$$

(c)

$$\begin{bmatrix} 0 & 1 & 1 & 0 & 1 \\ 1 & 0 & 1 & 0 & 0 \\ 0 & 0 & 0 & 0 & 0 \\ 1 & 0 & 2 & 0 & 0 \\ 0 & 0 & 0 & 1 & 0 \end{bmatrix}$$

(d)

$$\begin{bmatrix} 1 & 1 & 0 & 1 & 0 \\ 1 & 0 & 1 & 0 & 0 \\ 0 & 0 & 0 & 0 & 0 \\ 0 & 0 & 1 & 0 & 1 \\ 0 & 0 & 0 & 0 & 0 \end{bmatrix}$$

1.6 Find the adjacency matrix for each of the digraphs or tournaments given below.

(a)

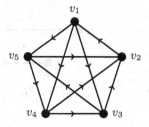

(b)

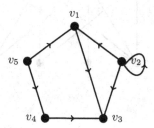

(c)

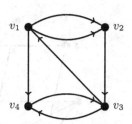

(d)

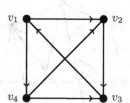

1.7 For each of the problems below, determine if the given pair of graphs are isomorphic. For those that are isomorphic, explicitly give the vertex correspondence and check that edge relationships are maintained. Otherwise, provide reasoning for why the pair of graphs are not isomorphic.

(a)

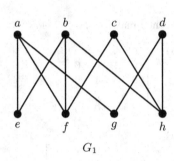
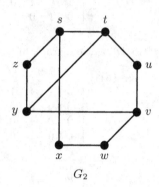

(b)

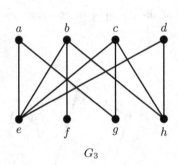
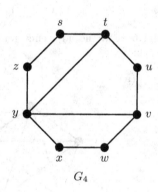

(c)

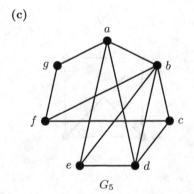
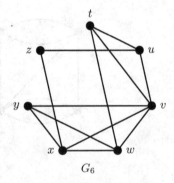

(d)

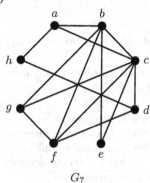

G_7

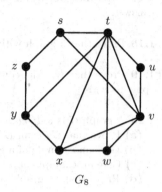

G_8

1.8 There are four possible tournaments on 4 vertices. Draw them and give their score sequences.

1.9 Determine if each of the following is a score sequence of a tournament.
  (a) $1, 1, 2, 3, 3$
  (b) $1, 1, 2, 2, 4$
  (c) $0, 2, 2, 2, 3$
  (d) $1, 2, 2, 2, 3$
  (e) $0, 2, 2, 3, 3, 5$
  (f) $1, 1, 2, 3, 4, 4$
  (g) $1, 2, 2, 2, 2, 5$
  (h) $1, 1, 1, 3, 4, 5$

1.10 For each of those from Exercise 1.9 that is a score sequence, draw the tournament that it represents.

1.11 For each of those from Exercise 1.9 that is a score sequence, determine if the tournament is strong.

1.12 What matrix operations could you perform to the adjacency matrix and still have the same underlying graph?

1.13 Recall that a graph is simple if it has no multi-edges or loops. What condition on the adjacency matrix can you make to ensure the graph is simple? Write your answer as a theorem and provide a proof.

1.14 Find the complements of $K_{2,2}$, $K_{2,3}$, $K_{2,2,3}$, and $K_{2,3,4}$. What type of structure is $\overline{K}_{m,n}$? $\overline{K}_{n_1,n_2,...,n_m}$? Explain your answer.

1.15 Use the Handshaking Lemma to prove Corollary 1.22: The sum of the degrees of any graph is always even.

1.16 Prove a k-regular graph on n vertices has $\frac{nk}{2}$ edges.

1.17 Determine the degree sum for a complete bipartite graph $K_{m,n}$. Prove your answer.

1.18 Prove that a graph with 15 vertices and 57 vertices cannot be bipartite.

1.19 Prove that K_n, the complete graph on n vertices has n subgraphs isomorphic to K_{n-1}.

1.20 A graph G is *self-complementary* if G is isomorphic to $\overline{G}$.
  (a) Prove that G_2 from Example 1.10 on page 14 is self-complementary.
  (b) Give an example of a graph on 4 vertices that is self-complementary.
  (c) Give an example of a graph on 5 vertices that is self-complementary.
  (d) Prove that no graph on 6 vertices is self-complementary.

1.21 Prove that if G and H are isomorphic then $\overline{G}$ and $\overline{H}$ are also isomorphic.

1.22 Recall that $G \vee H$ denotes the join of G and H.
  (a) Draw $K_3 \vee K_2$ and $K_3 \vee K_3$.
  (b) Prove that $K_3 \vee K_2 \cong K_5$ and $K_3 \vee K_3 \cong K_6$.
  (c) Prove the for any $m, n \geq 1$, $K_m \vee K_n \cong K_{m+n}$.

1.23 Perhaps the most famous round robin tournament is the FIFA World Cup Final Tournament. In the group stage of the tournament, 32 teams are split into eight groups of 4 teams so that each team plays three matches against the other teams in its group.
  (a) Assuming there are no ties, model all possible outcomes in a 4 team group with an unlabeled graph.
  (b) Give the degree sequence of each scenario from part (a).
  (c) Only the top two teams in each group advance to the elimination stage of FIFA World Cup Final Tournament, where 3 points are awarded for a win, one point for a tie, and no points for a loss. Again, assuming no ties, find determine who advances from each scenario in part (a), or explain why additional criteria are needed to determine who advances. Would allowing ties make this easier or harder to determine?

1.24 Similar to the Handshaking Lemma (Theorem 1.21), the sum of the degrees in a digraph has a relationship to the number of arcs. Prove the theorem below.

Theorem 1.33 Let $G = (V, A)$ be a digraph and $|A|$ denote the number of arcs in G. Then both the sum of the in-degrees of the vertices and the sum of the out-degrees equals the number of arcs; that is, if $V = \{v_1, v_2, \ldots, v_n\}$, then

$$\deg^-(v_1) + \cdots + \deg^-(v_n) = |A|$$
$$= \deg^+(v_1) + \cdots + \deg^+(v_n)$$

2

Graph Routes

This chapter will explore three main topics in Graph Theory related to routing problems. A route within a graph is simply a path, trail, circuit, or cycle, and the sections will explore how various requirements change the type of route needed. Each of these topics have a rich historical component, which will be introduced at the start of the section.

2.1 Eulerian Circuits

Most mathematical subfields arose through a natural observation of the physical world (such as geometry) or through a progression of ideas that are eventually compiled into a specific category (such as calculus). Few areas of mathematics have an origin story (akin to those of comic superheros). Probability owes its origins to letters between Pascal and Fermat regarding splitting the wager of a dice game ended before its completion (see [20]). The origin story for statistics begins with Ronald Fisher's experiment to test Muriel Bristol's claim that she could taste the difference as to whether milk was added to a tea cup before or after the tea (see [72]). Similarly, Graph Theory can trace its origins to a singular problem and the resulting publication of the answer. What is unique in this situation, however, is that the techniques used were novel and the applications of a graph were not widely understood until the advent of the computer.

2.1.1 Königsberg Bridge Problem

You arrive in a new city and hear of an intriguing puzzle captivating the population: can you leave your home, travel across each of the bridges in the city exactly once and then return home? Upon one look at the map, you claim "Of course it can't be done!" You describe the requirements needed for such a walk to take place and note this city fails those requirements. Easy enough!

The puzzle above is sometimes called the birth of graph theory. In 1736, Leonhard Euler, one of the greatest mathematicians of all time, published a short paper on the bridges of Königsberg, a city in Eastern Europe (see the following map).

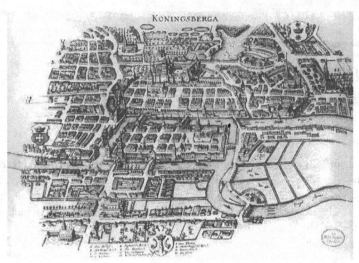

Map of Königsberg

Euler translated the problem into one of the "geometry of location" (*geometris situs*) and determined that only certain configurations would allow a solution to be possible. His publication set in motion an entirely new branch of mathematics, one that has profound impact in modern mathematics, computer science, management science, counterterrorism, ... and the list continues. Without worrying about the technicalities, see if you can find a solution to the Königsberg Bridge Problem! To aid in your analysis, note that Königsberg contains seven brides and four distinct landmasses.

2.1.2 Touring a Graph

Part of Euler's brilliance is how he figured out how to model a real world question with a mathematical object that we now call a graph. A part of the modeling process is determining what the answer we are searching for looks like in graph form. What type of structure or operation are we looking for? In our original search for a solution to the Königsberg Bridge Problem, we discuss traveling through the city. What would traveling through a graph mean? What types of restrictions might we place on such travel? Below are additional definitions needed to answer these questions, after which we will describe how to use a graph model to solve the Königsberg Bridge Problem.

Definition 2.1 Let G be a graph.

- A **walk** is a sequence of vertices so that there is an edge between consecutive vertices. A walk can repeat vertices and edges.

- A *trail* is a walk with no repeated edges. A trail can repeat vertices but not edges.

- A *path* is a trail with no repeated vertex (or edges). A path on n vertices is denoted P_n.

- A *closed walk* is a walk that starts and ends at the same vertex.

- A *circuit* is a closed trail; that is, a trail that starts and ends at the same vertex with no repeated edges though vertices may be repeated.

- A *cycle* is a closed path; that is, a path that starts and ends at the same vertex. Thus cycles cannot repeat edges or vertices. Note: we do not consider the starting and ending vertex as being repeated since each vertex is entered and exited exactly once. A cycle on n vertices is denoted C_n.

The *length* of any of these tours is defined in terms of the number of edges. For example, P_n has length $n - 1$ and C_n has length n.

Technically, since a path is a more restrictive version of a trail and a trail is a more restrictive form of a walk, any path can also be viewed as a trail and as a walk. However, a walk might not be a trail or a path (for example, if it repeats vertices or edges). Similarly, a cycle is a circuit and a closed walk. Unless otherwise noted, when we use any of the terms from Definition 2.1 we are referring to the most restrictive case possible; for example, if we ask for a walk in a graph then we want a walk that is not also a trail or a path.

In practice, it is often necessary to label the edges of a tour of a graph in the sequential order in which they are traveled. This is especially important when the graph is not simple.

Example 2.1 Given the graph below, find a trail (that is not a path) from a to c, a path from a to c, a circuit (that is not a cycle) starting at b, and a cycle starting at b.

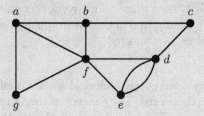

Solution: The order in which the edges will be traveled are noted in the following routes.

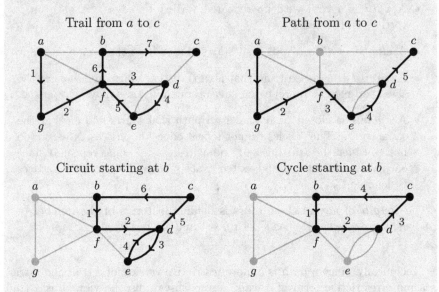

Trail from a to c Path from a to c

Circuit starting at b Cycle starting at b

Note that the trail from a to c is not a path since the vertex f is repeated and the circuit starting at b is not a cycle since the vertex d is repeated. Moreover, the examples given above are not the only solutions but rather an option among many possible solutions.

Earlier we defined what it meant for two vertices to be *adjacent*, namely x and y are adjacent if xy is an edge in the graph. Notice that in common language we may have wanted to say that x and y are connected since there is a line that connects these dots on the page. However, in graph theory the term *connected* refers to a different, though related, concept.

Definition 2.2 Let G be a graph. Two vertices x and y are **connected** if there exists a path from x to y in G. The graph G is **connected** if every pair of distinct vertices is connected.

This definition may seem overly technical when visually it is often easy to determine if a graph is connected. The concept of connectedness is surprisingly important in applications. The following example illustrates the importance of connectedness for the Königsberg Bridge Problem.

Example 2.2 Island City has two islands, a peninsula, and left and right banks, as shown on the left below. Modeling the relationship between the landmasses and bridges of Island City gives us the graph below on the right.

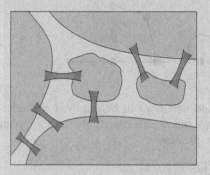 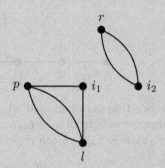

If the same puzzle were posed to the citizens of Island City as the one for Königsberg, it would be impossible to travel every bridge and then return home. In particular, there is no way to even travel from one side of the river to the other! Thus, if we even have a hope of finding a solution, the resulting graph must, at a *minimum*, be connected.

The Königsberg Bridge Problem is not just looking for a circuit, but rather a special type that hits every edge. In honor of Leonhard Euler, these special circuits are named for him. A similar term is used when allowing differing starting and ending location of the tour. The next section will discuss when a graph has an eulerian circuit, but for now we continue with some results related to touring a graph.

Theorem 2.3 Every $x - y$ walk contains an $x - y$ path.

Proof: Argue by strong induction on the length l of an $x - y$ walk W. For the base case, if $l = 0$ then the walk does not have any edges and so contains a singe vertex, that is $x = y$ and so W is a path of length 0.

Next assume $l \geq 1$ and suppose the claim holds for all walks of length $k < l$. If W has no repeated vertex, then it cannot have any repeated edges and so W is an $x - y$ path. Otherwise W has a repeated vertex, call it u. We can remove all edges and vertices between two appearances of u in the walk, and leave only one copy of u (see an example below). This will produce a shorter $x - y$ walk W' which is contained in W. Thus, by the induction hypothesis we know W' contains an $x - y$ path P, which is also contained in W. Thus W contains an $x - y$ path P.

Consider a walk as shown below. Using the technique from the proof above, we begin by identifying a repeated vertex, namely v_4.

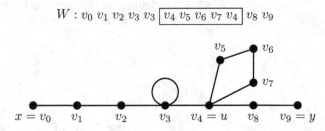

$$W : v_0 \ v_1 \ v_2 \ v_3 \ v_3 \ \boxed{v_4 \ v_5 \ v_6 \ v_7 \ v_4} \ v_8 \ v_9$$

Notice that these selected edges and vertices can be removed without destroying the walk from v_0 to v_9, as shown below. Also, we find another repeated vertex, namely v_3, and can repeat the process.

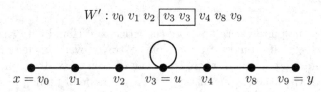

$$W' : v_0 \ v_1 \ v_2 \ \boxed{v_3 \ v_3} \ v_4 \ v_8 \ v_9$$

The final graph below does not contain any repeated vertices and so can be considered a path from v_0 to v_9, all of whose vertices and edges were contained in the original walk W.

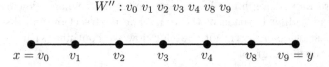

$$W'' : v_0 \ v_1 \ v_2 \ v_3 \ v_4 \ v_8 \ v_9$$

The theorem above shows that the underlying nature of graphs lend themselves to induction proofs. This is in part because we often want to remove a portion of the graph, whether it is a vertex, edge, or something more complex, and consider how that effects the graph.

The next theorem provides a condition guaranteeing a cycle within a graph. The proof does not utilize induction, but rather relies on a path that is as long as possible. We will use two related terms throughout this book, (for paths, matchings, colorings, etc.) that relate to the size of an object.

Definition 2.4 An object X is *maximum* if it is the largest among all objects under consideration; that is, $|X| \geq |A|$ for all $A \in \mathcal{U}$.

An object X is *maximal* if it cannot be made larger.

In certain scenarios, the adjectives maximum and maximal may be applied to the same object; however, this need not be true. Consider the graphs shown below. The path $b\ c\ d\ e\ f$ (highlighted on the left) is maximal because we cannot add any additional vertices and keep it a path. However, it is not maximum since a longer path can be found, namely $h\ g\ c\ d\ e\ f$ shown on the right. Note that this second path is a maximum path within the graph shown.

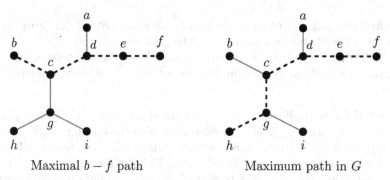

Maximal $b - f$ path   Maximum path in G

Theorem 2.5 If every vertex of a graph has degree at least 2 then G contains a cycle.

Proof: Let P be a maximal path in G and let x be an endpoint of P. Since P is maximal, it cannot be extended and so every neighbor of x must already be a vertex of P. Since x has degree at least 2, we know there must be another neighbor y of x in $V(P)$ via an edge not a part of the path.

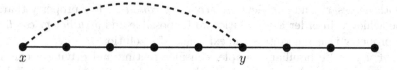

Then the edge xy completes a cycle with the portion of P from x to y.

Paths and cycles serve as the building blocks to many structures we wish to find within a larger graph. As seen above, the presence of a path between any two vertices indicates the graph is connected. What about circuits or cycles? Can a cycle contain every vertex? every edge? The remainder of this section will focus on the edges (through eulerian circuits) whereas the next section will explore vertices (through hamiltonian cycles).

2.1.3 Eulerian Graphs

Looking back at the Königsberg Bridge Problem, we now have the necessary pieces to describe the question in graph theoretic terminology, namely an exhaustive circuit that includes all vertices and edges of the graph. In honor of Euler's solution, these special types of circuits bear his name.

> **Definition 2.6** Let G be a graph. An *eulerian circuit* (or *trail*) is a circuit (or trail) that contains every edge and every vertex of G.
>
> If G contains an eulerian circuit it is called *eulerian* and if G contains an eulerian trail but not an eulerian circuit it is called *semi-eulerian*.

Part of Euler's brilliance was not only his ability to quickly solve a puzzle, such as the Königsberg Bridge Problem, but also the foresight to expand on that puzzle. What *makes* a graph eulerian? Under what conditions will a city have the proper tour? In his original paper, Euler laid out the conditions for such a solution, though as was typical of the time, he only proved a portion of the statement (see [6] or [33]).

> **Theorem 2.7** A graph G is eulerian if and only if
>
> (i) G is connected and
>
> (ii) every vertex has even degree.

The theorem above is of a special type in mathematics. It is written as an *"if and only if"* statement, which indicates that the conditions laid out are both necessary and sufficient. A *necessary condition* is a property that must be achieved in order for a solution to be possible and a *sufficient condition* is a property that guarantees the existence of a solution.

For a more familiar example, consider renting and driving a car. If you want to rent a car, a necessary condition would be having a driver's license; but this condition may not be sufficient since some companies will only rent a car to a person of at least 25 years of age. In contrast, having a driver's license is sufficient to be able to drive a car, but is not necessary since you can drive a car with a learner's permit as long as a guardian is present.

Mathematicians often search for a property (or collection of properties) that is both necessary and sufficient (such as a number is even if and only if it is divisible by 2). The theorem above gives both necessary and sufficient conditions for a graph to be eulerian. It should be clear why connectedness must be achieved if every vertex is to be reached in a single tour. Can you explain the degree condition? When traveling through a graph, we need to pair each entry edge with an exit edge. If a vertex is odd, then there is no pairing available and we would eventually get stuck at that vertex. A formal proof for the eulerian condition is given below.

> **Proof:** Suppose G is eulerian. Then there exists a circuit C that contains every vertex and every edge of G. Since two edges can be in the same circuit only if they are in the same component of G, we know G must be connected. Now, for any vertex v of G, every entry edge to v from the circuit must be paired with an exit edge at v, and so v must have even degree. Thus G is connected and all vertices have even degree.
>
> Conversely, suppose G is connected and every vertex has even degree. We will use induction on m, the number of edges in G to prove that G has an eulerian circuit. For a base case, consider if $m = 0$. Then since G is connected we know that it contains exactly one vertex, which is a trivial circuit.
>
> Now suppose G has $m > 0$ edges and that for all k, with $m > k \geq 0$, that any connected graph G' with k edges and all vertices of even degree has an eulerian circuit. Since G has at least one edge and G is connected with all vertices of even degree, we know all vertices must have degree at least 2. By Theorem 2.5, we know G contains a cycle C. Let G' be the graph obtained by deleting all the edges of C.

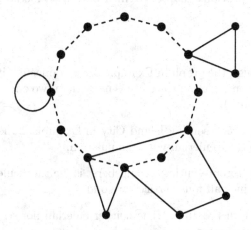

> Since C has 0 or 2 edges at each vertex of G, we know every vertex of G' is even. Thus every component of G' satisfies the induction hypothesis and has fewer than m edges, and therefore must each contain an eulerian circuit. To combine these into an eulerian circuit for G, we traverse C but when a component of G' is encountered for the first time, we detour along its eulerian circuit. This detour will start and end at the same vertex, and so we can continue traversing C until all edges have been traveled, thus completing an eulerian circuit of G.

The proof that the conditions above were indeed both necessary and sufficient was not published until 1873 (almost 140 years after Euler's initial publication!). The work was completed by German mathematician Carl Hierholzer, who unfortunately died too young to see his work in print. His contribution was recognized through the naming of one procedure for finding eulerian circuits that is discussed in the next section.

A similar result exists for semi-eulerian graphs. Since the starting and ending locations are different, then exactly two vertices must be odd since the first edge out of the starting vertex does not need to be paired with a return edge and the last edge to the ending vertex does not need to be paired with an exit edge. The proof appears in Exercise 2.11.

Corollary 2.8 A graph G is semi-eulerian if and only if

(i) G is connected and

(ii) exactly two vertices have odd degree.

Example 2.3 Consider the graphs appearing in the examples from this and the previous chapter. Which ones are eulerian? semi-eulerian? neither?

Solution:

- Even though the graph in Example 2.1 is connected, it is neither eulerian nor semi-eulerian since it has more than two odd vertices (namely, $a, b, e,$ and f).

- The graph representing Island City in Example 2.2 is not connected, so it is neither eulerian nor semi-eulerian.

- The graph representing Königsberg is neither eulerian nor semi-eulerian since all four vertices are odd.

- The graph in Example 1.1 is neither eulerian nor semi-eulerian since it is not connected.

- The graph in Example 1.7 is eulerian since it is connected and all the vertices have degree 4.

- The graph in Example 1.11 is semi-eulerian since it is connected and exactly two vertices are odd (namely, 372 and 271).

Note that the solution discussed in the third bullet above definitively answers the Königsberg Bridge Problem—there is no way to leave your home, travel across every bridge in the city exactly once, and return home!

In addition to the solution to the Königsberg Bridge Problem, Euler also determined some basic properties of graphs in his seminal paper, one of which (the Handshaking Lemma) we discussed in the previous chapter (see Theorem 1.21).

2.1.4 Algorithms

As we now know when a graph will be eulerian or semi-eulerian, the next obvious question is how do we find one. There are numerous methods for finding an eulerian circuit (or trail), though we will focus on only two of these. Each of these algorithms will be described in terms of the input, steps to perform, and output, so it is clear how to apply the algorithm in various scenarios. See Appendix E for the pseudocode for various algorithms appearing in this book. For more a more technical discussion of the algorithms, the reader is encouraged to explore [8] or [52].

The first method for finding an eulerian circuit that we discuss is Fleury's Algorithm. Although Fleury's solution was not the first in print, it is one of the easiest to walk through (no pun intended) [35]. As with all future algorithms presented in this book, an example will immediately follow the description of the algorithm and further examples are available in the Exercises. Note that Fleury's Algorithm will produce either an eulerian circuit or an eulerian trail depending on which solution is possible.

Fleury's Algorithm

Input: Connected graph G where zero or two vertices are odd.

Steps:

1. Choose a starting vertex, call it v. If G has no odd vertices, then any vertex can be the starting point. If G has exactly two odd vertices, then v must be one of the odd vertices.

2. Choose an edge incident to v that is unlabeled and label it with the number in which it was chosen, ensuring that the graph consisting of unlabeled edges remains connected.

3. Travel along the edge to its other endpoint.

4. Repeat Steps (2) and (3) until all edges have been labeled.

Output: Labeled eulerian circuit or trail.

The intention behind Fleury's Algorithm is that you are prevented from getting stuck at a vertex with no edges left to travel. In practice, it may be helpful to use two copies of the graph—one to keep track of the route and the other where labeled edges are removed. This second copy makes it easier to see which edges are unavailable to be chosen. In the example below, the vertex under consideration during a step of the algorithm will be highlighted and edges will be labeled in the order in which they are chosen.

Example 2.4 Input: A connected graph (shown below) where every vertex has even degree. We are looking for an eulerian circuit.

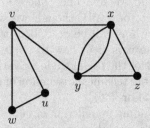

Step 1: Since no starting vertex is explicitly stated, we choose vertex v to be the starting vertex.

Step 2: We can choose any edge incident to v. Here we chose vx. The labeled graph is on the left and the unlabeled portions are shown on the right with edges removed that have already been chosen.

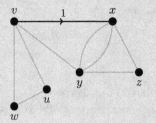
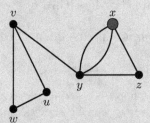

Step 3: Looking at the graph to the right, we can choose any edge out of x. Here we chose xy. The labeled and unlabeled graphs have been updated below.

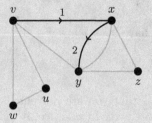
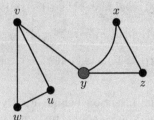

Step 4: At this point we cannot choose yv, as its removal would disconnect the unlabeled graph shown on the right in Step 3. However, yx and yz are both valid choices. Here we chose yx.

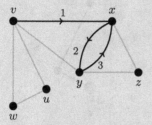 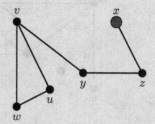

Step 5: There is only one available edge xz.

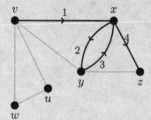 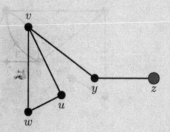

Step 6: There is only one available edge zy.

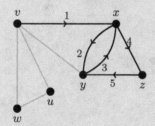 

Step 7: There is only one available edge yv.

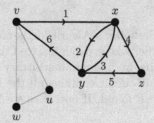 

Step 8: Both vw and vu are valid choices for the next edge. Here we chose vw.

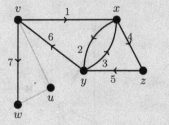 

Step 9: There is only one available edge wu.

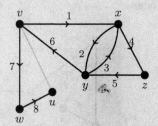 

Step 10: There is only one available edge uv.

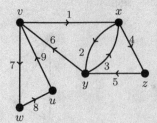

Output: The graph above on the left has an eulerian circuit labeled, starting and ending at vertex v.

The second algorithm we study, Hierholzer's Algorithm, is named for the German mathematician mentioned earlier whose paper inspired the procedure to follow. This efficient algorithm begins by finding an arbitrary circuit originating from the starting vertex. If this circuit contains all the edges of the graph, then an eulerian circuit has been found. If not, then we join another circuit to the existing one.

Hierholzer's Algorithm

Input: Connected graph G where all vertices are even.

Steps:

1. Choose a starting vertex, call it v. Find a circuit C originating at v.

2. If any vertex x on C has edges not appearing in C, find a circuit C' originating at x that uses two of these edges.

3. Combine C and C' into a single circuit C^*.

4. Repeat Steps (2) and (3) until all edges of G are used.

Output: Labeled eulerian circuit.

Note that Hierholzer's Algorithm requires the graph to be eulerian, whereas Fleury's Algorithm allows for the graph to be eulerian or semi-eulerian. In the implementation of Hierholzer's Algorithm shown below, a new circuit will be highlighted in bold with other edges in gray. As with Fleury's Algorithm, the edges will be labeled in the order in which they are traveled.

Example 2.5 Input: A connected graph where every vertex has even degree. We are looking for an eulerian circuit.

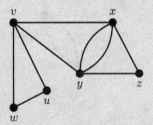

Step 1: Since no starting vertex is explicitly stated, we choose v and find a circuit originating at v. One such option is highlighted below.

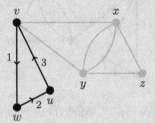

Step 2: As $\deg(v) = 4$ and two edges remain for v (shown in gray in the previous figure), a second circuit starting at v is needed. One option is shown below.

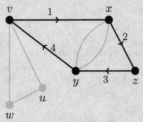

Step 3: Combine the two circuits from Step 1 and Step 2. There are multiple ways to combine two circuits, but it is customary to travel the first circuit created and then travel the second.

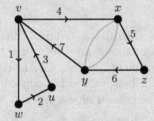

Step 4: As $\deg(x) = 4$ and two edges remain for x (shown in gray above), a circuit starting at x is needed. It is shown below.

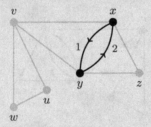

Step 5: Combine the two circuits from Step 3 and Step 4.

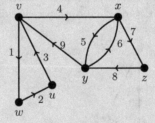

Output: The graph above gives a labeled eulerian circuit originating at v.

There are advantages and disadvantages for these two methods; in particular, both Fleury's and Hierholzer's Algorithms will find an eulerian circuit when one exists, whereas only Fleury's can be used to find an eulerian trail (see Exercise 2.10 for a modification of Hierholzer's that will find an eulerian trail). Try a few more examples and make additional comparisons. In practice, either algorithm is a good choice for finding an eulerian circuit—pick the method that works best for you.

Barring these differences, the two eulerian algorithms are fairly *efficient*. Algorithm efficiency will be discussed in Section 2.2 and Appendix D, but for now think of algorithm efficiency as measuring the time needed for an algorithm to find a solution. Efficiency takes into account the number of calculations needed to run the algorithm as the size of the problem grows. An algorithm is considered efficient if the run time grows at roughly the same speed as the size of the graph. For example, applying Fleury's Algorithm to a graph with 25 vertices is not that much more difficult than a graph on 10 vertices. An inefficient algorithm is one in which the run time grows much faster than the size of the graph; we will see one such algorithm in our discussion of the Traveling Salesman in Section 2.2.

2.1.5 Applications

The Königsberg Bridge Problem can be viewed as the first application of graph theory, and in particular eulerian circuits. The remainder of this section will be devoted to a variety of problems whose solution can be found using an eulerian circuit. The examples chosen are by no means an exhaustive list, but rather a sampling of the wide-ranging applications of graph theory.

Snowplow Routes

Having grown up in the northeast, when a snowplow came down our road could have a major effect on morning commutes. If a city wants to optimize their snowplows, they would want to plan the routes so no plow would need to travel over a road more than once (assuming it takes only one pass to clear the street). This is a little different than finding a singular eulerian circuit in the graph, since we do not need one snowplow to cover all the streets, but rather we want to split a city into sectors that each snowplow would then travel using an eulerian circuit or trail.

For simplicity, assume the town of Crystal Spring has already determined the sectors for their plows; the following map shows one such sector. To turn this into a problem of finding an eulerian circuit, we first model the town as a graph where the edges represent street blocks and the vertices the street intersections.

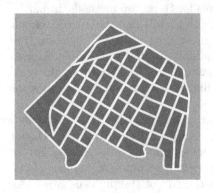

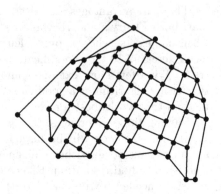

Crystal Spring Map   Crystal Spring Graph G_1

Notice that this graph is not eulerian since there are odd vertices, in fact there are quite a lot of them! At this point, we must adjust the graph so that a snowplow route can still be found. This is called an *eulerization* or *semi-eulerization*.

Definition 2.9 Given a connected graph $G = (V, E)$, an ***eulerization*** of G is the graph $G' = (V, E')$ so that

(i) G' is obtained by duplicating edges of G, and

(ii) every vertex of G' is even.

A ***semi-eulerization*** of G results in a graph G' so that

(i) G' is obtained by duplicating edges of G, and

(ii) exactly two vertices of G' are odd.

Our goal with an eulerization or semi-eulerization is to determine which edges should be traveled twice in order to find an optimal exhaustive route, that is a route that visits each edge at least once. The difficulty in this problem is in determining which edges should be duplicated. There is an additional issue we need to consider when choosing which edges to duplicate: are the edges equivalent or would some cost more to duplicate? Note cost could represent mileage, time, or dollars spent. For this example, we restrict our discussion to the scenario when the edges have equal cost. The next section discusses a variable cost example.

To begin, we must first identify which vertices are odd and determine if we want an eulerization or semi-eulerization. For the snow plow example we will assume we want an eulerization (though it is reasonable to look for a semi-eulerization so long as the starting and ending vertices are on the edge of the city sector, see Exercise 2.12). The graph below displays the odd vertices in bold.

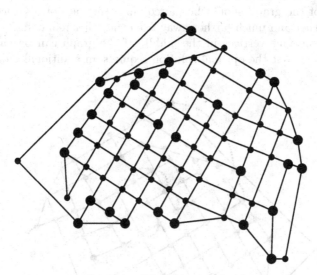

G_1 with odd vertices in bold

In the eulerization process, we are essentially pairing up the odd vertices and then duplicating the edges along a path between them. It should come as no surprise that we would like to pair up adjacent vertices as much as possible, since the path between them would consist of only one edge. The complexity of this problem arises when odd vertices are not adjacent and we must choose the best way to pair these vertices.

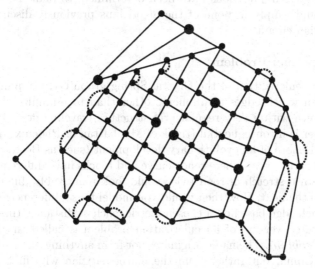

G_1 with edge duplications between adjacent odd vertices

In the previous graph, most of the odd vertices can be split into adjacent pairs, except the ones shown in bold. It should be clear that the two vertices at the top of the graph should be paired since they are of distance 2 away from each other and much farther away from the other remaining 4 vertices.

Finally, the four vertices in the middle of the graph can be paired in a variety of ways, but the optimal solution requires an additional 7 duplicated edges.

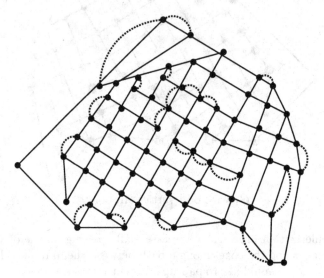

Optimal eulerization of G_1

Once the edge duplications have been determined, to plan the snowplow route we would simply use one of the algorithms previously discussed that find an eulerian circuit.

Chinese Postman Problem

In finding the eulerization of the Crystal Spring graph G_1, we didn't distinguish between which edges to duplicate other than to minimize the overall total. In a town with a very regular grid structure traveling down one block versus another is inconsequential (think of Manhattan or Phoenix). However, for cities with more of an evolutionary development (such as Boston or Providence) or in rural towns where roads curve and blocks have different lengths, traveling down a stretch of road twice could look remarkably different from one choice to the next. How then would you model these differences? We add weights to each edge based on a chosen metric, such as distance, time or cost.

The weighted version of an eulerization problem is called the ***Chinese Postman Problem***. The name originates not from anything particular about postmen in China, but rather from the mathematician who first proposed the problem—the Chinese mathematician Guan Meigu[42]. This problem first appeared in 1960, more than two centuries after Euler's original paper! The full solution was published about a decade later, where the main idea is that a Postman delivering mail in a rural neighborhood should repeat the shortest stretches of road (provided any duplications are necessary). We will discuss the process for a small example, since we can usually find the best duplications by inspection. A more complete solution will be given in Section 5.2.2.

Example 2.6 The citizens of the small island town of Sunset Island has hired a night patrol during the busy summer tourist season. Even though the patrolman on Sunset Island enjoys his evening strolls, he would like to complete a circuit through the town in as little time as possible. The weights on the graph below represent the average time it takes the patrolman to travel that stretch of road. Find an optimal eulerization taking into account these weights.

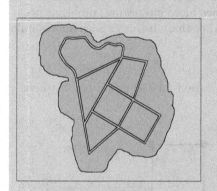 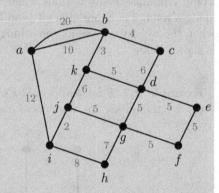

Solution: The optimal eulerization without taking into account the edge weights is shown below on the left. Note that the two duplicated edges total 18. A better eulerization is on the right which duplicates three edges for a total of 15.

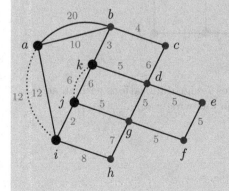 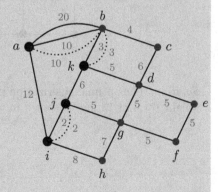

Be careful when duplicating edges in a graph with multi-edges, indicating which edge has been duplicated. The graph on the right duplicated one of the edges between a and b, and including the edge weight clarifies which option was used.

On a general graph, solving the Chinese Postman Problem can be quite challenging. However, most small examples can be solved by inspection since there are relatively few choices for duplicating edges. Would you duplicate 3 edges of weight 1 or one edge of weight 10? The choice should be obvious. In addition, if the weight of an edge represents distance, then we can rely on the real world properties of distance. For example, the shortest path between two points is a straight line and no one side of a triangle is longer than the sum of the other two (this is called the "triangle inequality"). These two properties would eliminate many options when the weight of an edge models distance along a road. The more difficult (and hence more interesting) problems occur when the weight represents something other than distance. Such an example is shown below.

Example 2.7 Find an optimal Eulerization for the graph below where the weights given are in terms of cost.

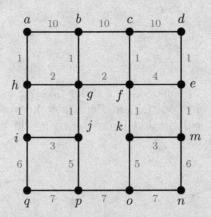

Solution: The optimal eulerization uses 7 edge duplications with an added weight total of 13 and is shown in the following graph.

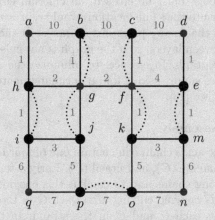

The eulerization for the unweighted graph would only use 5 edge duplications (try it!). On a weighted graph, we attempt to pair vertices along paths using edges of weight 1 as much as possible. We are able to do this for all vertices except o and p, and we duplicate edge op of weight 7. Any attempt to avoid using this edge would still require both o and p to be paired with another vertex (since they are both odd vertices) and would require the use of edges ko and jp, both of which have weight 5. There is no way to pair the remaining odd vertices while maintaining a total increase in weight of 13.

The choice of which edges to duplicate when working with a weighted graph relies in part on shortest paths between two vertices. The difficulty is in choosing which vertices to pair. This will be revisited in Chapter 5 with an algorithm for finding an optimal weighted-eulerization.

3D Printing

There are many different types of 3D printers, with wide ranging abilities to create items of varying size, material, and mode of creation. One mode of creation is Fused Deposition Modeling (FDM), which is a type of additive manufacturing where an object is built layer by layer using a plastic filament that is warmed, extruded onto previous layers and hardened as it cools. This process often makes use of a mesh approximation of the object, where finer meshes creates stronger objects but requiring more material. An important aspect of the printing process is determining the mesh, which will dictate the motion of the printing nozzle. Many programs have been developed to aid in this critical aspect of 3D printing.

Once a mesh has been created, the computer program must determine the path for the printer nozzle to create the object. If we view the lines of a mesh as the edges of a graph then the vertices would represent where the mesh crosses. Since each edge must be crossed, an eulerian circuit or trail (if the nozzle need not return to the same location) would represent the motion of the printer nozzle. Note that many programs do not stop and start the filament extrusion except between layers, so if the graph is not eulerian then we would need to eulerize the graph. In essence, to minimize the amount of filament being extruded we are simply asking for an optimal weighted-eulerization! For further information, see [26].

RNA Fragment Assembly

An RNA (ribonucleic acid) chain can contain one of four nucleotides: adenine (A), cytosine (C), guanine (G), and uracil (U). Early attempts to study RNA were hampered by long chains and to combat this enzymes were used to cut the strands, with a G-enzyme cutting after each G in the strand and a UC-enzyme cutting after each U and C in the strand. In the example shown below, if these enzymes are applied to the same RNA chain then two different collections of fragments will be produced.

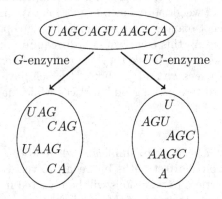

Some fragments obtained in this way will not end in the required nucleotides (such as CA on the left and A on the right) since they were at the end of the original RNA chain. If we were to attempt to reconstruct the original RNA chain, we would only know the ending fragment and not the arrangement of the other fragments. However, any such arrangement must be obtainable from both sets of fragments. For example, the chain $UAAGCAGUAGCA$ would also produce these two sets. In 1969, George Hutchinson devised a way to use this information to determine the possible starting RNA chains based upon an eulerian circuit in a digraph constructed to show the interactions of the fragments [50]. While there have been many advances in the study of RNA and DNA since the 1960's, Hutchinson's use of graph theory in the early study of genetics and has led to further collaborations between mathematics and biology.

2.2 Hamiltonian Cycles

Think back to the city of Königsberg. The previous
section determined when a graph would contain an
eulerian circuit, a special type of circuit that must
travel through every edge and vertex. This concept
arose from a desire to cross every bridge in the city.

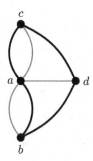

What if we change the requirements ever so
slightly so that we are only concerned with the landmasses? This could model a delivery service with
customers in every sector of the city. In graph theoretic terms, we are looking for a tour through the
graph that hits every vertex exactly once. An example of such a tour on the
graph representing Königsberg is shown above. What type of tour is this? If
we need to start and end at the same location, we are searching for a cycle. If
the starting and ending points can differ, we are searching for a path.

> **Definition 2.10** A cycle in a graph G that contains every vertex of G is
> called a ***hamiltonian cycle***. A path that contains every vertex is called
> a ***hamiltonian path***. A graph that contains a hamiltonian cycle is called
> ***hamiltonian***.

Recall that a cycle or a path can only pass through a vertex once, so
the hamiltonian cycles and paths travel through *every* vertex *exactly once*.
Moreover, using the language of Definition 1.5, we could describe hamiltonian
cycles and paths as spanning cycles and paths since they must include all
vertices of the graph.

As with eulerian circuits, these specific cycles (or paths) are named for the
mathematician who first formalized them, Sir William Hamilton. Hamilton
posed this idea in 1856 in terms of a puzzle, which he later sold to a game
dealer. The "Icosian Game" was a wooden puzzle with numbered ivory pegs
where the player was tasked with inserting the pegs so that following them in
order would traverse the entire board (shown on the following page). Perhaps
not too surprisingly, this game was not a big money maker.

It should be noted that T.P. Kirkman, a contemporary of Hamilton's,
did much of the early work in the study of hamiltonian circuits. Whereas
Hamilton primarily focused on one graph, Kirkman was concerned with the
conditions that will guarantee a graph has a hamiltonian cycle. However,
Hamilton deserves credit for publicizing the concept of a cycle that hits every
vertex exactly once. This section will explore when a graph has a hamiltonian
cycle and how to find an optimal, or near optimal, hamiltonian cycle.

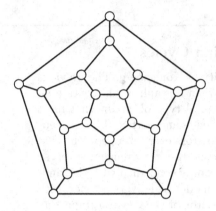

Icosian Game

When comparing an eulerian circuit with a hamiltonian cycle, only one requirement has been lifted: instead of a tour containing every edge and every vertex, we are now only concerned with the vertices. However, as often happens in mathematics, when restrictions are relaxed, the solution either does not exist or finding a solution becomes more difficult.

For the past one hundred sixty years, numerous mathematicians have searched for a solution to the hamiltonian cycle problem; that is, what are the necessary and sufficient conditions for a graph to contain a hamiltonian cycle? Recall that a *necessary condition* is a property that must be achieved in order for a solution to be possible and a *sufficient condition* is a property that guarantees the existence of a solution. As we saw in the previous section, necessary and sufficient conditions were found for eulerian circuits—a graph simply needs to be connected with all even vertices in order for an eulerian circuit to exist. The same is not true for hamiltonian cycles.

Example 2.8 For each of the graphs below, determine if they have hamiltonian cycles (and paths) and eulerian circuits (and trails).

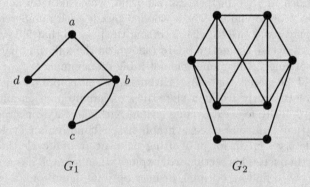

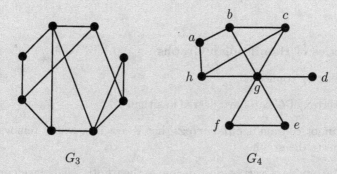

G_3 G_4

Solution: Since G_1 is connected and all vertices are even, we know it has an eulerian circuit. There is no hamiltonian cycle since we need to include c in the cycle and by doing so we have already passed through b twice, making it impossible to visit a and d.

Since G_2 is connected and all vertices are even, we know it is eulerian. Hamiltonian cycles and hamiltonian paths also exist. To find one such path, remove any one of edges from a hamiltonian cycle.

Some vertices of G_3 are odd, so we know it is not eulerian. Moreover, since more than two vertices are odd, the graph is not semi-eulerian. However, this graph does have a hamiltonian cycle (and so also a hamiltonian path). Can you find it?

Four vertices of G_4 are odd, we know it is neither eulerian nor semi-eulerian. This graph does not have a hamiltonian cycle since d cannot be a part of any cycle. Moreover, this graph does not have a hamiltonian path since any traversal of every vertex would need to travel through g multiple times.

Recall from Section 2.1.3 that if a graph has an eulerian circuit, then it cannot have an eulerian trail, and vice versa. The same is not true for the hamiltonian version:

- If a graph has a hamiltonian cycle, it automatically has a hamiltonian path (just leave off the last edge of the cycle to obtain a path).

- If a graph has a hamiltonian path, it may or may not have a hamiltonian cycle.

Before we wade into more complex theoretical results, we will focus on determining if a graph is hamiltonian. In particular, given small enough graphs and our reasoning skills, we can either find a hamiltonian cycle or explain why no such cycle exists. As should be obvious from the above examples, a few necessary conditions can be placed on the graph to ensure a hamiltonian cycle is possible. In addition, we can use what we know about cycles to show when a hamiltonian cycle is not possible. These include, but are not limited to, the following:

Properties of Hamiltonian Graphs

(1) G must be connected.

(2) No vertex of G can have degree less than 2.

(3) G cannot contain a **cut-vertex**, that is a vertex whose removal disconnects the graph.

(4) If G contains a vertex x of degree 2 then both edges incident to x must be included in the cycle.

(5) If two edges incident to a vertex x must be included in the cycle, then all other edges incident to x cannot be used in the cycle.

(6) If in the process of attempting to build a hamiltonian cycle, a cycle is formed that does not span G, then G cannot be hamiltonian.

One final note of caution: even with these properties and rules, determining if a graph is hamiltonian in nontrivial. If we are trying to prove a graph is not hamiltonian, we may need to also use any symmetries within the graph and break our argument into cases based on which edge is chosen. The example below works through two graphs and how we can prove they are not hamiltonian.

Example 2.9 Use the properties listed above to show that the graphs below are not hamiltonian.

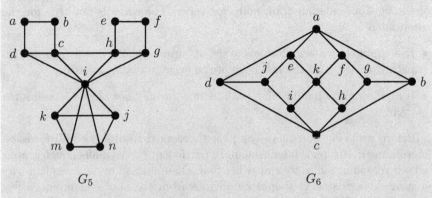

G_5 G_6

Solution: For G_5, notice that vertices $a, b, e,$ and f all have degree 2 and so all the edges incident to these vertices must be included in the cycle. But then edges cd and hg cannot be a part of a cycle since they would create smaller cycles that do not include all of the vertices in G_5.

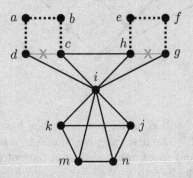

Since only one edge remains incident to d and g, namely di and gi, then these must be a part of the cycle. But in doing so, we would be forced to use ch since the other edges incident to i could not be chosen. This creates a cycle that does not span G_5, and so G_5 is not hamiltonian.

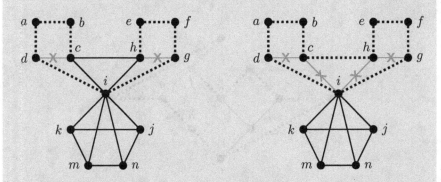

For G_6, we know we cannot use all of the edges $ab, ad, bc,$ and bd, as these four edges together create a cycle that does not span the graph. Since either b or d must have an edge not from this list, by symmetry we will choose bg to be a part of the hamiltonian cycle C we are attempting to build. Then either ab or bc must be the other edge incident to b in the cycle C, and again using symmetry we can choose ab. Now, we cannot use both af and fg, since together with ab and bg we would have a cycle that does not span G_6. Thus we must choose fk to be a part of the cycle C.

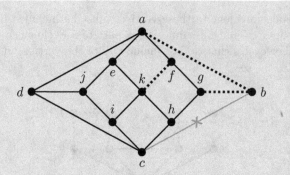

At this point we will consider two scenarios based on the last edge incident to f in C. For the first case, we choose to add af to C. Then we cannot include fg and so gh must also be a part of the cycle since it is the only edge left incident to g. We cannot use hk as it would close the cycle before it spans G_6, and so ch must also be a part of the cycle C. Since a already has two incident edges in the cycle, we cannot use either of ae or ad. Thus the other edges incident to d and e must be a part of the cycle, namely dj, dc, ej, and ek. But this closes the cycle without including vertex i.

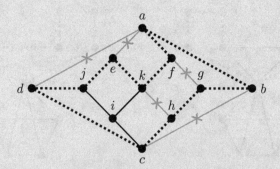

Thus af cannot be a part of a hamiltonian cycle C (if such a cycle exists). Thus fg must be chosen and we cannot use gh since g already has two incident edges as a part of the cycle. Thus the other edges incident to h must be used, namely hk and hc, and so the other edges incident to k cannot be, namely ke and ki. Based on the edges remaining, all of ci, ij, ej, and ea would be required, which closes a cycle that does not include d.

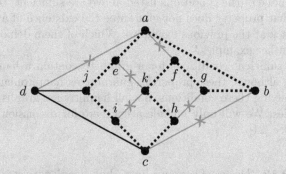

Since neither option produces a spanning cycle of G_6, we know it is not hamiltonian.

Perhaps a less obvious necessary condition is the one proven below, which relates the number of vertices to the number of components created with their removal.

Proposition 2.11 If G is a graph with a hamiltonian cycle, then $G - S$ has at most $|S|$ components for every nonempty set $S \subseteq V$.

Proof: Assume G has a hamiltonian cycle C and S is a nonempty subset of vertices. Then as we traverse C we will leave and return to S from distinct vertices in S since no vertex can be used more than once in a cycle. Since C must span $V(G)$, we know S must have at least as many vertices as the number of components of $G - S$.

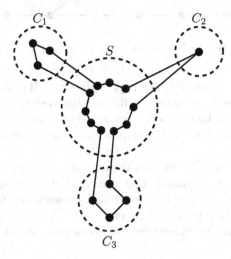

Note that none of the conditions listed above are sufficient, that is simply maintaining that property does not guarantee the existence of a hamiltonian cycle. Look back at the previous examples. Which of them demonstrate this? Can you find other examples?

There are sufficient conditions for a graph to contain a hamiltonian cycle; but again, although these properties guarantee a hamiltonian cycle exists for graph satisfying these conditions, not all hamiltonian graphs must satisfy these properties. We will only include a few here for discussion. For further information, see [21].

Theorem 2.12 (Dirac's Theorem) Let G be a graph with $n \geq 3$ vertices. If every vertex of G satisfies $\deg(v) \geq \frac{n}{2}$, then G has a hamiltonian cycle.

Proof: First note that G is connected (see Exercise 2.19). Let $P = x_0 x_1 \cdots x_k$ be a longest path of G. Then every neighbor of both x_0 and x_k must be included in the path. Let S be the set of vertices from P that are adjacent to x_k; that is, $S = \{x_i : x_i x_k \in E(G)\}$. Let T be the set of vertices whose immediate neighbor on P is adjacent to x_0; that is $T = \{x_i : x_{i+1} x_0 \in E(G)\}$. Since both x_0 and x_k have degree at least $\frac{n}{2}$ we know $|S| + |T| \geq n$. In addition, we know that neither S nor T can contain the vertex x_k and combining S and T cannot yield more vertices than all those in the graph, and so $|S \cup T| < n$. But as $|S \cup T| = |S| + |T| - |S \cap T|$, we have that $|S \cap T| \geq 1$, and so there exists some vertex x_i such that $x_0 x_{i+1}, x_i x_k \in E(G)$ (see the picture below).

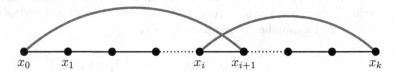

Consider the cycle $C = x_0 \underset{P}{-} x_i \; x_k \underset{P}{-} x_{i+1} \; x_0$, where $x \underset{P}{-} y$ indicates traveling from x to y along path P. It remains to show that C spans G, that is every vertex of G is a part of the cycle. If C does not span G then there must be some vertex v with an edge to some vertex of C. But then we can form a longer path than P using v and a spanning path of C. Thus C spans G and is therefore a hamiltonian cycle for G.

Dirac's Theorem [25] relies on the high degree at every vertex of a graph in order to guarantee a hamiltonian cycle, yet this isn't quite necessary. The main concern is that there is a way to get between two nonadjacent vertices without needing to repeat an edge or vertex. The following result is a slight weakening on Dirac's sufficient conditions to guarantee a hamiltonian cycle.

Theorem 2.13 (Ore's Theorem) Let G be a graph with $n \geq 3$ vertices. If $\deg(x) + \deg(y) \geq n$ for all distinct nonadjacent vertices, then G has a hamiltonian cycle.

Notice that Ore's Theorem [68] is only concerned with the sum of the degrees of nonadjacent pairs of vertices: if this sum is large enough then we can guarantee a hamiltonian cycle, but if the sum is too small we cannot make any conclusion about the graph. If G is known to be hamiltonian, then adding edges to G cannot destroy the existence of a hamiltonian cycle, it can only make it easier to find one. So if G is hamiltonian, then so must be $G + e$ for some edge e. However the converse is not true: removing an edge from a hamiltonian graph can create a graph that is not hamiltonian. What we will see below is that if we are careful with which edges we choose to add to G, and if the resulting graph is hamiltonian, then G itself will be hamiltonian.

Definition 2.14 The *hamiltonian closure* of a graph G, denoted $cl(G)$, is the graph obtained from G by iteratively adding edges between pairs of nonadjacent vertices whose degree sum is at least n, that is we add an edge between x and y if $\deg(x) + \deg(y) \geq n$, until no such pair remains.

Example 2.10 Find a hamiltonian closure for the graph below.

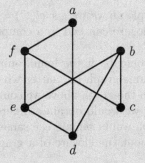

Solution: First note that the degrees of the vertices are $2, 3, 2, 3, 3, 3$ (in alphabetic order). Since the closure of G is formed by putting an edge between nonadjacent vertices whose degree sum is at least 6, we must begin with edge bf or df. We chose to start with bf. In the sequence of graphs shown on the next page, the edge added at each stage is a thick dashed line.

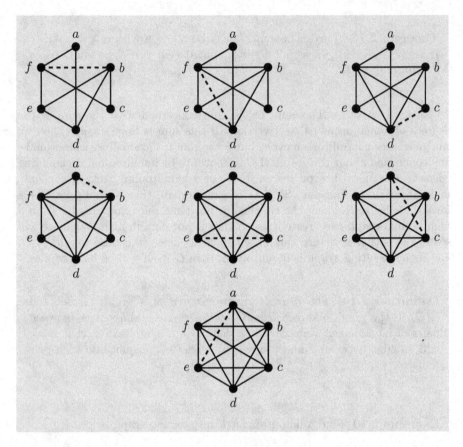

Note that in the example above we get $cl(G) = K_6$. While the process of finding the hamiltonian closure can end in a complete graph, it need not to (see Exercise 2.7).

An obvious concern would be if the hamiltonian closure is well-defined, that is, if we obtain the same graph no matter what order we choose to add new edges. For example, in the graph above we could have added ab at step 2 instead of df. In doing so, we would obtain a different graph in the next few steps, but the ending graph would remain the same (see Exercise 2.13). If we want to use a property about the closure of a graph, we need to ensure the closure is unique.

Theorem 2.15 The closure of G is well-defined.

Proof: Let $e_1, e_2, \ldots, e_j$ and $f_1, f_2, \ldots, f_k$ be the sequence of edges added to G to form $cl(G)$, creating graphs G_1 and G_2, respectively. We will prove that $G_1 = G_2$ by showing that both G_1 and G_2 contain all the edges

$(e_1, \ldots, e_j, f_1, \ldots, f_k)$. Suppose for a contradiction that this is not true. Then there exists some first $e_i = xy$ such that $e_i \in E(G_1)$ but $e_i \notin E(G_2)$. Let G_3 be the graph obtained by adding $e_1, e_2, \ldots e_{i-1}$ to G. Then G_3 is a subgraph of both G_1 and G_2.

Since e_i was a valid edge in the creation of G_1, we know that $\deg_{G_1}(x) + \deg_{G_1}(y) \geq n$, but then this is also true for G_3, that is $\deg_{G_3}(x) + \deg_{G_3}(y) \geq n$. Moreover, since e_i was a valid edge we know that x and y were not adjacent until the addition of e_i. Thus x and y are not adjacent in G_3, nor in G_2 since e_i is not an edge of G_2. But then x and y would remain nonadjacent vertices in the creation of G_2 that satisfy the degree condition, which contradicts the formation of the closure of G. Thus $G_1 = G_2$ and the closure of G is well-defined.

Bondy and Chvátal [7] proved that a graph and its closure have the same conclusion to the hamiltonian question. In particular, if you can prove that the closure of a graph has a certain property known to guarantee the existence of a hamiltonian cycle, then the original graph must also be hamiltonian.

Theorem 2.16 A graph G is hamiltonian if and only if its closure $cl(G)$ is hamiltonian.

Lemma 2.17 If G is a graph with at least 3 vertices such that its closure $cl(G)$ is complete, then G is hamiltonian.

Our final result related to determining if a graph is hamiltonian comes from Chvátal [15] in which he further relaxes the degree condition, which allows for vertices of small degree as long as there are enough vertices of large degree, whether or not the vertices are adjacent.

Theorem 2.18 Let G be a simple graph where the vertices have degree $d_1, d_2, \ldots, d_n$ such that $n \geq 3$ and the degrees are listed in increasing order. If for any $i < \frac{n}{2}$ either $d_i > i$ or $d_{n-i} \geq n-i$, then G is hamiltonian.

Proof: Assume G is a graph in which for any $i < \frac{n}{2}$ either $d_i > i$ or $d_{n-i} \geq n - i$. We will prove that $cl(G)$ is a complete graph, and so by Lemma 2.17 we would know that both $cl(G)$ and G are hamiltonian. Suppose for a contradiction that $cl(G)$ is not complete. Then there exist nonadjacent vertices x and y in $cl(G)$, and $\deg_{cl(G)}(x) + \deg_{cl(G)}(y) \leq n - 1$, as otherwise xy would be added to G in the formation of $cl(G)$. Choose x and y among all such nonadjacent pairs in $cl(G)$ so that the sum of their degrees in $cl(G)$ is as large as possible, and without loss of generality we may assume $\deg_{cl(G)}(x) \leq \deg_{cl(G)}(y)$.

Let $k = \deg_{cl(G)}(x)$. So $\deg_{cl(G)}(y) \leq n - 1 - k$ and so $k \leq \frac{n-1}{2} < \frac{n}{2}$.

Let A be the set of vertices, other than y, that are not adjacent to y in $cl(G)$. Thus

$$\begin{aligned} |A| &= n - 1 - \deg_{cl(G)}(y) \\ &\geq (n-1) - (n-1-k) \\ &= k \end{aligned}$$

Since we chose x and y so that their degree sum is as large as possible, we know every vertex $v \in A$ satisfies

$$\deg_G(v) \leq \deg_{cl(G)}(v) \leq \deg_{cl(G)}(x) = k.$$

Then every vertex in A has degree at most k in G, and so G has at least k vertices of degree at most k. Thus $d_k \leq k$.

Similarly, let B be the set of vertices, other than x, in $cl(G)$ that are not adjacent to x. Then

$$\begin{aligned} |B| &= n - 1 - \deg_{cl(G)}(x) \\ &= n - 1 - k \end{aligned}$$

and every vertex v in B satisfies

$$\deg_G(v) \leq \deg_{cl(G)}(v) \leq \deg_{cl(G)}(y) \leq n - 1 - k.$$

Thus there are $n - 1 - k$ vertices of degree at most $n - 1 - k$, implying $d_{n-k-1} \leq n - k - 1$. But then x is also a vertex of degree at most $n - 1 - k$ since

$$\begin{aligned} \deg_G(x) &\leq \deg_{cl(G)}(x) \\ &\leq \deg_{cl(G)}(y) \\ &\leq n - 1 - k \end{aligned}$$

Thus $d_{n-k} \leq n - k - 1$.

Thus $d_k \leq k$ and $d_{n-k} < n - k$, and since $k < \frac{n}{2}$ this contradicts the assumption that $d_k > k$ or $d_{n-k} \geq n - k$. Thus $cl(G)$ must be complete and so by Lemma 2.17 we know G is hamiltonian.

As should not be obvious, the question of whether a graph is hamiltonian is non-trivial and can be quite difficult to answer for even moderate-sized graphs. We now turn to a class of graphs known to contain (many!) hamiltonian cycles, where instead of questioning if the graph is hamiltonian, we focus on finding an optimal hamiltonian cycle.

2.2.1   The Traveling Salesman Problem

The discussion above should make clear the difficulty in determining if a graph is hamiltonian. But what if a graph is know to have a hamiltonian cycle? For example, every complete graph K_n (for $n \geq 3$) must contain a hamiltonian cycle since it satisfies the criteria of Dirac's Theorem. In this scenario, finding a hamiltonian cycle is quite elementary, and so, as mathematicians do, we generalize the problem to one in which the edges are no longer equivalent and have a weight associated to them. Then instead of asking whether a graph simply *has* a hamiltonian cycle, we can now ask how do we find the *best* hamiltonian cycle.

Historically, the extensive study of hamiltonian circuits arose in part from a simple question: A traveling salesman has customers in numerous cities; he must visit each of them and return home, but wishes to do this with the least total cost; determine the cheapest route possible for the salesman. In fact, Proctor and Gamble can be credited with the modern study of hamiltonian circuits when they sponsored a seemingly innocent competition in the 1960s asking for a shortest hamiltonian circuit visiting 33 cities across the United States. Mathematicians were intrigued and an entire branch of mathematics and computer science developed. For over half a century, some of the brightest minds have tackled the Traveling Salesman Problem (my graph theory professor in college called it "the disease") and numerous books and websites are devoted to finding an optimal solution to both the general question and to specific instances (such as a cycle through all cities in Sweden). A full discussion of the problem is beyond the scope of this book, though you are encouraged to peruse [16] or [17]. You could honestly spend a semester just discussing the various algorithms, so we restrict ourselves to just a handful of these, with plenty of examples and exercises.

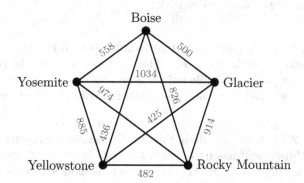

The graph that models the general Traveling Salesman Problem (TSP) is a *weighted complete graph*, such as the one shown above from Example 1.7. Recall from Definition 1.8 that a weighted graph is one in which each edge is assigned a weight, which usually represents either distance, time, or cost. It is standard to use a complete graph since theoretically it should be possible to travel between any two cities, such as the previous graph indicating the driving distance between a home city and various national parks.

Brute Force Algorithm

To find the hamiltonian cycle of least total weight, one obvious method is to find all possible hamiltonian cycles and pick the cycle with the smallest total. The method of trying every possibility to find an optimal solution is referred to as an *exhaustive search*, or use of the *Brute Force Algorithm*. This method can be used for any number of problems, not just the Traveling Salesman Problem, though the description below is only for finding hamiltonian cycles. Knowing this, you might be asking yourself why this problem is still being studied. If we have an algorithm that will find the optimal hamiltonian cycle, why are mathematicians still interested? The answer will soon become clear.

Brute Force Algorithm

Input: Weighted complete graph K_n.

Steps:

1. Choose a starting vertex, call it v.

2. Find all hamiltonian cycles starting at v. Calculate the total weight of each cycle.

3. Compare all $(n-1)!$ cycles. Pick one with the least total weight. (Note: there should be at least two options).

Output: Minimum hamiltonian cycle.

Although both K_3 and K_4 contain hamiltonian cycles (try it!), the first graph with some complexity is K_5. The example below walks through the process of finding all 24 possible weighted cycles. You are encouraged to find these on your own and then check your solution with the chart provided.

Example 2.11 Sam is planning his next business trip from his hometown of Addison and has determined the cost for travel between any of the five cities he must visit. This information is modeled in the weighted complete graph on the next page, where the weight is given in terms of dollars. Use Brute Force to find all possible routes for his trip.

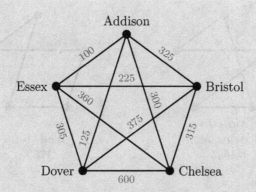

Solution: One method for finding all hamiltonian cycles, and ensuring you indeed have all 24, is to use alphabetical or lexicographic ordering of the cycles. Note that all cycles must start and end at Addison and we will abbreviate all cities with their first letter. For example, the first cycle is *abcdea* and appears first in the list below. Below each cycle is its reversal and total weight.

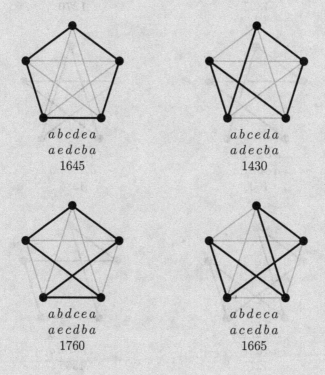

abcdea
aedcba
1645

abceda
adecba
1430

abdcea
aecdba
1760

abdeca
acedba
1665

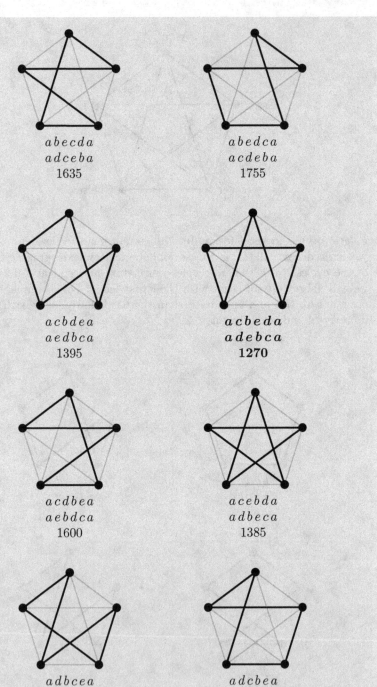

abecda
adceba
1635

abedca
acdeba
1755

acbdea
aedbca
1395

acbeda
adebca
1270

acdbea
aebdca
1600

acebda
adbeca
1385

adbcea
aecbda
1275

adcbea
aebcda
1365

From the data above we can identify the optimal cycle as $acbeda$, which provides Sam with the optimal route of Addison to Chelsea to Bristol to Essex to Dover and back to Addison, for a total at a cost of \$1270.

If you attempted to find all 24 cycles by hand, how long did it take you? You may have gotten into a rhythm and improved after the first few, but it still took some time to complete. What if you tried K_6? K_7? K_{15}? How many cycles would you need to check? Ignoring reversals you have 120, 720, and roughly 87 billion cycles to find.

The Traveling Salesman Problem is of interest to mathematicians in part because of how quickly the size of the problem grows. As the size of the input grows (say from K_5 to K_{15}) the number of calculations needed to obtain an output explodes (from 24 to about 87 billion; see Appendix D for further discussion). You may be thinking "Yeah...but we have these things called computers that can work faster than any human, so that shouldn't be a problem."

Computer performance is often measured in FLOPS, an acronym for floating-point operations per second. Roughly speaking, a floating-point operation can consist of arithmetic calculations (such as adding or subtracting two numbers). Top of the line desktop processors have performance ratings of roughly 175 billion FLOPS, better known as 175 GFLOPS [51]. At the time of publication, the best supercomputer in the world had a performance rating of about 33 million GFLOPS and the sum of the top 500 supercomputers was 308 million GFLOPS [78].

To determine how quickly these computers would complete the Brute Force Algorithm, we first need to determine the number of FLOPS required. Given a specified starting vertex, we know there are essentially $(n - 1)!/2$ possible hamiltonian cycles for the complete graph K_n. For each of these cycles we need to perform n additions to find its total weight. Once a cycle and its weight have been computed, we must compare them to the previously computed cycle, only keeping in memory the one of least total weight, requiring another $(n - 1)!/2$ calculations. Altogether, we estimate the time required to fully implement the Brute Force Algorithm on K_n is

$$n \cdot \frac{(n-1)!}{2} + \frac{(n-1)!}{2} = \frac{(n+1) \cdot ((n-1)!)}{2}$$

$$= \frac{(n+1)!}{2n} \text{ FLOPS.}$$

Suppose you have been given access to the highest rated supercomputer (and also the top 500) in the world and would like to find the optimal hamiltonian cycle on K_n for each n from 5 to 50. How long will this take? The table below gives you an estimate of the time requirements for increasing values of n. To put some of these numbers into context, scientists believe the earth is about 4.54 billion years old, that is 4.54×10^9 years! Using Brute Force is computationally impractical for graphs with more than 24 vertices, even when using the top 500 supercomputers in the entire world!

	Supercomputers	
n	**Best**	**Top 500**
5	2×10^{-15} seconds	2×10^{-16} seconds
15	2×10^{-5} seconds	2×10^{-6} seconds
20	40 seconds	4 seconds
21	14 minutes	2 minutes
22	5 hours	32 minutes
23	4.5 days	12 hours
24	16 weeks	12 days
25	7.5 years	10 months
26	2 centuries	2 decades
30	132 million years	14 million years
40	4.1×10^{23} years	4.3×10^{22} years
50	1.4×10^{40} years	1.5×10^{39} years

The discussion above illustrates how ineffective Brute Force is when trying to solve an instance of the Traveling Salesman Problem. Although it will eventually find a solution, the time necessary to finish all the computations can be quite unreasonable. Mathematicians have been searching for algorithms that will find the optimal cycle in a relatively short time span; that is, an algorithm that is both efficient and optimal. Not only has no such algorithm been found for the Traveling Salesman Problem, but some mathematicians believe no such algorithm even exists. Further discussion of the difference between efficient and optimal can be found in Appendix D.

The next few sections discuss ***approximate algorithms,*** which are efficient but not optimal. This means they can find a good hamiltonian cycle without taking too much computational time. In some instances these algorithms may in fact find the optimal cycle, but there is no guarantee that this will always occur. This will become evident through the examples and exercises.

Nearest Neighbor

If you do not have time to run Brute Force, how would you find a good hamiltonian cycle? You could begin by taking the edge to the "closest" vertex from your starting location, that is the edge of the least weight. And then? Maybe move to the closest vertex from your new location? This strategy is called the ***Nearest Neighbor Algorithm.***

Nearest Neighbor Algorithm

Input: Weighted complete graph K_n.

Steps:

1. Choose a starting vertex, call it v. Highlight v.

2. Among all edges incident to v, pick the one with the smallest weight. If two possible choices have the same weight, you may randomly pick one.

3. Highlight the edge and move to its other endpoint u. Highlight u.

4. Repeat Steps (2) and (3), where only edges to unhighlighted vertices are considered.

5. Close the cycle by adding the edge to v from the last vertex highlighted. Calculate the total weight.

Output: hamiltonian cycle.

Example 2.12 Apply the Nearest Neighbor Algorithm to the graph from Example 2.11.

Solution: At each step we will show two copies of the graph. One will indicate the edges under consideration and the other traces the route under construction.

Step 1: The starting vertex is a.

Step 2: The edge of smallest weight incident to a is ae with weight 100.

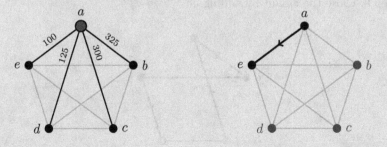

Step 3: From e we only consider edges to $b, c,$ or d. Choose edge eb with weight 225.

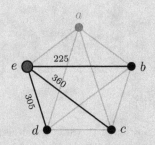
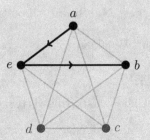

Step 4: From b we consider the edges to c or d. The edge of smallest weight is bc with weight 315.

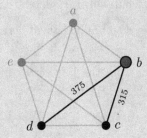
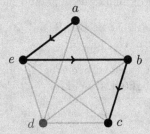

Step 5: Even though cd does not have the smallest weight among all edges incident to c, it is the only choice available.

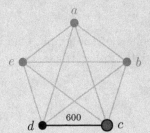
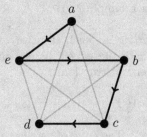

Step 6: Close the circuit by adding da.

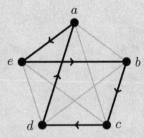

Output: The circuit is $a\,e\,b\,c\,d\,a$ with total weight 1365.

In the previous example, the final circuit was Addison to Essex to Bristol to Chelsea to Dover and back to Addison. Although Nearest Neighbor did not find the optimal circuit of total cost $1270, it did produce a fairly good circuit with total cost $1365. Perhaps most important was the speed with which Nearest Neighbor found this circuit.

Example 2.12 also illustrates some drawbacks for Nearest Neighbor. First, the last two edges are completely determined since we cannot travel back to vertices that have already been chosen, possibly forcing us to use the heaviest edges in the graph, as happened above. Second, the arbitrary choice of a starting vertex could eliminate lighter weight edges from consideration.

Though we cannot do anything about the former concern, we can address the latter. By using a different starting vertex, the Nearest Neighbor Algorithm may identify a new hamiltonian cycle, which may be better or worse than the initial cycle. Instead of only considering the circuits starting at the chosen vertex (for example, Sam's hometown of Addison), we will run Nearest Neighbor with each of the vertices as a starting point. This is called **Repetitive Nearest Neighbor**.

Repetitive Nearest Neighbor Algorithm

Input: Weighted complete graph K_n.

Steps:

1. Choose a starting vertex, call it v.

2. Apply the Nearest Neighbor Algorithm.

3. Repeat Steps (1) and (2) so each vertex of K_n serves as the starting vertex.

4. Choose the cycle of least total weight. Rewrite it with the desired reference point.

Output: hamiltonian cycle.

The last step of this algorithm calls for a cycle to be rewritten. Looking back at the table of cycles in Example 2.11, if you did not know the starting vertex was a, you could not discern this from the cycle highlighted in the graph. In fact, any vertex in a cycle could be the reference point, though there is often a designated reference point based on the scenario being modeled.

Example 2.13 Apply the Repetitive Nearest Neighbor Algorithm to the graph from Example 2.11.

Solution: The five cycles are shown below, with the original name, the rewritten form with *a* as the reference point, and the total weight of the cycle. You should notice that the cycle starting at *d* is the same as the one starting at *a*, and the cycle starting at *b* is their reversal.

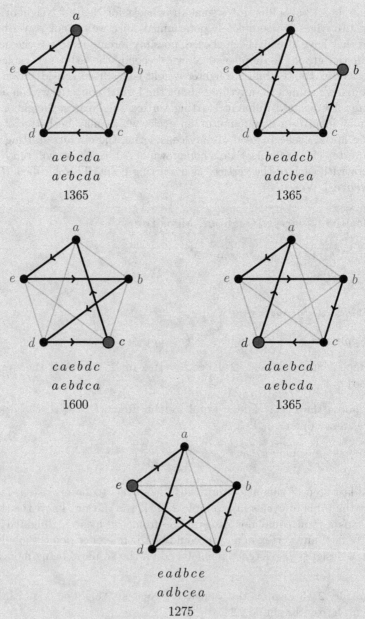

aebcda
aebcda
1365

beadcb
adcbea
1365

caebdc
aebdca
1600

daebcd
aebcda
1365

eadbce
adbcea
1275

It should come as no surprise that Repetitive Nearest Neighbor performs better than Nearest Neighbor; however, there is no guarantee that this improvement will produce the optimal cycle. In Example 2.13, the algorithm found the second best cycle, which only differed from the optimal by $5! This small increase in cost would clearly be worth the time savings over Brute Force.

Cheapest Link

Even though Repetitive Nearest Neighbor addressed the concern of missing small weight edges, it is still possible that some of these will be bypassed as we travel a cycle. The *Cheapest Link Algorithm* attempts to fix this by choosing edges in order of weight as opposed to edges along a tour. Unlike either version of Nearest Neighbor, Cheapest Link does not follow a path that eventually closes into a cycle, but rather chooses edges by weight in such a way that they eventually form a cycle.

Cheapest Link Algorithm

Input: Weighted complete graph K_n.

Steps:

1. Among all edges in the graph pick the one with the smallest weight. If two possible choices have the same weight, you may randomly pick one. Highlight the edge.

2. Repeat Step (1) with the added conditions:

   (a) no vertex has three highlighted edges incident to it; and

   (b) no edge is chosen so that a cycle closes before hitting all the vertices.

3. Calculate the total weight.

Output: hamiltonian cycle.

Example 2.14 Apply the Cheapest Link Algorithm to the graph from Example 2.11.

Solution: In each step shown, unchosen edges are shown in gray, previously chosen edges are in black, and the newly chosen edge in bold. An edge that is skipped will be marked with an X.

Step 1: The smallest weight is 100 for edge *ae*.

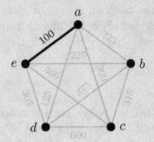

Step 2: The next smallest weight is 125 with edge *ad*.

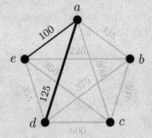

Step 3: The next smallest weight is 225 for edge *be*.

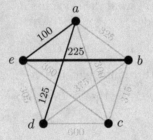

Step 4: Even though *ac* has weight 300, we must bypass it as it would force *a* to have three incident edges that are highlighted.

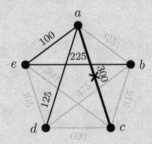

Step 5: The next smallest weight is 305 for edge *ed*, but again we must bypass it as it would close a cycle too early as well as force *e* to have three incident edges that are highlighted.

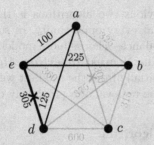

Step 6: The next available is *bc* with weight 315.

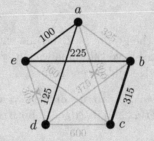

Step 7: At this point, we must close the cycle with the only one choice of *cd*.

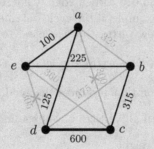

Output: The resulting cycle is *a e b c d a* with total weight 1365.

In the example above, Cheapest Link ran into the same trouble as the initial cycle created using Nearest Neighbor; although the lightest edges were chosen, the heaviest also had to be included due to the outcome of the previous steps. However, Cheapest Link will generally perform quite well (in the example above, it found the same cycle as Nearest Neighbor) and is an efficient algorithm.

Nearest Insertion

The downfall of the previous two algorithms is the focus on choosing the smallest weighted edges at every opportunity. In some instances, it may be beneficial to work for a balance—choose more moderate weight edges to avoid later using the heaviest. The ***Nearest Insertion Algorithm*** does just this by forming a small circuit initially from the smallest weighted edge and balances adding new edges with the removal of a previously chosen edge.

Nearest Insertion Algorithm

Input: Weighted complete graph K_n.

Steps:

1. Among all edges in the graph, pick the one with the smallest weight. If two possible choices have the same weight, you may randomly pick one. Highlight the edge and its endpoints.

2. Pick a vertex that is closest to one of the two already chosen vertices. Highlight the new vertex and its edges to both of the previously chosen vertices.

3. Pick a vertex that is closest to one of the three already chosen vertices. Calculate the increase in weight obtained by adding two new edges and deleting a previously chosen edge. Choose the scenario with the smallest total. For example, if the cycle obtained from (2) was $a - b - c - a$ and d is the new vertex that is closest to c, we calculate:

$$w(dc) + w(db) - w(cb) \text{ and } w(dc) + w(da) - w(ca)$$

and choose the option that produces the smaller total.

4. Repeat Step (3) until all vertices have been included in the cycle.

5. Calculate the total weight.

Output: hamiltonian cycle.

Example 2.15 Apply the Nearest Insertion Algorithm to the graph from Example 2.11.

Solution: At each step shown, the graph on the left highlights the edge being added and the graph on the right shows how the cycle is built.

Step 1: The smallest weight edge is *ae* at 100.

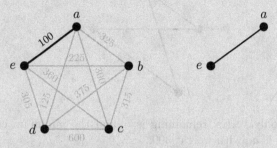

Step 2: The closest vertex to either *a* or *e* is *d* through the edge *ad* of weight 125. Form a cycle by adding *ad* and *de*.

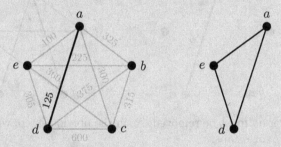

Step 3: The closest vertex to any of *a, d,* or *e* is *b* through the edge *be* with weight 225.

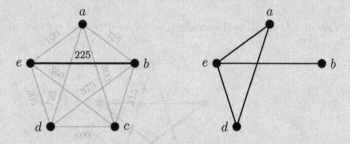

In adding edge *be*, either *ae* or *de* must be removed so that only two edges are incident to *e*. To determine which is the better choice, compute the following expressions:

$$be + ba - ea = 225 + 325 - 100 = 450$$
$$be + bd - ed = 225 + 375 - 305 = 295$$

Since the second total is smaller, we create a larger cycle by adding edge *bd* and removing *ed*.

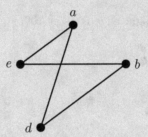

Step 4: The only vertex remaining is *c*, and the minimum edge to the other vertices is *ac* with weight 300.

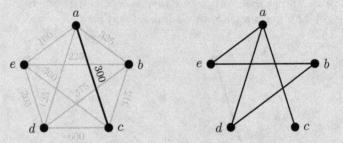

Either *ae* or *ad* must be removed. As in the previous step, we compute the following expressions:

$$ca + cd - ad = 300 + 600 - 125 = 775$$
$$ca + ce - ae = 300 + 360 - 100 = 560$$

The second total is again smaller, so we add *ce* and remove *ae*.

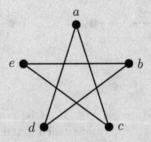

Output: The cycle is *a c e b d a* with total weight 1385.

In the example above, Nearest Insertion performed slightly worse than

Cheapest Link and Repetitive Nearest Neighbor. Among all three algorithms, Repetitive Nearest Neighbor found the cycle closest to optimal. In general, when comparing algorithm performance we focus less on absolute error $(1275 - 1270 = 5)$ but rather on *relative error*.

Definition 2.19 The ***relative error*** for a solution is given by

$$\epsilon_r = \frac{Solution - Optimal}{Optimal}$$

Absolute error gives the exact measure away from optimal, but can be misleading if the weights themselves are either very large or very small. Using relative error allows us to compare the performance of an algorithm across multiple examples where the scale may vary. It is important to remember that relative error (and absolute error) can only be calculated when the optimal solution is known.

Example 2.16 Find the relative error for each of the algorithms performed on the graph from Example 2.11.

Solution:

- Repetitive Nearest Neighbor: $\epsilon_r = \dfrac{1275 - 1270}{1270} = 0.003937 \approx 0.39\%$

- Cheapest Link: $\epsilon_r = \dfrac{1365 - 1270}{1270} = 0.074803 \approx 7.48\%$

- Nearest Insertion: $\epsilon_r = \dfrac{1385 - 1270}{1270} = 0.090551 \approx 9.05\%$

It is possible for an approximation algorithm to find the optimal solution. However, this is unlikely and any of these algorithms could in fact find the absolute worst choice. In addition, no one approximation algorithm will always perform better than the others. There are instances in which one of the algorithms performs better than the others and instances where it performs worse. More practice with these approximation algorithms can be found in the Exercises.

2.2.2 Tournaments Revisited

We began Chapter 1 by introducing tournaments as an example of graph modeling. Recall that a tournament is a complete graph where each edge has been assigned a specific direction. As we have just spent considerable energy investigating methods for finding hamiltonian cycles on complete graphs, it is natural to wonder if these same ideas can be applied to tournaments.

First, consider the two tournaments shown below. Their underlying graph is K_5, which we know has $4! = 24$ unique hamiltonian cycles with a specific reference point. But if we are now required to follow the direction of an arc in the tournament, could we still find a hamiltonian cycle?

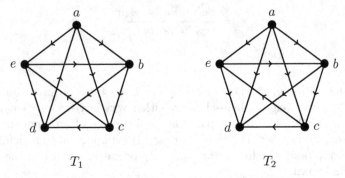

$$T_1 \qquad\qquad T_2$$

The tournament T_1 on the left has a hamiltonian cycle, given by $a\,e\,b\,c\,d\,a$, whereas the tournament T_2 on the right cannot have a hamiltonian cycle since a has in-degree $\deg^-(a) = 0$ and every vertex along a cycle must have nonzero in-degree and out-degree. But is this condition enough? Hopefully we have seen enough of the complexity surrounding hamiltonian graphs to suspect there is far more to it than just nonzero degrees. In fact, the tournament T_3 shown on the left below has degree sequence $1, 1, 1, 4, 4, 4$ and yet no hamiltonian cycle can exist since if a cycle must exit vertices c, d, and e then arcs ce, dc, and ed must all be included, which creates a subcycle, as shown on the right.

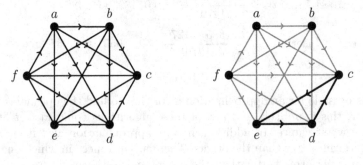

$$T_3 \qquad\qquad\qquad T_3 \text{ with subcycle}$$

But examining the tournament T_3 above yields something interesting— even though no hamiltonian cycle exists we can find a hamiltonian path, one of which is $f\,a\,b\,e\,d\,c$. In fact, if you look back at any tournament appearing in this book, you would find a hamiltonian path within each one! There is nothing special about these tournaments though, as the following result proves that all tournaments have hamiltonian paths.

Theorem 2.20 Every tournament has a hamiltonian path.

Proof: Let T be a tournament on n vertices. We will argue by induction on n that T has a hamiltonian path. If $n \le 2$ then T either has a trivial path of no edges or T has a path represented by the only edge of the tournament.

Suppose that for some $n \ge 3$ that for all $1 \le k < n$ that any tournament on k vertices has a hamiltonian path. Let T be a tournament on n vertices and suppose x is a vertex with nonzero out-degree in T. Let $T' = T - x$. Then by the induction hypothesis there must exist a hamiltonian path $P = v_1 v_2 \cdots v_{n-1}$ in T'. Let v_i be the first vertex in P where $x \to v_i$. Note that since x has nonzero out-degree in T we know that such a v_i exists.

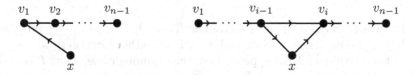

If $i = 1$, then adjoin x to the start of P to obtain the path $P' = x\, v_1 v_2 \cdots v_{n-1}$, which is a hamiltonian path for T. Otherwise $i \ge 2$ and so $v_{i-1} \to x$ and $x \to v_i$. We can insert x between v_{i-1} and v_i to obtain $P' = v_1 \cdots v_{i-1}\, x\, v_i \cdots v_n - 1$, which is a hamiltonian path for T.

Thus by induction every tournament has a hamiltonian path.

So what type of tournament will have a hamiltonian cycle? As we have seen above, having a vertex with in-degree or out-degree 0 prevents a hamiltonian cycle (and so the transitive tournaments are not hamiltonian), but that can't be the only deciding factor as the tournament with degree sequence $1, 1, 1, 4, 4, 4$ also is not hamiltonian. We need an additional constraint on tournaments to guarantee hamiltonian cycles.

Definition 2.21 A digraph G is ***strong*** if for any pair of distinct vertices x and y there exists an $x - y$ path and a $y - x$ path in G.

Look back on the digraphs appearing in this section—which ones would be classified as strong? If we tried to examine all pairs of vertices and looked for possible paths between them, we could be working for quite a long time (though with small tournaments this wouldn't be to bad). Luckily there is a technique from Chapter 1 on score sequences that can be modified to determine if a tournament is strong.

Theorem 2.22 An increasing sequence $S : s_1, s_2, \ldots, s_n$ (for $n \geq 2$) of nonnegative integers is a score sequence of a strong tournament if and only if

$$s_1 + s_2 + \cdots + s_k > \frac{k(k-1)}{2}$$

for each k with $1 \leq k \leq n - 1$ and with equality holding at $k = n$.

One note of caution: this works only for tournaments, whereas we can talk about more general digraphs as strong or not strong. For our purposes, however, we can apply this to show T_1 is strong, whereas T_3 is not (see Exercise 2.17). The theorem below concludes our hamiltonian tournament section.

Theorem 2.23 A tournament is strong if and only if it is hamiltonian.

Proof: Suppose T is hamiltonian. Then there exists a spanning cycle $v_1 v_2 \cdots v_n$. Then we can find a path (in either direction) between any two vertices of T using portions of the spanning cycle. Thus T is strong.

Conversely, suppose T is strong and let $C = v_1 v_2 \cdots v_k v_1$ be the longest cycle in T. If C spans T, then it is a hamiltonian cycle. Otherwise, there exists at least one vertex x that is not part of the cycle. If there exist some i so that $v_i \to x \to v_{i+1}$, then x could be inserted into C to create a longer cycle, contradicting the assumption that C is the longest cycle of T. Thus if there is an edge from x to any vertex of C, then there must be an edge from x to every vertex of C. Similarly, if there is an edge to x from any vertex of C, then there is an edge to x from every edge of C. Thus we can partition the vertices not on C into two sets:

$$N^+(C) = \{x \in V(T) \mid v_i \to x \text{ for all } 1 \leq i \leq k\}$$
$$N^-(C) = \{x \in V(T) \mid x \to v_i \text{ for all } 1 \leq i \leq k\}$$

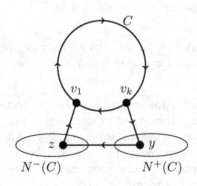

Note that $N^+(C)$ and $N^-(C)$ cannot both be empty. Moreover, since T is strong if $x \in N^+(C)$ then any path from x to C must travel through $N^-(C)$ and if $x \in N^-(C)$ then any path from C to x must travel through $N^+(C)$. Thus there must exist some $y \in N^+(C)$ and some $z \in N^-(C)$ so that $y \to z$ (as shown in the previous figure). By definition, $v_k \to y$ and $z \to v_1$ and so $C' = v_1 v_2 \cdots v_k\, y\, z\, v_1$ is a longer cycle in T. This is a contradiction and so C must span T, making T a hamiltonian tournament.

2.3   Shortest Paths

The shortest way to travel between two locations is perhaps one of the oldest questions. As any mathematics student knows, the answer to this question is a line. But this relies on an xy-plane with no barriers to traveling in a straight line. What happens when you must restrict yourself to an existing structure, such as roadways or rail lines? This problem can be described in graph theoretic terms as the search for a *shortest path* on a weighted graph. Recall that a path is a sequence of vertices in which there is an edge between consecutive vertices and no vertex is repeated. As with the algorithms for the Traveling Salesman Problem, the weight associated to an edge may represent more than just distance (e.g., cost or time) and the shortest path really indicates the path of least total weight.

As with the previous two topics in this chapter, our study of shortest paths can be traced to a specific moment of time. In 1956 Edsger W. Dijkstra proposed the algorithm we are about to study not out of necessity for finding a shortest route, but rather as a demonstration of the power of a new "automatic computer" at the Mathematical Centre in Amsterdam. The goal was to have a question easily understood by a general audience while also allowing for audience participation in determining the inputs of the algorithm. In Dijkstra's own words "the demonstration was a great success" [23]. Perhaps more surprising is how important this algorithm would become to modern society— almost every GIS (Geographic Information System, or mapping software) uses a modification of Dijkstra's Algorithm to provide directions. In addition, Dijkstra's Algorithm provides the backbone of many routing systems and some studies in epidemiology.

Note, we will only investigate *how* to find a shortest path since determining if a shortest path exists is quickly answered by simply knowing if the graph is connected. The following section will consider implications of shortest paths.

2.3.1   Dijkstra's Algorithm

Numerous versions of Dijkstra's Algorithm exist, though two basic descriptions adhere to Dijkstra's original design (see [22]). In one, a shortest path from your chosen starting and ending vertex is found. Though useful in its own right, we will study the more general version that finds the shortest path from a specific vertex to all other vertices in the graph (since if we only cared for the shortest path from a to b, we could halt the algorithm once b is reached).

Dijkstra's Algorithm is a bit more complex than the algorithms we have studied so far. Each vertex is given a two-part label $L(v) = (x, (w(v))$. The first portion of the label is the name of the vertex used to travel to v. The second part is the weight of the path that was used to get to v from the designated starting vertex. At each stage of the algorithm, we will consider a set of *free* vertices, denoted by an F below. Free vertices are the neighbors of previously visited vertices that are themselves not yet visited.

Dijkstra's Algorithm

Input: Weighted connected simple graph $G = (V, E, w)$ and designated *Start* vertex.

Steps:

1. For each vertex x of G, assign a label $L(x)$ so that $L(x) = (-, 0)$ if $x = Start$ and $L(x) = (-, \infty)$ otherwise. Highlight *Start*.

2. Let $u = Start$ and define F to be the neighbors of u. Update the labels for each vertex v in F as follows:

   if $w(u) + w(uv) < w(v)$, then redefine $L(v) = (u, w(u) + w(uv))$

   otherwise do not change $L(v)$

3. Highlight the vertex with lowest weight as well as the edge uv used to update the label. Redefine $u = v$.

4. Repeat Steps (2) and (3) until each vertex has been reached. In all future iterations, F consists of the un-highlighted neighbors of all previously highlighted vertices and the labels are updated only for those vertices that are adjacent to the last vertex that was highlighted.

5. The shortest path from *Start* to any other vertex is found by tracing back using the first component of the labels. The total weight of the path is the weight given in the second component of the ending vertex.

Output: Highlighted path from *Start* to any vertex x of weight $w(x)$.

Perhaps the most complex portion of this algorithm is the labeling of the vertices and how they are updated with iterations of Step (2) and Step (3). In the initial step of Dijkstra's Algorithm, all vertices have no entry in the first part of the label and the second part is 0 for the starting vertex and ∞ for all others. Note that the set F of free vertices consists of all neighbors of highlighted vertices and all are under consideration for becoming the next highlighted vertex. It is important that we do not only consider the neighbors of the last vertex highlighted, as a path from a previously chosen vertex may in fact lead to the shortest path. The example below provides a detailed explanation in the updating of the vertex labels and how to use them to find a shortest path.

Example 2.17 Apply Dijkstra's Algorithm to the graph below where $Start = g$.

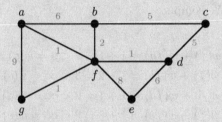

Solution: In each step, the label of a vertex will be shown in the table on the right.

Step 1: Highlight g. Define $L(g) = (-,0)$ and $L(x) = (-,\infty)$ for all $x = a, \cdots, f$.

$a_{(-,\infty)}$ $b_{(-,\infty)}$ $c_{(-,\infty)}$

$d_{(-,\infty)}$

$f_{(-,\infty)}$

$g_{(-,0)}$ $e_{(-,\infty)}$

$F = \{\}$

a	$(-,\infty)$
b	$(-,\infty)$
c	$(-,\infty)$
d	$(-,\infty)$
e	$(-,\infty)$
f	$(-,\infty)$
g	$(-,0)$

Step 2: Let $u = g$. Then the neighbors of g comprise $F = \{a, f\}$. We compute

$$w(g) + w(ga) = 0 + 9 = 9 < \infty = w(a)$$
$$w(g) + w(gf) = 0 + 1 = 1 < \infty = w(f)$$

Update $L(a) = (g, 9)$ and $L(f) = (g, 1)$. Since the minimum weight for all vertices in F is that of f, we highlight the edge gf and the vertex f.

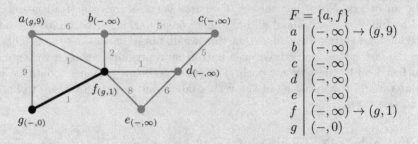

$$
\begin{array}{l}
F = \{a, f\} \\
\begin{array}{c|l}
a & (-, \infty) \to (g, 9) \\
b & (-, \infty) \\
c & (-, \infty) \\
d & (-, \infty) \\
e & (-, \infty) \\
f & (-, \infty) \to (g, 1) \\
g & (-, 0)
\end{array}
\end{array}
$$

Step 3: Let $u = f$. Then the neighbors of all highlighted vertices are $F = \{a, b, d, e\}$. We compute

$$
\begin{aligned}
w(f) + w(fa) &= 1 + 1 = 2 < 9 = w(a) \\
w(f) + w(fb) &= 1 + 2 = 3 < \infty = w(b) \\
w(f) + w(fd) &= 1 + 1 = 2 < \infty = w(d) \\
w(f) + w(fe) &= 1 + 8 = 9 < \infty = w(e)
\end{aligned}
$$

Update $L(a) = (f, 2)$, $L(b) = (f, 3)$, $L(d) = (f, 2)$ and $L(e) = (f, 9)$. Since the minimum weight for all vertices in F is that of a or d, we choose to highlight the edge fa and the vertex a.

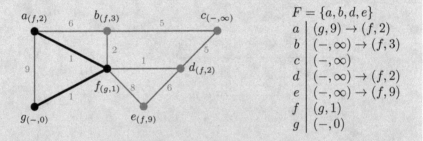

$$
\begin{array}{l}
F = \{a, b, d, e\} \\
\begin{array}{c|l}
a & (g, 9) \to (f, 2) \\
b & (-, \infty) \to (f, 3) \\
c & (-, \infty) \\
d & (-, \infty) \to (f, 2) \\
e & (-, \infty) \to (f, 9) \\
f & (g, 1) \\
g & (-, 0)
\end{array}
\end{array}
$$

Step 4: Let $u = a$. Then the neighbors of all highlighted vertices are $F = \{b, d, e\}$. Note, we only consider updating the label for b since this is the only vertex adjacent to a, the vertex highlighted in the previous step.

$$
w(a) + w(ba) = 2 + 6 = 8 \not< 2 = w(b)
$$

We do not update the label for b since the computation above is not less than the current weight of b. The minimum weight for all vertices in F is that of d, and so we highlight the edge fd and the vertex d.

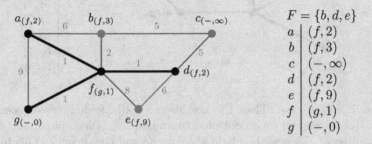

$$F = \{b, d, e\}$$

a	$(f, 2)$
b	$(f, 3)$
c	$(-, \infty)$
d	$(f, 2)$
e	$(f, 9)$
f	$(g, 1)$
g	$(-, 0)$

Step 5: Let $u = d$. Then the neighbors of all highlighted vertices are $F = \{b, c, e\}$. We compute

$$w(d) + w(dc) = 2 + 5 = 7 < \infty = w(c)$$
$$w(d) + w(de) = 2 + 6 = 8 < 9 = w(e)$$

Update $L(c) = (d, 7)$ and $L(e) = (d, 8)$. Since the minimum weight for all vertices in F is that of b, we highlight the edge bf and the vertex b.

$$F = \{b, c, e\}$$

a	$(f, 2)$
b	$(f, 3)$
c	$(-, \infty) \rightarrow (d, 7)$
d	$(f, 2)$
e	$(f, 9) \rightarrow (d, 8)$
f	$(g, 1)$
g	$(-, 0)$

Step 6: Let $u = b$. Then the neighbors of all highlighted vertices are $F = \{c, e\}$. However, we only consider updating the label of c since e is not adjacent to b. Since

$$w(b) + w(bc) = 3 + 5 = 8 \not< 7 = w(c)$$

we do not update the labels of any vertices. Since the minimum weight for all vertices in F is that of c we highlight the edge dc and the vertex c. This terminates the iterations of the algorithm since our ending vertex has been reached.

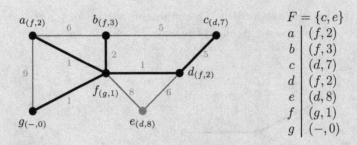

Step 7: Let $u = c$. Then the neighbors of all highlighted vertices are $F = \{e\}$. However, we do not need to update any labels since c and e are not adjacent. Thus we highlight the edge de and the vertex e. This terminates the iterations of the algorithm since all vertices are now highlighted.

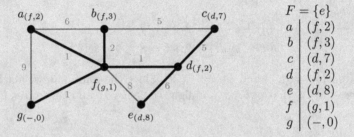

Output: The shortest paths from g to all other vertices can be found highlighted above. For example the shortest path from g to c is $g\,f\,d\,c$ and has a total weight 7, as shown by the label of c.

2.3.2   Walks Using Matrices

Recall in Section 1.4 we saw how to model a graph using an adjacency matrix. Matrix representations of graphs are useful when using a computer program to investigate certain features or processes on a graph. Another use for the adjacency matrix is to count the number of walks between two vertices within a graph. For review of matrix operations, see Appendix C.

Consider the graph shown below with its adjacency matrix A on the right.

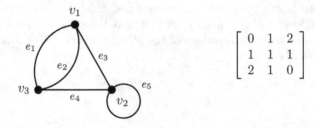

If we want a walk of length 1, we are in essence asking for an edge between two vertices. So to count the number of walks of length 1 from v_1 to v_3, we need only to count the number of edges (namely 2) between these vertices. What if we want the walks of length 2? By inspection, we can see there is only one, which is

$$v_1 \underset{e_3}{\to} v_2 \underset{e_4}{\to} v_3$$

Now consider the walks from v_1 to v_2. There is only one walk of length 1, and yet three of length 2:

$$v_1 \underset{e_3}{\to} v_2 \underset{e_5}{\to} v_2$$
$$v_1 \underset{e_1}{\to} v_3 \underset{e_4}{\to} v_2$$
$$v_1 \underset{e_2}{\to} v_3 \underset{e_4}{\to} v_2$$

How could we count this? If we know how many walks there are from v_1 to v_2 (1) and then the number from v_2 to itself (1), we can get one type of walk from v_1 to v_2. Also, we could count the number of walks from v_1 to v_3 (2) and then the number of walks from v_3 to v_2 (1). In total we have $1*1+2*1=3$ walks from v_1 to v_2. Note that we did not include any walks of the form $v_1v_1v_2$ since there are no edges from v_1 to itself.

Viewing this as a multiplication of vectors, we have

$$\begin{bmatrix} 0 & 1 & 2 \end{bmatrix} \cdot \begin{bmatrix} 1 \\ 1 \\ 1 \end{bmatrix} = 0*1+1*1+2*1 = 3$$

If we do this for the entire adjacency matrix, we have

$$A^2 = \begin{bmatrix} 5 & 3 & 1 \\ 3 & 3 & 3 \\ 1 & 3 & 5 \end{bmatrix}$$

Thus the entry a_{ij} in A^2 represents the number of walks between vertex v_i and v_j of length 2. If we multiplied this new matrix by A again, we would simply be counting the number of ways to get from v_i to v_j using 3 edges. The theorem below summarizes this for walks of any length n.

Theorem 2.24 Let G be a graph with adjacency matrix A. Then for any integer $n > 0$ the entry a_{ij} in A^n counts the number of walks from v_i to v_j.

2.3.3 Distance, Diameter, and Radius

Dijkstra's Algorithm provides the method for determining the shortest path between two points on a graph, which we define as the *distance* between those vertices. There are many theoretical implications for this distance. We will investigate a few of these below; further discussion will occur throughout this book, most notably in Chapter 3 when discussing trees and in Chapter 4 when discussing connectivity. In particular, we will begin with defining the diameter and radius of a graph and the eccentricity of a vertex.

Definition 2.25 Given two vertices x, y in a graph G, define the *distance* $d(x, y)$ as the length of the shortest path from x to y. The *eccentricity* of a vertex x is the maximum distance from x to any other vertex in G; that is $\epsilon(x) = \max_{y \in V(G)} d(x, y)$.

The *diameter* of G is the maximum eccentricity among all vertices, and so measures the maximum distance between any two vertices; that is $diam(G) = \max_{x, y \in V(G)} d(x, y)$. The *radius* of a graph is the minimum eccentricity among all vertices; that is $rad(G) = \min_{x \in V(G)} \epsilon(x)$.

If a graph is connected, all of these parameters will be nonnegative integers. What happens if the graph is disconnected? If x and y are in separate components of G then there is no shortest path between them and $d(x, y) = \infty$. This would then make $diam(G) = rad(G) = \infty$ since $\epsilon(v) = \infty$ for all vertices in G. Conceptually, you can think of the diameter as the longest path you can travel between any two points on a graph and the radius as the shortest distance among all pairs of vertices.

Example 2.18 Find the diameter and radius for the graph below.

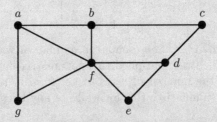

Solution: Note that f is adjacent to all vertices except c, but there is a path of length 2 from f to c. As no vertex is adjacent to all other vertices, we know the radius is 2. The longest path between two vertices is from g to c, and is of length 3, so the diameter is 3.

While graphs can exist with arbitrarily large diameter and radius, there is a direct relationship between the diameter and radius of a graph and that of its complement. The first result below is for disconnected graphs, though it can be further generalized (see Exercise 2.30) for connected graphs.

Theorem 2.26 If G is disconnected then $\overline{G}$ is connected and $diam(\overline{G}) \leq 2$.

Proof: Assume G is disconnected. To prove $\overline{G}$ is connected, we must show there is an $x - y$ path for any pair of vertices x, y. If x and y are not adjacent in G then $xy \in E(\overline{G})$. Thus the edge xy is itself a $x - y$ path. Otherwise, $xy \in E(G)$ and so x and y are in the same component of G and not adjacent in $\overline{G}$. Since G is not connected, there must exist some vertex z in a different component from x and y, and so z cannot be adjacent to either of x or y. This implies $xz, yz \in E(\overline{G})$, and so xzy is a $x - y$ path in $\overline{G}$. Note that every pair of vertices in $\overline{G}$ fall into one of these two cases and so satisfy $d(x, y) \leq 2$. Thus $\overline{G}$ is connected and $diam(\overline{G}) \leq 2$.

A similar result can be found in terms of the radius of a graph and its complement. Note that the result below does not assume connectivity of G.

Theorem 2.27 For a simple graph G if $rad(G) \geq 3$ then $rad(\overline{G}) \leq 2$.

Proof: Since G is simple and $rad(G) = r \geq 3$, we know that $\epsilon(v) \geq r$ for all vertices in G. In particular, there exists some path $wxyz$ such that w is not adjacent to y and z, and x is not adjacent to z. Since $r \geq 3$ we know $\epsilon(x) \geq 3$ and so there must exist another vertex v (not one of w, y, z) such that $d(x, v) \geq 3$. Thus v is not adjacent to either of x or y. Moreover, v cannot be adjacent to both w and z since otherwise $d(w, z) < 3$. Thus at least one of the edges vw or vz, but possibly both, cannot exist in G.

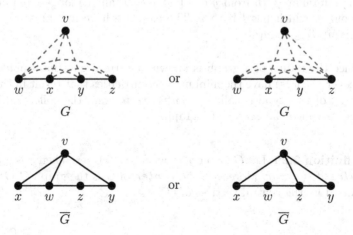

In either case, we have that the distance in $\overline{G}$ between any two of these vertices is at most 2, and since this holds for any collection of vertices in G, we see that $rad(\overline{G}) \leq 2$.

Since the radius measures the shortest distance between two vertices and the diameter the longest, we should expect that these quantities cannot be too far away from one another. In geometry when we study circles, we define the diameter to be twice the radius. But is the same true in graph theory? As we have already seen above, the diameter need not be equal to twice the radius, but as the result below proves it cannot be larger than twice the radius.

Theorem 2.28 For any simple graph G, $rad(G) \leq diam(G) \leq 2rad(G)$.

Proof: First note that if the radius of a graph G is r then there is some vertex v with $\epsilon(v) = r$. Thus there cannot be a longest path from v to any other vertex that is shorter than r. Thus $rad(G) \leq diam(G)$.

Next, suppose x, y, and z are vertices of G. Then the shortest path from x to z may or may not travel along the shortest paths from x to y and then from y to z.

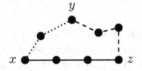

In either case, we know that $d(x, y) + d(y, z) \geq d(x, z)$. So suppose x and z are chosen so that $d(x, z) = diam(G)$. Let y be a vertex so that $\epsilon(y) = rad(G) = r$. Then we know that $d(x, y) \leq r$ and $d(y, z) \leq r$ since no minimum path from y can be longer than the longest minimum path from y, which has length r. Therefore we have $d(x, z) \leq r + r$ and so $diam(G) \leq 2rad(G)$.

Once the radius of a graph is known, a natural question would be which vertex (or vertices) have the minimum eccentricities that result in that radius. A vertex of this type is called a *central vertex* and the collection of central vertices is called the *center* of a graph.

Definition 2.29 Let G be a graph with $rad(G) = r$. Then x is a **central vertex** if $\epsilon(x) = r$. Moreover, the **center** of G is the graph $C(G)$ that is induced by the central vertices of G.

Example 2.19 Find the center of the graph from Example 2.18.

Solution: Vertices a, b, d, e, and f all have eccentricities of 2. The center of G is the graph induced by these vertices, as shown below.

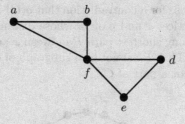

Radius and diameter rely on (shortest) paths in their definition. If we turn instead to cycles, we can ask a similar question as to the length of both the shortest and longest cycles within a graph.

Definition 2.30 Given a graph G, the ***girth*** of G, denoted $g(G)$, is the minimum length of a cycle in G. The ***circumference*** of G is the maximum length of a cycle.

Note that the circumference of any graph is at most n, the number of vertices since no vertex can be repeated in a cycle. Moreover, if a graph does not have any cycles (what we will call a tree in Chapter 3) then we define $g(G) = \infty$ and the circumference to be 0.

Example 2.20 Find the girth and circumference for the graph from Example 2.18.

Solution: Since the graph is simple, we know the girth must be at least 3, and since we can find triangles within the graph we know $g(G) = 3$. Moreover, the circumference is 7 since we can find a cycle containing all the vertices (try it!).

It shouldn't be too surprising that the diameter and girth of a graph are related, since removing one edge xy from a cycle creates a path and would increase the distance between x and y.

Theorem 2.31 If G is a graph with at least one cycle then $g(G) \leq 2diam(G) + 1$.

Proof: Assume C us a cycle of minimum length in G, namely C is of length $g(G)$. Suppose for a contradiction that $g(G) \geq 2diam(G)+2$. Then there exist two vertices x and y on C whose distance along C us at least $diam(G) + 1$. But any shortest path P between x and y must have length at most $diam(G)$. Thus P cannot be a subgraph of C, i.e. P must contain edges or vertices not a part of C.

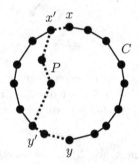

Let x' be the last vertex for which C and P agree and y' be the next vertex on P and C. Then the cycle C' formed using the portions of C and P between x' and y' is shorter than cycle C. This is a contradiction to C being a cycle of minimum length. Thus $g(G) \leq 2diam(G) + 1$.

While girth, radius, and diameter have clear relationships to one another, the same cannot be said about their relationship to degree measures of a graph. In fact, we can find connected graphs with large diameter when the minimum degree is quite small or quite large (see Exercise 2.15); however, in order to maintain a small minimum degree with a large diameter, we will need to have a large number of vertices in the graph. Conversely, if a graph has a small diameter and maximum degree then it cannot have too many vertices. The proof below uses the closed form of a geometric series as follows:

$$\sum_{i=0}^{k-1} x^i = \frac{x^k - 1}{x - 1}$$

We will use this result below with $x = d - 1$.

Theorem 2.32 Let G be a graph with n vertices, radius at most k, and maximum degree at most d, with $d \geq 3$. Then $n < \dfrac{d}{d-2}(d-1)^k$.

Proof: Assume v is a central vertex of a graph G with maximum degree at most d (with $d \geq 3$). Define V_i to be the vertices whose distance from v is i. Then $V_0 = \{v\}$, V_1 contains all the neighbors of v, and V_2 contains the vertices of distance 2 from v. Thus $|V_0| = 1$ and $|V_1| \leq d$ since d is the maximum degree of G.

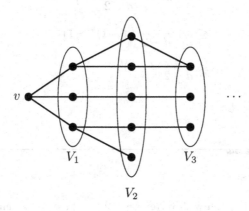

Every vertex in V_{i+1} (for $i > 0$) has an edge to a vertex in V_i, and every vertex in V_i has at most $d - 1$ neighbors in V_{i+1} since each vertex in V_i also has an edge to a vertex in V_{i-1}. Thus $|V_{i+1}| \leq (d-1)|V_i|$ and $|V_i| \leq (d-1)|V_{i-1}|$. This gives

$$
\begin{aligned}
|V_{i+1}| &\leq (d-1)|V_i| \\
&\leq (d-1)(d-1)|V_{i-1}| \\
&\;\;\vdots \\
&\leq \underbrace{(d-1)\cdots(d-1)}_{i \text{ times}}|V_1| \\
&\leq (d-1)^i d
\end{aligned}
$$

Since G has radius k we know that $V(G)$ is the union of these sets, that is $V(G) = \bigcup\limits_{i=0}^{k} V_i$. Moreover, these sets are disjoint. Thus we can calculate n, the number of vertices in G, as

$$n = |V_0| + |V_1| + \cdots + |V_k|$$
$$\leq 1 + d + d(d-1) + \cdots + d(d-1)^{k-1}$$
$$\leq 1 + d \sum_{i=0}^{k-1} (d-1)^i$$
$$\leq 1 + d \left(\frac{(d-1)^k - 1}{d-2} \right)$$
$$\leq 1 + \frac{d}{d-2}((d-1)^k - 1)$$
$$< \frac{d}{d-2}(d-1)^k.$$

2.4   Exercises

2.1 Let G be a graph with vertex set $V(G) = \{a, b, c, d, e\}$ and edge set $E(G) = \{ab, ae, bc, cd, de, ea, eb\}$.
  (a) Draw G.
  (b) Is G connected?
  (c) Is G simple?
  (d) List the degrees of every vertex.
  (e) Find all edges incident to b.
  (f) List all the neighbors of a.
  (g) Find a walk, trail, and path in G, each of which has length 3.
  (h) Find a closed walk, circuit, and cycle in G, each of which starts at e.
  (i) Is G eulerian, semi-eulerian, or neither? Explain your answer.

2.2 Which of the following scenarios could be modeled using (i) an eulerian circuit or trail? (ii) hamiltonian cycle or path? Explain your answer.
  (a) A photographer wishes to visit each of the seven bridges in a city, take photos, then return to his hotel.
  (b) Salem Public Works must repave all the streets in the downtown area.
  (c) Frank's Flowers needs to deliver bouquets to 6 customers throughout the city, starting and ending at the flower shop.
  (d) Richmond Water Authority must read all the water meters throughout the town. One worker is tasked with this job.
  (e) Sam works in sales for a Fortune 500 company. He spends each day visiting his clients around southwest Virginia and must plan his route to avoid backtracking as much as possible.

2.3 For each of the graphs below (i) find the degree of each vertex (ii) use your results from (i) to determine if the graph is eulerian, semi-eulerian, or neither, and (iii) find an eulerian circuit or eulerian trail if it exists. Explain your answer.

(a)

(b)

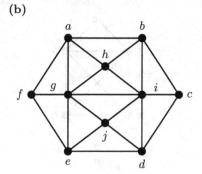

(c)

(d)

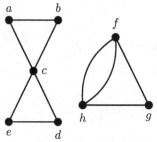

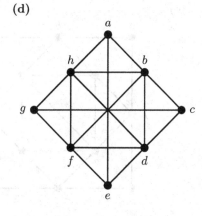

(e)

(f)

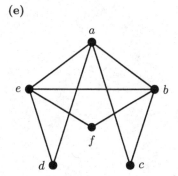

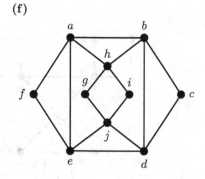

2.4 Find an eulerian circuit or eulerian trail for each of the graphs below.

(a)

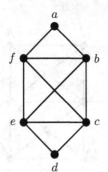

(b)

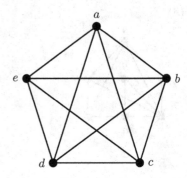

(c)

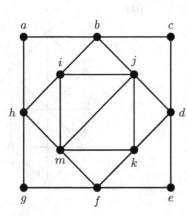

(d)

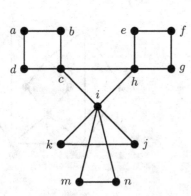

2.5 Determine if each of the graphs below are hamiltonian. For those that are, find a hamiltonian cycle. Otherwise, provide a clear and concise argument as to why the graph is not hamiltonian.

(a) **(b)**

(c) **(d)**

(e)

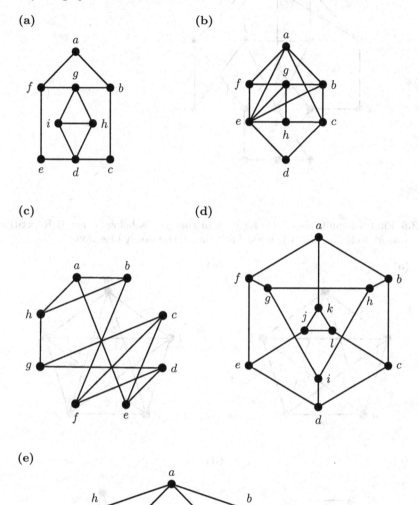

(f)

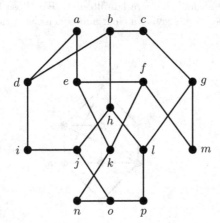

2.6 Find a hamiltonian cycle for each of the graphs below using (i) Repetitive Nearest Neighbor, (ii) Cheapest Link, and (iii) Nearest Insertion.

(a)

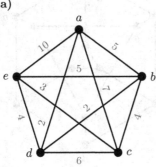

(b)

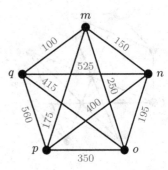

(c)

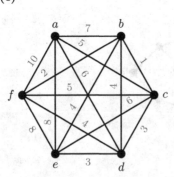

(d)

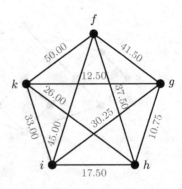

(e)

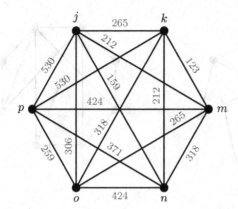

(f)

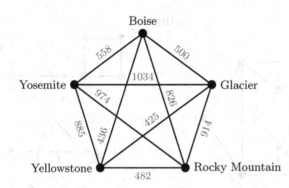

2.7 Find the hamiltonian closure of each of the graphs below.

(a)   **(b)**

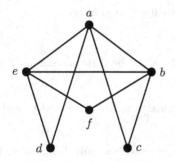   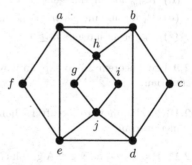

(c)   **(d)**

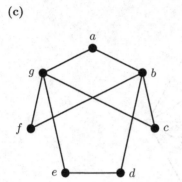 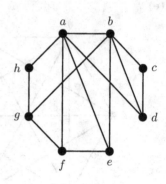

2.8 Determine the diameter, radius, and center for the graphs or class of graphs described below.

(a)   **(b)**

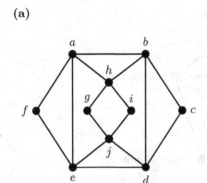 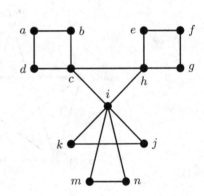

    (c) K_n for any integer $n \geq 1$

    (d) $K_{m,n}$ for any integers $m, n \geq 1$

    (e) P_n for any integer $n \geq 1$ (Hint: consider if n is even or odd)

    (f) C_n for all integers $n \geq 3$

2.9 Use Theorem 2.7 to explain why a graph cannot be both eulerian and semi-eulerian.

2.10 Explain how to modify Hieholzer's Algorithm so it can be used to find an eulerian trail.

2.11 Prove Corollary 2.8: A graph G is semi-eulerian if and only if G is connected and exactly two vertices have odd degree.

2.12 Determine an optimal semi-eulerization for the town of Crystal Spring on page 62 if the two odd vertices must be on the edge of the graph.

2.13 Show the steps to calculate the hamiltonian closure of G from Example 2.10 on page 77 when the first edge added is fd instead of bf.

2.14 Find a graph G for which both G and $\overline{G}$ are hamiltonian.

2.15 Find a graph (G) with
  (a) diameter k and $\delta(G) = 1$.
  (b) diameter k and $\delta(G) = k$.

2.16 Let G be a graph with n vertices and circumference n. Explain why G must be hamiltonian.

2.17 Use Theorem 2.22 to show T_1 from page 98 is strong but T_3 from page 98 is not.

2.18 Assume G is a connected simple graph with n vertices and m edges. Prove $n \leq m + 1$. (Hint: Argue by induction on m.)

2.19 Assume that every vertex in a simple graph G has degree at least $\dfrac{n}{2}$. Prove that G is connected.

2.20 Recall that $\delta(G)$ is the minimum degree of the graph G.
  (a) Prove that a simple graph G is connected if $\delta(G) \geq \dfrac{n-1}{2}$.
  (b) Show that this bound is best possible; that is, show there exists a graph G with $\delta(G) = \dfrac{n-2}{2}$ that is disconnected.

2.21 Recall that a graph G is regular if each vertex has the same degree. Suppose G is a connected regular graph with n vertices and $n \geq 3$.
  (a) Prove that G must be eulerian if n is odd.
  (b) Give examples to demonstrate that if n is even then G may or may not be eulerian.

2.22 Suppose x and y are distinct nonadjacent vertices of G such that $\deg(x) + \deg(y) \geq n$. Prove $G + xy$ is hamiltonian if and only if G is hamiltonian.

2.23 Under what conditions will $K_{m,n}$ be hamiltonian? Prove your answer.

2.24 Prove that $K_{n,2n,3n}$ is hamiltonian for all positive n.

2.25 Prove that $K_{n,2n,3n+1}$ is not hamiltonian for all positive n.

2.26 Prove that a graph is bipartite if and only if it does not contain any odd cycles.

2.27 Prove a graph G has radius 1 if and only if G contains a vertex adjacent to all other vertices of G.

2.28 Let G be a graph with adjacent vertices x and y. Prove $\epsilon(x)$ and $\epsilon(y)$ differ by at most 1.

2.29 Prove that every graph contains a path of length at least $\delta(G)$ and if $\delta(G) \geq 2$ then G also has a cycle of length at least $\delta(G) + 1$.

2.30 If G is a simple graph with $diam(G) \geq 3$ then $diam(\overline{G}) \leq 3$.

2.31 Prove that if G is a simple graph with $diam(G) \geq 4$ then $diam(\overline{G}) \leq 2$.

2.32 Prove that if G is a regular graph with $diam(G) = 3$ then $diam(\overline{G}) = 2$.

2.33 Prove that a transitive tournament has exactly one hamiltonian path.

2.34 Prove that every regular tournament (with at least 3 vertices) is strong.

3

Trees

The previous chapter was focused on the routes within a graph, whether they be exhaustive (such as an eulerian circuit or hamiltonian cycle) or optimal (such as the shortest path or cheapest hamiltonian cycle). In this chapter we switch focus to an underlying structure that guarantees connectivity. We will revisit the subject of connectivity in Chapter 4 when the question switches to how connected a graph is and various measures for evaluating the notion of connectedness. For now, consider the following scenario:

> Franklin Homes is building a new housing development and will be laying down fiber optic cable. As there is a cost for each foot of cable used, they need to determine the optimal system of cables while ensuring each house is connected to the junction box at the entrance.

If the only concern is connectivity, we could just lay cable between every house and the junction box, but this would be extremely inefficient since the cable to one house may run directly over that of another home. Franklin Homes should instead build out a structure that maintains connectivity, but eliminates any cycles and minimizes the total length of cable used; in graph theoretic terms, they want a *tree*.

Definition 3.1 A graph G is

- *acyclic* if there are no cycles or circuits in the graph.

- a *tree* if it is both acyclic and connected.

- a *forest* if it is an acyclic graph.

In addition, a vertex of degree 1 is called a *leaf*.

Note that a forest is simply a graph in which every component is itself a tree.

Trees arise in many seemingly unrelated disciplines, including probability, chemistry, and computer science. In Chapter 1, Example 1.8 discussed how to model a game with a probability tree. Additional applications of trees will be discussed in Sections 3.3 and 3.4. We will begin with a discussion of finding a tree within a larger graph, followed by some theoretical results about trees.

3.1   Spanning Trees

In Chapter 1 we introduced the notion of a spanning subgraph and in Chapter 2 we described hamiltonian cycles as spanning cycles in a graph. Here we will discuss the tree version, where we want a spanning subgraph that is itself a tree.

> **Definition 3.2** A *spanning tree* is a spanning subgraph that is also a tree.

Recall that if an edge appears in a subgraph, then both endpoints must also be included in the subgraph. However, if a vertex appears in a subgraph, any number of its incident edges may be included.

Example 3.1 For each of the graphs below, find a spanning tree and a subgraph that does not span.

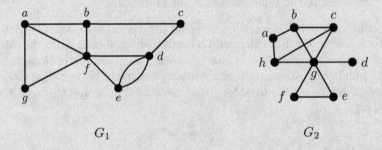

G_1 G_2

Solution: To find a spanning tree, we must form a subgraph that is connected, acyclic, and includes every vertex from the original graph. The graphs T_1 and T_2 below are two examples of spanning trees for their respective graphs; other examples exist.

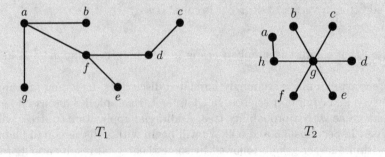

T_1 T_2

The subgraph H_1 below is neither spanning nor a tree since some vertices from G_1 are missing and there is a multi-edge (and hence a circuit) between d and e. The subgraph H_2 below is not spanning since it does not contain vertex a, but it is a tree since no circuits or cycles exist. As above, these are merely examples and other non-spanning subgraphs exist.

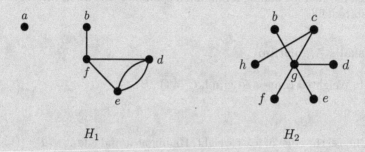

$$H_1 \qquad\qquad H_2$$

Under what conditions will a graph have a spanning tree? Clearly the more difficult criteria is the tree, not spanning, since every graph contains a spanning subgraph. The trick then is to ensure the spanning subgraph is both connected and acyclic. If the original graph is not connected, then we have no hope of finding a spanning tree; however, if the graph has cycles, we only need to remove enough edges to ensure the result is connected yet no cycles remain. Thus every connected graph contains a spanning tree. How then can we find a best one? And how to we determine what is best?

3.1.1 Minimum Spanning Trees

In Chapter 2 we not only asked if a graph had a hamiltonian cycle, but also how to find the optimal one within a weighted graph. Just as knowing if a spanning tree exists within a graph was an easy question to answer, finding an optimal one is also straightforward.

Definition 3.3 Given a weighted graph $G = (V, E, w)$, T is a *minimum spanning tree*, or MST, of G if it is a spanning tree with the least total weight.

There are many algorithms that can find a MST, but we will focus on only two. Additional algorithms will appear in the Exercises.

Kruskal's Algorithm

Similar to Dijkstra's Algorithm studied in Section 2.3, Kruskal's Algorithm is fairly modern, first published in 1956 [60]. Joseph Kruskal was an American mathematician best known for his work in statistics and computer science. This algorithm is unique in that it is both efficient and optimal while still easily implemented and understandable for a non-scientist. In fact, it is the preferred method for finding a minimum spanning tree when the edges can be easily sorted[40].

Kruskal's Algorithm

Input: Weighted connected graph $G = (V, E)$.

Steps:

1. Choose the edge of least weight. Highlight it and add it to $T = (V, E')$.

2. Repeat Step (1) so long as no circuit is created. That is, keep picking the edges of least weight but skip over any that would create a cycle in T.

Output: Minimum spanning tree T of G.

Kruskal's Algorithm does not distinguish between two edges of the same weight, in part because it does not influence the outcome. If at any point there are two edges to choose from of the same weight, you can pick either one. In addition, at each step of the algorithm we are building a forest subgraph that will eventually result in a spanning tree.

Example 3.2 Find the minimum spanning tree of the graph G below using Kruskal's Algorithm.

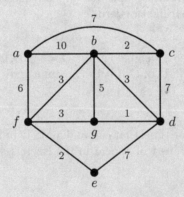

Solution: At each step, the newest edge under consideration will be shown as a dotted line and the previously chosen edges will be in black. Unchosen edges will be shown in gray.

Step 1: Pick the smallest edge, gd and highlight it.

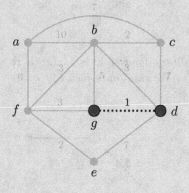

Step 2: Pick the next smallest edge. There are two edges of weight 2 (bc and ef). Either is a valid choice. We choose bc.

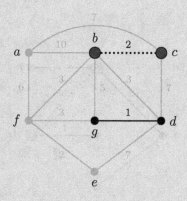

Step 3: The other edge of weight 2, ef, is still a valid choice. Add it to T.

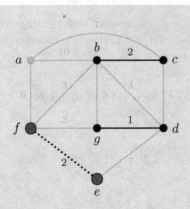

Step 4: The next smallest edge weight is 3, and there are 3 edges to choose from $(bd, bf,$ and $fg)$. We randomly pick bf.

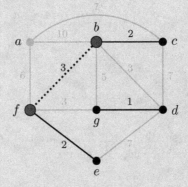

Step 5: Both of the other edges of weight 3 are still available. We choose bd.

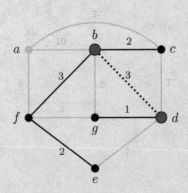

Step 6: At this point, we cannot choose the last edge of weight 3, fg, since it would create a circuit (namely, $b\,d\,g\,f\,b$).

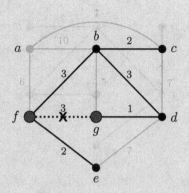

The next smallest edge is *bg* of weight 5. Again, we cannot choose this
edge since it would create a circuit (*b d g b*).

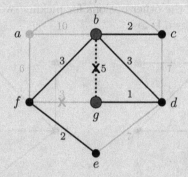

The next available edge is *af* of weight 6. This is also the last edge
needed since we now have a tree containing all the vertices of *G*.

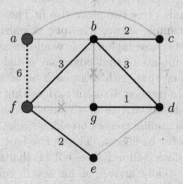

Output: The following tree with total weight 17.

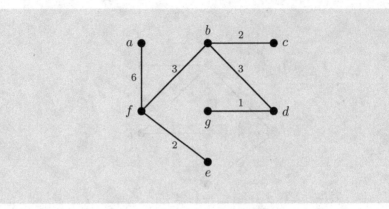

In Steps 4 and 5 above, we made a choice of which edge of weight 3 to add to the subgraph (that would eventually become a spanning tree). There are two other possible minimum spanning trees (each of which has total weight 17) that correspond to the other options for picking two of the three edges of weight 3. These are shown below.

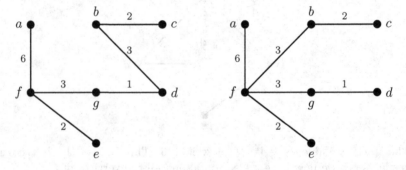

Perhaps the most surprising aspect of Kruskal's Algorithm is the process you would like to take (namely picking the smallest weight edges) also guarantees a minimum spanning tree (we prove this in Theorem 3.12 on page 138). Looking back at the example above, when we skipped over an edge (say *bg* of weight 5), we did so because including it would create a cycle. This means a path between the endpoints of that edge (say *b* and *g*) must already exist and the other edges along that path must each be of weight no greater than the edge we skip over. In essence, Kruskal's Algorithm is focused on avoiding cycle creation and eventually arrives at a connected subgraph. Conversely, if you think of finding a spanning tree as breaking cycles, then the largest edge on that cycle should never be chosen. This is the basis behind another algorithm, Reverse Delete, described in Exercise 3.11, that focuses on maintaining connectedness and eventually arrives at an acyclic subgraph. We conclude this section with another MST algorithm (Prim's Algorithm) that focuses on building an MST out from a specific vertex and has a similar feel to Dijsktra's Algorithm from Section 2.3.1.

Prim's Algorithm

The algorithm described below is widely known as Prim's Algorithm, named for the American mathematician and computer scientist Robert C. Prim. Prim worked closely with Kruskal at Bell Laboratories and published this algorithm in 1957 (one year after the publication of Kruskal's Algorithm). However, it was originally discovered in 1930 by the Czech mathematician Vojtěch Jarnik and also republished in 1959 by Dijkstra (who should be familiar from Section 2.3)[40].

Prim's Algorithm contrasts from Kruskal's in that the structure obtained in each step is itself a tree. By the end of the process, a spanning tree will be found. It begins by denoting a starting vertex for the tree, similar to that of rooted trees that appear in Section 3.3.

Prim's Algorithm

Input: Weighted connected graph $G = (V, E)$.

Steps:

1. Let v be the root. If no root is specified, choose a vertex at random. Highlight it and add it to $T = (V', E')$.

2. Among all edges incident to v, choose the one of minimum weight. Highlight it. Add the edge and its other endpoint to T.

3. Let S be the set of all edges with exactly endpoint from $V(T)$. Choose the edge of minimum weight from S. Add it and its other endpoint to T.

4. Repeat Step (3) until T contains all vertices of G, that is $V(T) = V(G)$.

Output: Minimum spanning tree T of G.

Similar to Dijkstra's Algorithm from Section 2.3, we will consider vertices adjacent to previously chosen vertices. Unlike Dijkstra's Algorithm, however, we are not concerned with weights along a path but rather the total weight of all edges chosen.

Example 3.3 Use Prim's algorithm to find a minimum spanning tree for the graph given in Example 3.2.

Solution: As before, the newest edges under consideration will be shown as dotted lines and the previously chosen edges will be in black. Unchosen edges will be shown in gray. The vertices in T at each stage will be highlighted as well.

Step 1: Since no root was specified, we choose a as the starting vertex.

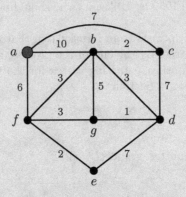

Step 2: We consider the edges incident to a, namely ab, ac, and af. These are shown as dotted lines in the graph on the left. The edge of least weight is af. This is added to the tree, shown in bold on the right.

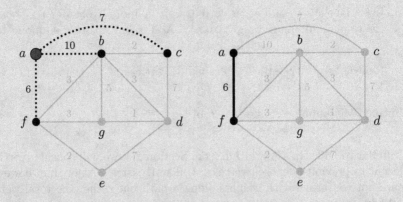

Step 3: The set S consists of edges with one endpoint as a or f, as shown in the graph to the left. The edge of minimum weight from these is ef. This is added to the tree, as shown on the right.

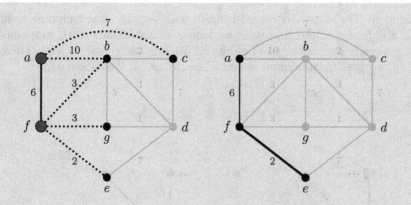

Step 4: The new set S consists of edges with one endpoint as a, e, or f. The next edge added to the tree could either be fg or fb. We choose fg.

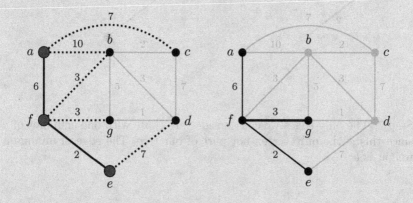

Step 5: We consider the edges where exactly one endpoint is from a, e, f, or g. The next edge to add to the tree is dg.

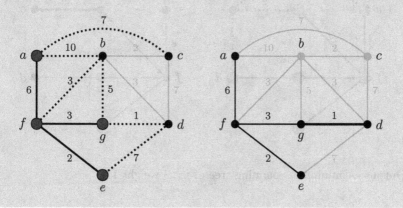

Step 6: The edges to consider must have exactly one endpoint from $a, d, e, f,$ or g. Note that de is no longer available since both endpoints are already part of the tree (and its addition would create a cycle). There are two possible minimum weight edges, bf or bd. We choose bf.

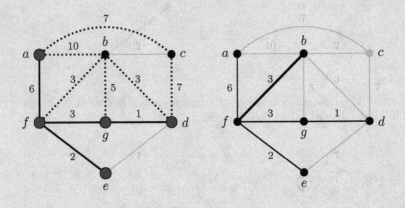

Step 7: The only edges we can consider are those with one endpoint of c since this is the only vertex not part of our tree. The edge of minimum weight is bc.

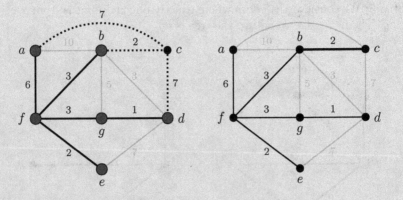

Output: A minimum spanning tree of total weight 17.

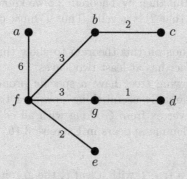

Both minimum spanning tree algorithms described in this chapter are efficient and optimal, and result in roughly the same computation requirements. It should be noted that many other algorithms exist and the study of minimum spanning trees did not originate with Kruskal and Prim. In fact, both mathematicians cited the work of Otakar Borůvka, a Czech mathematician who is credited with the first minimum spanning tree algorithm in 1926. Borůvka's Algorithm, which can be found in Exercise 3.29, is slightly more complex and requires all edge weights to be unique, though is better suited for parallel computing.

3.2 Tree Properties

As finding a minimum spanning tree is (in graph theoretic terms) quick and easy, we focus our attention on the properties of trees and what a spanning tree can tell us about its graph. As is common in mathematics, the things with the simplest definitions provide an abundance of material to study in depth. Trees in particular provide ample opportunities to strengthen our proof writing skills, specifically induction and contradiction methods. We begin with some results that arise through counting techniques. Recall that a vertex of degree 1 is called a leaf.

Theorem 3.4 Every tree with at least two vertices has a leaf.

Proof: Suppose for a contradiction that there exists a tree T with at least two vertices that does not contain a leaf. Since T must be connected, we know no vertex has degree 0, and therefore every vertex of T must have degree at least 2. But then by Theorem 2.5 we know T must have a cycle, which contradicts that T is acyclic. Thus T must contain a leaf.

Exercise 3.18 expands on this theorem to show that every tree (with at least two vertices) in fact has at least two leaves.

Beyond simply showing trees have a specific property, the result above allows us to do a remarkably useful thing—prune a tree! Recall that $G - v$ denotes removing the vertex from G along with all edges incident to v. The proof of the following lemma appears in Exercise 3.16.

Lemma 3.5 Given a tree T with a leaf v, the graph $T - v$ is still a tree.

Removing a leaf from a tree will always remove exactly one vertex and one edge, creating a tree with a smaller size. This technique naturally lends itself to induction arguments.

Theorem 3.6 A tree with n vertices has $n - 1$ edges for all $n \geq 1$.

Proof: Argue by induction on n, the number of vertices in T. If $n = 1$ then T cannot have any edges since any edge in a graph with one vertex must be a loop, which would create a cycle.

Suppose for some $k \geq 1$ that any tree with k vertices has $k - 1$ edges and let T be a tree with $k + 1$ vertices. Since $k + 1 \geq 2$, by Theorem 3.4 we know T must have at least one leaf, call it v. Then by Lemma 3.5 we know $T - v$ is a tree with k vertices. Thus by the induction hypothesis $T - v$ must have $k - 1$ edges. Since v is a leaf, its removal from T deleted exactly one edge and so T must have k edges.

Therefore, for all $n \geq 1$ any tree with n vertices must have $n - 1$ edges.

The corollary below on the total degree of a tree is an immediate consequence of Theorem 3.6, the proof of which appears in Exercise 3.17.

Corollary 3.7 The total degree of a tree on n vertices is $2n - 2$.

The following two results are often written as a singular theorem, since they can each be shown to be equivalent characterizations of trees. The proof of the second result appears in Exercise 3.20. Note that a common proof technique is demonstrated below: when we want to show there is a unique object of a certain type, we assume two exist and find a contradiction.

Proposition 3.8 Let T be a tree. Then for every pair of distinct vertices x and y there exists a unique $x - y$ path.

Proof: First note that there must be at least one $x - y$ path since T is connected. Now suppose this path is not unique. Then there exist distinct vertices x and y such that there are at least two $x - y$ paths in T, call them P_1 and P_2.

Let $v_0 v_1$ be the first edge in P_1 not in P_2. Then there must exist some u_1 such that $v_0 u_1$ is an edge on P_2 that is not on P_1. Let v_n be the next vertex on both P_1 and P_2. The dotted lines below show then edges in P_1 not in P_2 and the dashed lines show the edges on P_2 not in P_1.

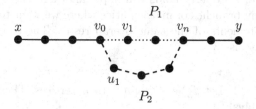

Then $C = v_0 \underset{P_1}{-} v_n \underset{P_2}{-} v_0$ is a cycle since no edges are repeated in a path and no vertex is repeated since then v_n would be that earlier vertex. But this contradicts that T is acyclic, and so there can only be one $x - y$ path in T.

Proposition 3.9 Every tree is minimally connected, that is the removal of any edge disconnects the graph.

We now know that removing an edge from a tree disconnects the graph, which can be restated as every edge of a tree is a *bridge*, that is an edge whose removal will disconnect the graph (these will be studied in more detail in Chapter 4). But what happens if we add an edge to a tree? We know the resulting graph cannot be minimally connected, but what does that really mean? The result below shows that this graph will contain exactly one cycle.

Proposition 3.10 If any edge e is added to a tree T then $T + e$ contains exactly one cycle.

Proof: Let T be a tree with vertices x and y, and let T' be the graph created by adding $e = xy$. Then T' must contain at least one cycle since if x and y are already adjacent then e creates a multiedge, and otherwise there exists a path between x and y in T that results in a cycle when e is added.

It remains to show that this cycle is unique. So suppose T' has at least two cycles, say C_1 and C_2. We know that neither cycle can exist in T since T is a tree and so e must be an edge on both cycles. But then $C_1 - e$ and $C_2 - e$ must each be paths connecting x and y, which contradicts Proposition 3.8 that there is a unique path between x and y. Thus T' contains exactly one cycle.

The theorem below summarizes the previous results on trees, and shows that knowing one property holds will imply another. These types of theorems appear in many branches of mathematics where we want to indicate multiple equivalent statements of the same overall result. The proof of this theorem appears in Exercise 3.24.

Theorem 3.11 Let T be a graph with n vertices. The following conditions are equivalent:

(a) T is a tree.

(b) T is acyclic and contains $n - 1$ edges.

(c) T is connected and contains $n - 1$ edges.

(d) There is a unique path between every pair of distinct vertices in T.

(e) Every edge of T is a bridge.

(f) T is acyclic and for any edge e from T, $T + e$ contains exactly one cycle.

Using these results about the overall properties of trees, we can now prove why Kruskal's Algorithm is guaranteed to find a minimum spanning tree.

Theorem 3.12 Kruskal's Algorithm produces a minimum spanning tree.

Proof: Let G be a connected, weighted graph with n vertices. Let T be the spanning tree obtained by applying Kruskal's Algorithm to G and suppose the edges added to T were $e_1, e_2, \ldots, e_{n-1}$, in that order.

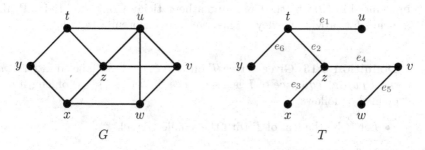

G T

Suppose for a contradiction that T is not a minimum spanning tree for G. Then among all minimum spanning trees, choose T' to be the one that agrees with the construction of T for the longest time. Then there is some $k < n - 1$ such that T' contains $e_1, e_2, \ldots, e_k$ and no minimum spanning tree of G contains all of $e_1, e_2, \ldots, e_k, e_{k+1}$.

Since T' is a spanning tree of G, we must have a cycle C in $T' + e_{k+1}$. But T does not have any cycles (since it is a tree) and so C must contain an edge not in T, call it e'. If we remove e' from $T' + e_{k+1}$ then the cycle C is broken and what remains is a spanning tree of G. Thus $T' + e_{k+1} - e'$ is a spanning tree of G containing edges $e_1, e_2, \ldots, e_k, e_{k+1}$.

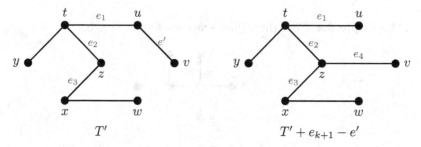

T' $T' + e_{k+1} - e'$

Since e' must have been available to be chosen when e_{k+1} was chosen by Kruskal's Algorithm, we know that $w(e_{k+1}) \leq w(e')$. Thus the total weight of $T' + e_{k+1} - e'$ is no bigger than the weight of T'. But $T' + e_{k+1} - e'$ has more edges in common with T than T', which is a contradiction to the way in which T' was chosen. Thus T must be a minimum spanning tree of G.

3.2.1   Tree Enumeration

In Section 1.6 we discussed how a degree sequence can help determine the structure of a graph, though it may not be unique. As it turns out, there is a similar result for trees, but one in which the tree can be uniquely determined from the sequence given. Instead of focusing on degrees, this sequence, called a *Prüfer sequence*, uses the location of leaves within a tree. These sequences were introduced in 1918 by the German mathematician Ernst Paul Heinz Prüfer as a means for proving Cayley's Theorem (see Theorem 3.14)[70].

Definition 3.13 Given a tree T on $n > 2$ vertices (labeled $1, 2, \ldots, n$), the ***Prüfer sequence*** of T is a sequence $(s_1, s_2, \ldots, s_{n-2})$ of length $n - 2$ defined as follows:

- Let l_1 be the leaf of T with the smallest label.

- Define T_1 to be $T - l_1$.

- For each $i \geq 1$, define $T_{i+1} = T_i - l_{i+1}$, where l_{i+1} is the leaf with the smallest label of T_i.

- Define s_i to be the neighbor of l_i.

While this definition of the Prüfer sequence is quite detailed, it is not too difficult to work with in practice. The main idea is to prune the leaf of the smallest index, while keeping track of its unique neighbor. At the end of the pruning, two adjacent vertices will remain.

Example 3.4 Find the Prüfer sequence for the tree below.

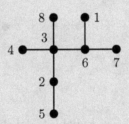

Solution: The pruning of leaves is shown below, with l_i and s_i listed for each step.

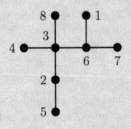

$T : l_1 = 1 \quad s_1 = 6$

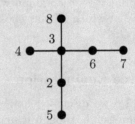

$T_1 : l_2 = 4 \quad s_2 = 3$

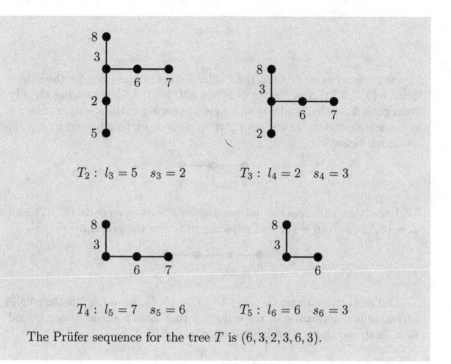

$T_2: l_3 = 5 \quad s_3 = 2$ $T_3: l_4 = 2 \quad s_4 = 3$

$T_4: l_5 = 7 \quad s_5 = 6$ $T_5: l_6 = 6 \quad s_6 = 3$

The Prüfer sequence for the tree T is $(6, 3, 2, 3, 6, 3)$.

Looking back at this example, we should see a few items stand out. First, the leaves of T are $1, 4, 5, 7,$ and 8, none of which appear in the Prüfer sequence. This is because we are always noting the neighbor of a leaf, not the leaf itself. You should also notice that a vertex appears in the sequence one less than its degree. Also, the vertex with label n will always be adjacent to the vertex appearing as s_{n-2}, which occurs since the last graph created (T_{n-2}) will consist of only two vertices and n will never be the smallest leaf remaining. Using these ideas, we can actually work backwards to build the tree from a given Prüfer sequence.

Example 3.5 Find the tree associated to the Prüfer sequence $(1, 5, 5, 3, 2)$.

Solution: First note that the sequence is of length 5, so the tree must have 7 vertices. At every stage we will consider the possible leaves of a subtree created by the earlier pruning. Initially our set of leaves is $L = \{4, 5, 7\}$, since these do not appear in the sequence and so must be the leaves of the full tree. To begin building the tree, we look for the smallest value not appearing in the sequence, namely 4. This must be adjacent to 1 by entry s_1, as shown on the next page.

Next, we remove s_1 from the Prüfer sequence and consider the subsequence $(5, 5, 3, 2)$. Now our set of leaves is $L = \{1, 6, 7\}$ since 4 has already been placed as a leaf and 1 is no longer appearing in the sequence. Since 1 is the smallest value in our set L, we know it must be adjacent to $s_2 = 5$, as shown below.

Repeating this process, we see that our subsequence is $(5, 3, 2)$ and $L = \{6, 7\}$. Thus 6 is the leaf adjacent to 5. See the graph below.

Our next subsequence is $(3, 2)$ and $L = \{5, 7\}$, meaning 5 is the smallest value not appearing in the sequence that hasn't already been placed as a "leaf," so there must be and edge from 5 to 3.

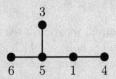

Finally we are left with the single entry (2). We know that $L = \{3, 7\}$ and both of these must be adjacent to 2.

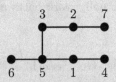

To verify we have the correct tree, we can use the process shown in Example 3.4 above to find the Prüfer sequence of our final tree and verify it matches the one given.

By having a method to determine a tree based upon one of these Prüfer sequences, we can now enumerate how many different trees exist. This result is named for the British mathematician Arthur Cayley, who is known for his contributions to pure mathematics, mainly in algebra. Cayley included this result on trees in his Collected Papers, a 14 volume collection of his work that was first published in 1889 [11]; however, the result below was first proven in 1860 (from the viewpoint of matrices) by the German mathematician Carl Wilhelm Borchardt [9]. Moreover, Cayley's orginal proof for this result was much more involved and difficult to follow. Prüfer's introduction of the sequences above make the proof of Cayley's Theorem fairly trivial.

Theorem 3.14 (Cayley's Theorem) There are n^{n-2} different labeled trees on n vertices.

Proof: First, note that there are n^{n-2} possible sequences of length $n-2$ since each spot on the sequence has n options. Each of these sequences can be viewed as a Prüfer sequence and will uniquely determine a tree as shown above. Thus there are n^{n-2} different labeled trees on n vertices.

3.3 Rooted Trees

When we look at the graph theory terminology used for trees, mathematicians have adopted many of the same terms that we use for biological trees (such as leaf and forest). So what then would we mean by a root of a tree? We can think of the root as the place from which a tree grows.

Definition 3.15 A *rooted tree* is a tree T with a special designated vertex r, called the *root*. The *level* of any vertex in T is defined as the length of its shortest path to r. The *height* of a rooted tree is the largest level for any vertex in T.

Example 3.6 Find the level of each vertex and the height of the rooted tree shown below.

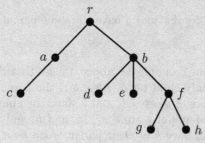

Solution: Vertices a and b are of level 1, c, d, e, and f of level 2, and g and h of level 3. The root r has level 0. The height of the tree is 3.

Most people have encountered a specific type of rooted tree: a family tree. In fact, much of the terminology for rooted trees comes not from a plant version of a tree but rather from genealogy and family trees. The root of a family tree would be the person for whom the descendants are being mapped and the level of a vertex would represent a generation; see the tree below. With this application in mind, the terminology below is used to describe how various vertices are related within a rooted tree.

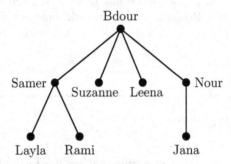

Definition 3.16 Let T be a tree with root r. Then for any vertices x and y

- x is a *descendant* of y if y is on the unique path from x to r;

- x is a *child* of y if x is a descendant of y and exactly one level below y;

- x is an *ancestor* of y if x is on the unique path from y to r;

- x is a *parent* of y if x is an ancestor of y and exactly one level above y;

- x is a *sibling* of y if x and y have the same parent.

Using the tree from Example 3.6 above, we see that the parent of a is the root r and c is the only child for a. Also, b is the parent of e, but e has no children. The ancestors of g are f, b, and r since the unique path from g to r is $g\,f\,b\,r$. The descendants of b are d, e, f, g, and h, and the siblings of d are e and f since they all have b as their parent. Also note that the leaves of a rooted tree are exactly those vertices without any children, namely $c, d, e, g,$ and h in the graph from Example 3.6.

A common type of tree, especially in computer science, adds the restriction that no vertex can have more than two children. You can think of this as a decision tree, where at every stage you can answer yes or no (or for the more technically inclined this would correspond to 1's and 0's). These trees are called binary trees.

Definition 3.17 A tree in which every vertex has at most two children is called a *binary tree*. If every parent has exactly two children we have a *full binary tree*. Similarly, if every vertex has at most k children then the tree is called a *k-nary tree*.

Example 3.7 Trees can be used to store information for quick access. Consider the following string of numbers:

$$4, 2, 7, 10, 1, 3, 5$$

We can form a tree by creating a vertex for each number in the list. As we move from one entry in the list to the next, we place an item below and to the left if it is less than the previously viewed vertex and below and to the right if it is greater. If we add the restriction that no vertex can have more than two edges coming down from it, then we are forming a binary tree.

For the string above, we start with a tree consisting of one vertex, labeled 4 (see T_1). The next item in the list is a 2, which is less than 4 and so its vertex is placed on the left and below the vertex for 4. The next item, 7, is larger than 4 and so its vertex is placed on the right and below the vertex for 4 (see T_2).

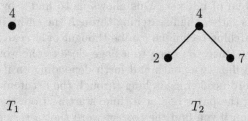

$$T_1 \qquad\qquad\qquad T_2$$

The next item in the list is 10. Since 4 already has two edges below it, we must attach the vertex for 10 to either 2 or 7. Since 10 is greater than 4, it must be placed to the right of 4 and since 10 is greater than 7, it must be placed to the right of 7 (see T_3). A similar reasoning places 1 to the left and below 2 (see T_4).

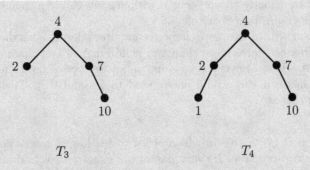

T_3 T_4

The next item is 3, which is less than 4 and so must be to the left of 4. Since 3 is greater than 2, it must be placed to the right of 2 (see T_5). The final item is 5, which is greater than 4 but less than 7, placing it to the right of 4 but to the left of 7 (see T_6).

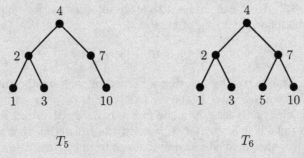

T_5 T_6

This final tree T_6 represents the items in our list. If we want to search for an item, then we only need to make comparisons with at most half of the items in the list. For example, if we want to find item 5, we first compare it to the vertex at the top of the tree. Since 5 is greater than 4, we move along the edge to the right of 4 and now compare 5 to this new vertex. Since 5 is less than 7 we move along the edge to the left of 7 and reach the item of interest. This allows us to find 5 by making only two comparisons rather than searching through the entire list.

This searching technique can be thought of as searching for a word in a dictionary. Once you open to a page close to the word you are searching for, you flip pages back and forth depending on if you are before or after the word needed. Searching through the list item-by-item would be like flipping the pages one at a time, starting from the beginning of the dictionary, until you find the correct word (not a very efficient method).

Due to the more regular nature of binary trees, they have some special properties in terms of the number of leaves and their maximum height. We list these below, but the proofs appear in the Exercise 3.27.

Theorem 3.18 Let T be a binary tree with height h and l leaves. Then

(i) $l \le 2^h$.

(ii) if T is a full binary tree and all leaves are at height h, then $l = 2^h$.

(iii) if T is a full binary tree, then $n = 2l - 1$.

Earlier in this chapter we were concerned with finding a minimum spanning tree, often for the use of a specific application needing to connect items at a minimum cost. Here we use search trees to find paths within a graph from a specified root. The applications of these are mainly still within the realm of graph theory, such as finding connected components or bridges within a graph or testing if a graph is planar (see Chapter 7). However, both search trees we discuss below arose, in part, as a way to solve a maze, and have applications into the study of artificial life. Depth-first search trees are credited to the nineteenth century French author and mathematician Charles Pierre Trémaux, and breadth-first search trees were introduced in the 1950s by the American mathematician E.F. Moore.

3.3.1 Depth-First Search Tree

The main idea behind a depth-first tree is to travel along a path as far as possible from the root of a given graph. If this path does not encompass the entire graph, then branches are built off this central path to create a tree. The formal description of this algorithm relies on an ordered listing of the neighbors of each vertex and uses this order when adding new vertices to the tree. For simplicity, we will always use an alphabetical order when considering neighbor lists.

Depth-First Search Tree

Input: Simple graph $G = (V, E)$ and a designated root vertex r.

Steps:

1. Choose the first neighbor x of r in G and add it to $T = (V, E')$.

2. Choose the first neighbor of x and add it to T. Continue in this fashion—picking the first neighbor of the previous vertex to create a path P. If P contains all the vertices of G, then P is the depth-first search tree. Otherwise continue to Step (3).

3. Backtrack along P until the first vertex is found that has neighbors
not in T. Use this as the root and return to Step (1).

Output: Depth-first search tree T.

In creating a depth-first search tree, we begin by building a central spine
from which all branches originate. These branches are as far down on this
path as possible. In doing so, the resulting rooted tree is often of large height
and is more likely to have more vertices at the lower levels.

Example 3.8 Find the depth-first search tree for the graph below with
the root a.

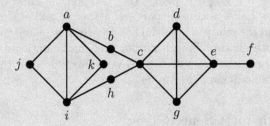

Solution:

Step 1: Since a is the root, we add b as it is the first neighbor of a.
Continuing in this manner produces the path shown below. Note this
path stops with f since f has no further neighbors in G.

Step 2: Backtracking along the path above, the first vertex with an unchosen neighbor is e. This adds the edge eg to T. No other edges from g
are added since the other neighbors of g are already part of the tree.

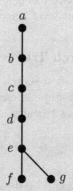

Step 3: Backtracking again along the path from Step 1, the next vertex with an unchosen neighbor is c. This adds the path $chij$ to T.

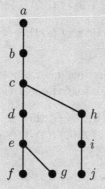

Step 4: Backtracking again along the path from Step 3, the next vertex with an unchosen neighbor is i. This adds the edge ik to T and completes the depth-first search tree as all the vertices of G are now included in T.

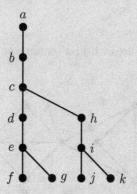

Output: The tree above is the depth-first search tree.

Note that the tree created above has height 5 and with one vertex each at level 1 and 2, two vertices each at level 3 and 4, and four vertices at level 5.

3.3.2 Breadth-First Search Tree

The main objective for a breadth-first search tree is to add as many neighbors of the root as possible in the first step. At each additional step, we are adding all available neighbors of the most recently added vertices.

Breadth-First Search Tree

Input: Simple graph $G = (V, E)$ and a designated root vertex r.

Steps:

1. Add all the neighbors of r in G to $T = (V, E')$.

2. If T contains all the vertices of G, then we are done. Otherwise continue to Step (3).

3. Beginning with x, the first neighbor of r that has neighbors not in T, add all the neighbors of x to T. Repeat this for all the neighbors of r.

4. If T contains all the vertices of G, then we are done. Otherwise repeat Step (3) with the vertices just previously added to T.

Output: Breadth-first tree T.

As with depth-first, we will use an alphabetical order when considering neighbor lists. At each stage we are adding a new level to the tree and visually we will place the vertices from left to right, thus aiding in the next stage of vertex additions.

Example 3.9 Find the breadth-first search tree for the graph below with the root a.

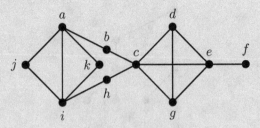

Solution:

Step 1: Since a is the root, we add all of the neighbors of a to T.

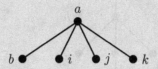

Step 2: We next add all the neighbors of b, i, j, and k, that are not already in T, beginning with b as it is the first neighbor of a that was added in Step 1. This adds the edge bc. Moving to i we add the edge ih. No other edges are added since j and k do not have any unchosen neighbors.

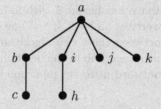

Step 3: We next add all the neighbors of c not in T, namely d, e, and g. No other vertices are added since all the neighbors of h are already part of the tree T.

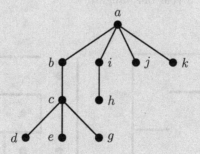

Step 4: Since d has no unchosen neighbors, we move ahead to adding the unchosen neighbors of e. This completes the breadth-first search tree as all the vertices of G are now included in T.

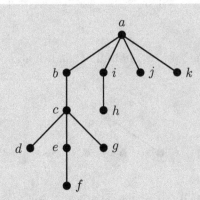

Output: The tree above is the breadth-first search tree.

It should come as no surprise that breadth-first search trees are likely to be of shorter height than their depth-first search counterpart. The breadth-first search tree created above has height 4 with four vertices on level 1, two vertices on level 2, three vertices on level 3, and one on level 4.

The main difference between these two algorithms is that depth-first focuses on traveling as far into the graph in the beginning, whereas the breadth-first focuses on building outward using neighborhoods.

3.3.3   Mazes

As previously stated, both breadth-first and depth-first search trees can be used to solve a maze. We conclude our section on Rooted Trees with a maze example because it allows us to more easily visualize the difference between the two types of search trees. On the left below is a quite simple maze (too easy according to my 8 year old daughter).

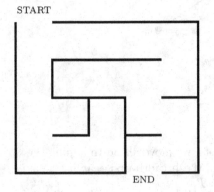

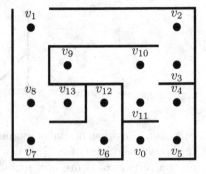

To convert it into a graph, we will place vertices at the starting and ending locations (v_1 and v_0, respectively), as well as at any location at which a turn, choice, or dead end occurs, as shown above (though we omit the edges of the graph). A solution to the maze would be a path from v_1 to v_0.

On the left below are the routes formed by using the Breadth-First Search. Notice that every corner of this maze is examined, and from this we can create the tree, shown below it. From the tree, we can find the maze solution by tracing the unique path from v_1 (START) to v_0 (END). On the right below we have the routes formed by the Depth-First Search. Notice that we reach END before all parts of the maze are examined. If we continued the DFS Algorithm further, we would obtain the DFS tree shown below the maze.

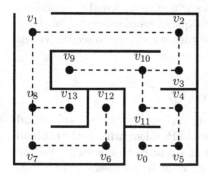

Breadth-First Search

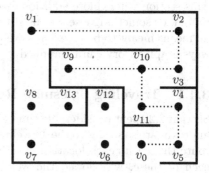

Depth-First Search

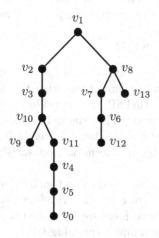

Maze Breadth-First Tree

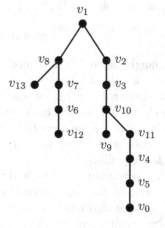

Maze Depth-First Tree

Some interesting things occurred in these trees. In fact, they are essentially the same tree, and only the order in which the vertices were added to the tree differ. This is quite different from the BFS and DFS trees seen in Examples 3.8 and 3.9, which occurs since there is only one solution for a maze and we did not have any cycles. Note that a full DFS tree was unnecessary to find the maze solution, yet we grew the entire BFS tree when looking for the maze solution. This may not always occur however, and is mainly an artifact from the structure of a maze.

3.4   Additional Applications

This section will explore additional examples where trees, whether spanning, rooted, or something else, are used to answer questions of interest. This is by no means meant to be an exhaustive list, but rather a chance to see the varied ways graph theory can be applied.

3.4.1   Traveling Salesman Revisited

Section 2.2 spent considerable time discussing various algorithms used to find approximate solutions to the Traveling Salesman Problem: what is the optimal route that visits each location exactly once and returns to the start? Within each of these, we allowed the weight on an edge to represent either cost, distance, or time. Here we present one additional approximate algorithm that can be used in a specific instance of the Traveling Salesman Problem when the weights assigned to the edges satisfy the ***triangle inequality***;that is, for a weighted graph $G = (V, E, w)$, given any three vertices x, y, z we have

$$w(xy) + w(yz) \geq w(xz)$$

The triangle inequality is named to reference a well-known fact in geometry that no one side of a triangle is longer than the sum of the other two sides.

The *metric Traveling Salesman Problem* (mTSP) only considers scenarios where the weights satisfy the triangle inequality. When the weight function is modeling distance, we are within the mTSP realm; when the weight function models cost or time, we may or may not be in a scenario that satisfies the triangle inequality.

Minimum spanning trees, and the algorithms used to find such subgraphs, can be used to find an approximate solution to a metric Traveling Salesman Problem. The algorithm below combines three ideas we have studied so far: eulerian circuits, hamiltonian cycles, and minimum spanning trees. A minimum spanning tree is modified by duplicating every edge, ensuring all vertices have even degree and allowing an eulerian circuit to be obtained. This circuit is then modified to create a hamiltonian cycle. Note that this procedure is guaranteed to work only when the underlying graph is complete. It may still find a proper hamiltonian cycle when the graph is not complete, but cannot be guaranteed to do so.

---

mTSP Algorithm

Input: Weighted complete graph K_n, where the weight function w satisfies the triangle inequality.

Steps:

1. Find a minimum spanning tree T for K_n.

2. Duplicate all the edges of T to obtain T^*.

3. Find an eulerian circuit for T^*.

4. Convert the eulerian circuit into a hamiltonian cycle by skipping any previously visited vertex (except for the starting and ending vertex).

5. Calculate the total weight.

Output: hamiltonian cycle for K_n.

The example below is similar to those from Section 2.2, except the distances shown satisfy the triangle inequality. Recall that to find an optimal hamiltonian cycle on a graph with 6 vertices, we would need to calculate all 60 possible hamiltonian cycles.

Example 3.10 Nour must visit clients in six cities next month and needs to minimize her driving mileage. The table below lists the driving distances between these cities. Use the mTSP Algorithm to find a good plan for her travels if she must start and end her trip in Philadelphia. Include the total distance.

	Boston	Charlotte	Memphis	New York	Philadelphia	D.C.
Boston	·	840	1316	216	310	440
Charlotte	840	·	619	628	540	400
Memphis	1316	619	·	1096	1016	876
New York City	216	628	1096	·	97	228
Philadelphia	310	540	1016	97	·	140
Washington, D.C.	440	400	876	228	140	·

Solution: The details for finding a minimum spanning tree and an eulerian circuit will be omitted (You are encouraged to work through these!). In addition, city names will be represented by their first letter.

Step 1: A minimum spanning tree for the six cities is given below.

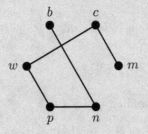

Step 2: Duplicate all the edges of the tree above.

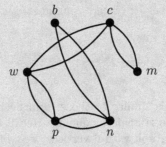

Step 3: Find an eulerian circuit starting at p. The circuit shown below is $p\,n\,b\,n\,p\,w\,c\,m\,c\,w\,p$.

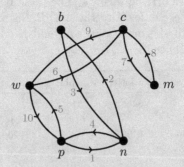

Step 4: We follow the eulerian circuit from Step 3 until we reach vertex b. Since we are looking for a hamiltonian cycle, we cannot repeat vertices and so we cannot return to n. The next vertex along the circuit that has not been previously visited is w.

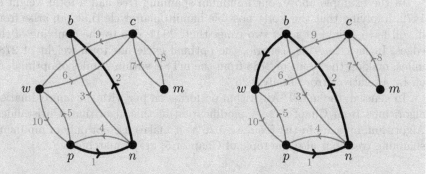

We follow the circuit again until m is reached. Again, we cannot return to c and at this point we must return to p since all other vertices have been visited.

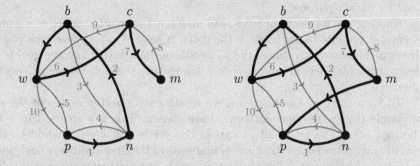

This solution can also be obtained by using the written form of the eulerian circuit by keeping a vertex the first time it appears and then returning to the starting vertex:

$$p\,n\,b\,n\,p\,w\,c\,m\,c\,w\,p \longrightarrow p\,n\,b\,w\,c\,m\,p$$

Output: The final hamiltonian cycle is shown below (Philadelphia→New York→Boston→Washington, D.C.→Charlotte→Memphis→Philadelphia) with a total weight of 2788 miles.

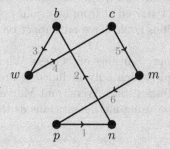

In the example above, our minimum spanning tree had a total weight of 1472, implying that the worst possible hamiltonian cycle that can arise from it will have weight at most two times that, 2944, due to the doubling of the edges. In fact, for this scenario, the optimal cycle has total weight of 2781 miles, making the result of 2788 from the mTSP within 7 miles of optimal, or off by a relative error of only 0.25%!

In general, the mTSP Algorithm performs on par with the approximation algorithms from Chapter 2. A modification of this algorithm, Christofides' Algorithm, appears in the Exercise 5.26 as it makes use not only of minimum spanning trees but also the topic of Chapter 5, graph matchings.

3.4.2   Decision Trees

Decision trees allow for options to be mapped out, where you can use a series of questions to arrive at the best solution. You have probably seen flow charts used for such a purpose.

Another use for decision trees is to map out the solution space of a game. A current favorite in my house is the game "Guess Who?" where each player draws a card matching one of the 24 people on their board and the goal is to determine which person their opponent has chosen, through a series of Yes/No questions.

To form the tree for this game, we would have a vertex represent the set of people that your opponent could have chosen. This is a rooted tree with the root being the set of all 24 cards. The vertices at level 1 would be the two distinct sets of cards that are available after the first question asked, such as "Does your person have blue eyes?" with one set being those with blue eyes and the other the people without blue eyes. The next level would be the sets formed after the second question, perhaps "Does your person have brown hair?" At this point, we would have 4 distinct sets of people.

Note that the tree formed by this game is a binary tree since at each stage we only have two options for the answer. Also, the order in which the questions are asked would create different trees. When playing the game, you also would not be looking for all possible paths from the root, only along the branch based on your opponents answer. For example, we could map out the four sets obtained from the two questions above, but in reality would only care about the unique set arrived at from how your opponent answered. Thus, the order of your questions could have an impact on how quickly you arrive at a solution.

There have been many editions of "Guess Who?" released by Hasbro games, from the original (which only included 5 women) to newer versions using characters from Disney, Star Wars, and Marvel Comics. The example below is a smaller version using mathematicians as the pool of cards.

Example 3.11 The junior and senior math majors are facing off on a Mathematician Guess Who game. Below is a list of the cards. The junior team picked Pierre de Fermat and asked the questions (in order) "Is your person a woman? From pre-1900's? Known for Number Theory?" The senior team picked Sophie Germain and asked "Is your person a woman? Known for Graph Theory? Known for Number Theory?" Use a decision tree to determine which team won.

Pierre de Fermat (PF)	Carl Gauss (CG)	Leonhard Euler (LE)
Srinivasa Ramanujan (SR)	Alan Turing (AT)	Emmy Noether (EN)
Maryam Mirzakhani (MM)	Georg Cantor (GC)	Isaac Newton (IN)
Sophie Germain (SG)	Paul Erdős (PE)	Ada Lovelace (AL)

Solution: Below are two trees representing the sequence of questions asked by each team. Since only the junior team has a single person in their final set, we know they won the game after 3 questions.

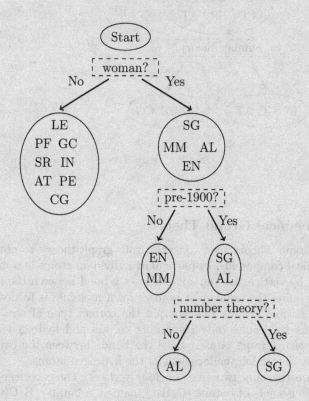

Junior Team Tree

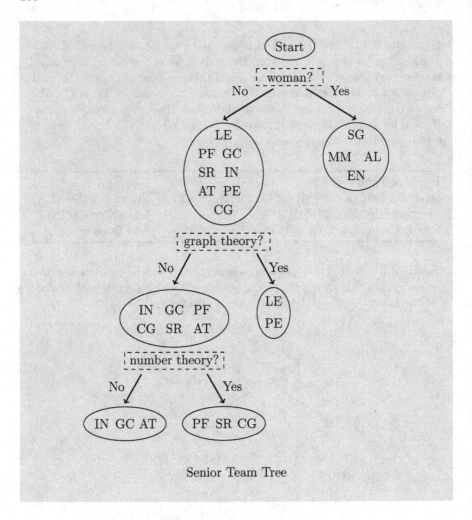

Senior Team Tree

3.4.3   Chemical Graph Theory

Chemical Graph Theory uses concepts from graph theory to obtain results about chemical compounds. In particular, individual atoms in a molecule are represented by vertices and an edge denotes a bond between the atoms. One way to determine the number of isomers for a molecule is to determine the number of distinct graphs that contain the correct type of each atom. For hydrocarbons (molecules only containing carbon and hydrogen atoms) the *hydrogen-depleted graph* is used since the bonds between the carbon atoms will uniquely determine the locations of the hydrogen atoms.

Below are the only two trees on four vertices. These correspond to the only possible isomers of butane (C_4H_{10}), namely n-butane ($H_3C(CH_2)_2CH_3$) and isobutane (($H_3C)_3CH$), whose full molecular forms are displayed below their respective hydrogen-depleted graph. By using graph theory, we can prove no other isomers of butane are possible since no other trees on four vertices exist [2].

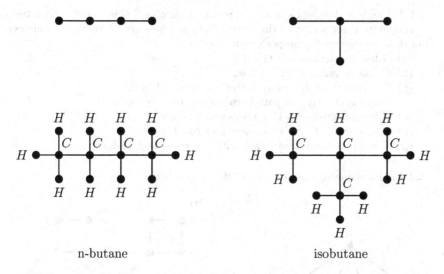

n-butane   isobutane

While molecules do not need to adhere to a tree structure, for those that do we can use our knowledge of graphs to determine all possible isomers. In doing so, we must keep in mind the chemical properties of the atoms involved. For example, if we restricted ourselves to the hydrogen-depleted graph for hexane (C_6H_{14}), we would begin by looking at the possible trees on 6 vertices. Below are two such trees on 6 vertices, but only the one on the left corresponds to an isomer for hexane (called neohaxe or 2,2-dimethylbutane) whereas the tree on the right cannot represent a hexane molecule since the central vertex has degree 5 and carbon atoms have at most 4 bonds. Thus we would need to restrict ourselves with finding the different tree structures with maximum degree 4 (see Exercise 3.28).

neohexane   not an isomer

3.5   Exercises

3.1 For each of the graphs described below, determine if G is (i) definitely a tree,
(ii) definitely not a tree, or (iii) may or may not be a tree. Explain your answer
or demonstrate with a proper graph.

  (a) G has 10 vertices and 11 edges.

  (b) G has 10 vertices and 9 edges.

  (c) G is connected and every vertex has degree 1 or 2.

  (d) There is exactly one path between any two vertices of G.

  (e) G is connected with 15 vertices and 14 edges.

  (f) G is connected with 15 vertices and 20 edges.

  (g) G has two components, each with 9 vertices and 8 edges.

3.2 Find a spanning tree for each of the graphs below.

(a)   **(b)**

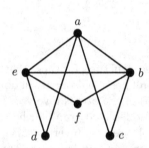

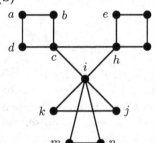

(c)   **(d)**

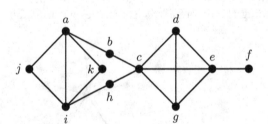

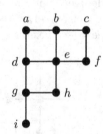

3.3 Find a minimum spanning tree for each of the graphs below using (i) Kruskal's Algorithm and (ii) Prim's Algorithm.

(a)

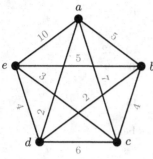

(b)

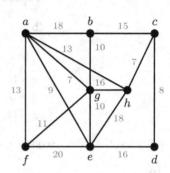

(c)

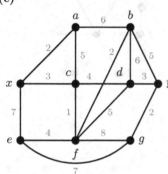

(d)

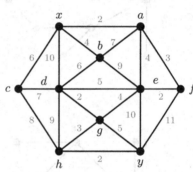

(e)

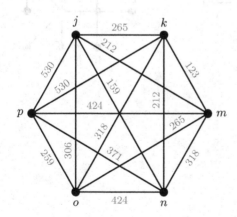

3.4 Find a minimum spanning tree for the graph represented by the table below.

	a	b	c	d	e	f	g
a	·	5	7	8	10	3	11
b	5	·	2	4	1	12	7
c	7	2	·	6	7	5	4
d	8	4	6	·	2	10	12
e	10	1	7	2	·	6	9
f	3	12	5	10	6	·	15
g	11	7	4	12	9	15	·

3.5 Find the Prüfer sequence for each of the trees below.

(a)

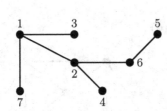

(b)

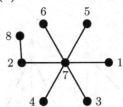

(c)

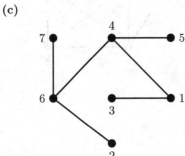

(d)

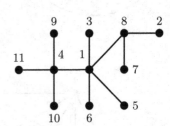

(e)

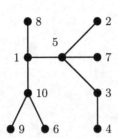

3.6 Find the tree associated with each of the Prüfer sequences listed below.
(a) $(2,2,2,2)$
(b) $(2,4,4,3)$
(c) $(2,2,1,1,3)$
(d) $(7,7,1,3,4)$

3.7 For the rooted tree T below, with root r, identify the following:
(a) Level of r, f, h, and k.
(b) The height of T.
(c) Parents of r, b, c, f, i.
(d) Children of r, b, c, f, i.
(e) Ancestors of r, a, d, h, k.
(f) Descendants of r, a, d, h, k.
(g) Siblings of a, f, h, i.

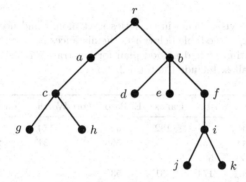

3.8 Complete each of the following on the two graphs shown below.
(a) Find the breadth-first search tree with root a.
(b) Find the breadth-first search tree with root i.
(c) Find the depth-first search tree with root a.
(d) Find the depth-first search tree with root i.

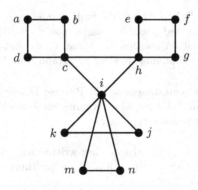

G_1

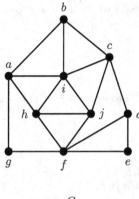

G_2

3.9 Nour must visit clients in six cities next month and needs to minimize her driving mileage. The table below lists the distances between these cities. Use the mTSP Algorithm to find a good plan for her travels if she must start and end her trip in Dallas. Include the total distance.

	Austin	Dallas	El Paso	Fort Worth	Houston	San Antonio
Austin	·	182	526	174	146	74
Dallas	182	·	568	31	225	253
El Paso	526	568	·	537	672	500
Fort Worth	174	31	537	·	237	241
Houston	146	225	672	237	·	189
San Antonio	74	253	500	241	189	·

3.10 The weight of the edges of the graph in Exercise 3.3(e) satisfies the triangle inequality. Apply the mTSP Algorithm to find a hamiltonian cycle and compare it to those found from Chapter 2 (Exercise 2.6(e)).

3.11 The Reverse Delete Algorithm finds a minimum spanning tree by deleting the largest weighted edges as long as you do not disconnect the graph. In essence, it is Kruskal's Algorithm in reverse. Verify that Reverse Delete produces the same minimum spanning trees for the graph from Example 3.2.

3.12 Under what circumstances would Reverse Delete be a better choice than Kruskal's Algorithm? Under what circumstance would Kruskal's be a better choice? Explain your answer.

3.13 The algorithms in this chapter are written with a connected graph as input. Determine the output of each of the algorithms below if the input was a disconnected graph.
   (a) Kruskal's Algorithm
   (b) Prim's Algorithm

(c) Reverse Delete Algorithm
(d) Depth-First Search
(e) Breadth-First Search

3.14 How would you modify Kruskal's Algorithm if a specific edge must be included in the spanning tree? Would the resulting tree be a minimum spanning tree? Explain your answer.

3.15 Let G be a forest with k components and n vertices.
(a) Determine bounds on the number of leaves for G.
(b) Prove G has $n - k$ edges.

3.16 Prove Lemma 3.5: Given a tree T with a leaf v, the graph $T - v$ is still a tree.

3.17 Prove Corollary 3.7: The total degree of a tree on n vertices is $2n - 2$.

3.18 Expand Theorem 3.4 to show that every tree with at least two vertices has at least two leaves.

3.19 Prove that every non-leaf in a tree is a cut-vertex. (Recall a cut-vertex is a vertex whose removal will disconnect the graph).

3.20 Prove Proposition 3.9: Every tree is minimally connected, that is every edge of a tree is a bridge.

3.21 Let T be a tree of maximum degree k. Prove that T has at least k leaves.

3.22 Let T be a tree with e edges.
(a) If e is even, prove that T has at least one even vertex.
(b) If e is odd, give examples to show that T may or may nor have any even vertices.

3.23 (a) Prove that every tree is bipartite.
(b) Let X, Y be a bipartition of the vertices of T where $|X| \geq |Y|$. Prove that T has a leaf in X, and T also has a leaf in Y if $|X| = |Y|$.

3.24 Prove Theorem 3.11 on the equivalent statements about a tree.

3.25 Recall that the eccentricity of a vertex x is the maximum distance from x to any other vertex in G (so $\epsilon(x) = \max_{y \in V(G)} d(x, y)$), and a vertex is central if it has the minimum eccentricity among all vertices in G. Assume T is a tree with at least three vertices.
(a) Prove that if x is a leaf of T with unique neighbor y then $\epsilon(x) = \epsilon(y) + 1$.
(b) Prove that if x is a central vertex of T then x is not a leaf.

3.26 Recall that the center of a graph, $C(G)$, is the subgraph induced by all the vertices of G that satisfy $\epsilon(v) = rad(v)$. Prove that $C(T)$ is a either a vertex or an edge for any tree T. (Hint: Argue by induction on the number of vertices.)

3.27 Prove Theorem 3.18 on page 147.

3.28 Following the process described in Section 3.4.3, determine the possible isomers of pentane (C_5H_{12}) and hexane (C_6H_{14}).

3.29 In Section 3.3.1, we studied two different algorithms for finding a minimum spanning tree. As mentioned earlier, both Kruskal and Prim cited the work of the Czech mathematician Otakar Borůvka. Below is a description of his algorithm, first published in 1926.

Borůvka's Algorithm

Input: Weighted connected graph $G = (V, E)$ where all the weights are distinct.

Steps:

1. Let T be the forest where each component consists of a single vertex.

2. For each vertex v of G, add the edge of least weight incident to v to T.

3. If T is connected, then it is a minimum spanning tree for G. Otherwise, for each component C of T, find the edge of least weight from a vertex in C to a vertex not in C. Add the edge to T.

4. Repeat Step (3) until T has only one component, making T a tree.

Output: A minimum spanning tree for G.

Apply Borůvka's Algorithm to the following two graphs. Use either Kruskal's Algorithm or Prim's Algorithm to verify that Borůvka's Algorithm found a minimum spanning tree.

(a)   **(b)**

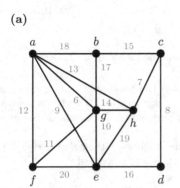
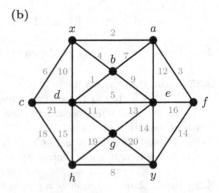

4

Connectivity and Flow

In each of the previous chapters, we have used connectivity in the context of other problems. For example, in Chapter 2 we needed to know if a graph was connected in order to determine if it is eulerian and in Chapter 3 we define trees as minimally connected graphs, since the removal of any edge would disconnect the graph. This chapter focuses on connectivity as its own topic, where we now consider *how* connected a graph is, and not just whether it is connected or not. Note that this chapter is more theoretical than the previous, though we tie in network flow and end with a section on applications of connectivity.

Consider graphs G_1, G_2, and G_3 below. It should be plain to see that they are all connected graphs. We could describe other features of these graphs (are they eulerian? hamiltonian? acyclic?) but one distinguishing factor between them should be the simple difference in edge count, with G_1 having fewer edges than the other two. Notice that G_2 and G_3 both contain 13 edges, but their underlying structure is quite different. Visually, is seems that the edges in G_3 are more clumped than in G_2. One way to describe this clumping is in how many edges or vertices would need to be removed before the graph is no longer connected, which is one way we measure connectivity.

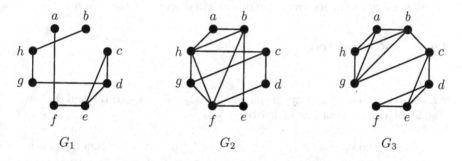

$$G_1 \qquad\qquad G_2 \qquad\qquad G_3$$

4.1   Connectivity Measures

When we define a graph to be connected, we refer to the existence of a way to move between any two vertices in a graph, specifically as the existence of a path between any pair of vertices. We will see that measuring how connected a graph is has a similar description, but we first use the standard notion described above in terms of vertex (or edge) removal.

> **Definition 4.1** A *cut-vertex* of a graph G is a vertex v whose removal disconnects the graph, that is, G is connected but $G - v$ is not. A set S of vertices within a graph G is a *cut-set* if $G - S$ is disconnected.

Note that any connected graph that is not complete has a cut-set, whereas K_n does not have a cut-set (see Exercise 4.13). Moreover, a graph can have many different cut-sets and of varying sizes. For example, two different cut-sets are shown below for graph G_1 above.

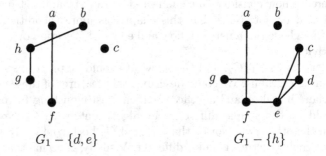

$$G_1 - \{d, e\} \qquad\qquad G_1 - \{h\}$$

Although we can find many different cut-sets for graph G_1, we may want to choose one over another based on some sense of optimality. In particular, when we evaluate how connected a graph is, we are really asking what is the fewest number of vertices whose removal will disconnect the graph.

4.1.1   *k*-Connected

> **Definition 4.2** For any graph G, we say G is *k*-connected if the smallest cut-set is of size at least k.
>
> Define the connectivity of G, $\kappa(G) = k$, to be the maximum k such that G is k-connected, that is there is a cut-set S of size k, yet no cut-set exists of size $k - 1$ or less. Define $\kappa(K_n) = n - 1$.

The distinction between k-connected and connectivity k is subtle yet important. For example if we say a graph is 3-connected, then we know there cannot be a cut-set of size 2 or less in the graph; however, we only know that its connectivity is at least 3 ($\kappa(G) \geq 3$).

Example 4.1 Find $\kappa(G)$ for each of the graphs shown above on page 169.

Solution: The removal of any one of $d, e, f, g,$ or h in G_1 will disconnect the graph, so $\kappa(G_1) = 1$. Similarly, $G_3 - c$ has two components and so $\kappa(G_3) = 1$. However, $\kappa(G_2) = 2$ since the removal of any one vertex will not disconnect the graph, yet $S = \{b, h\}$ is a cut-set. Note this means G_2 is both 1-connected and 2-connected, but not 3-connected.

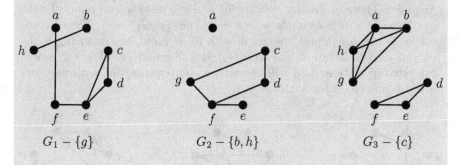

$$G_1 - \{g\} \qquad\qquad G_2 - \{b, h\} \qquad\qquad G_3 - \{c\}$$

The example above demonstrates that more than one minimal cut-set can exist within a graph. Moreover, any connected graph is 1-connected, but we are more interested in how large k can be before G fails to be k-connected.

4.1.2 k-Edge-Connected

A similar notion with regards to edges exists, where we now look at how many edges need to be removed before the graph is disconnected. Recall that when we remove an edge $e = xy$ from a graph, we are *not* removing the endpoints x and y.

Definition 4.3 A *bridge* in a graph $G = (V, E)$ is an edge e whose removal disconnects the graph, that is, G is connected but $G - e$ is not. An *edge-cut* is a set $F \subseteq E$ so that $G - F$ is disconnected.

Clearly every connected graph has an edge-cut since removing all the edges from a graph will result in just a collection of isolated vertices. As with the vertex version, we are more concerned with the smallest size of an edge-cut.

Definition 4.4 We say G is *k-edge-connected* if the smallest edge-cut is of size at least k.

Define $\kappa'(G) = k$ to be the maximum k such that G is k-edge-connected, that is there exists a edge-cut F of size k, yet no edge-cut exists of size $k - 1$.

Example 4.2 Find $\kappa'(G)$ for each of the graphs shown on page 169.

Solution: There are many options for a single edge whose removal will disconnect G_1 (for example af or dg). Thus $\kappa'(G_1) = 1$. For G_2, no one edge can disconnect the graph with its removal, yet removing both ab and ah will isolate a and so $\kappa(G_2) = 2$. Similarly $\kappa'(G_3) = 2$, since the removal of bc and cg will create two components, one with vertices a, b, g, h and the other with c, d, e, f.

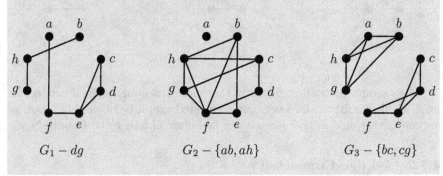

$$G_1 - dg \qquad\qquad G_2 - \{ab, ah\} \qquad\qquad G_3 - \{bc, cg\}$$

4.1.3   Whitney's Theorem

Can you discern any relationship between the vertex and edge connectivity measures? The examples above should demonstrate that these measures need not be equal, though they can be. How does the minimum degree of a graph play a role in these? Notice how in both G_2 and G_3 above we found an edge-cut by removing both edges incident to a specific vertex.

Theorem 4.5 (Whitney's Theorem) For any graph G, $\kappa(G) \leq \kappa'(G) \leq \delta(G)$.

Proof: Let G be a graph with n vertices and $\delta(G) = k$ and suppose x is a vertex with $\deg(x) = k$. Let F be the set of all edges incident to x. Then $G - F$ is disconnected, since x is now isolated, and so F is an edge-cut. Thus $\kappa'(G) \leq k$.

It remains to show that $\kappa(G) \leq \kappa'(G)$. If $G = K_n$, then $\delta(G) = n - 1$

and $\kappa(G) = n - 1$ by definition, and so $\kappa(G) = \kappa'(G)$. Otherwise, let F be a minimal edge-cut of G and define G_1 and G_2 to be the two components of $G - F$. We will consider how these components are related, and in both cases find a cut-set S of size less than that of F.

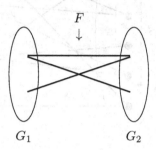

Case 1: Every vertex of G_1 is adjacent to every vertex of G_2. Then $|F| = |G_1| \cdot |G_2| \geq n - 1$ since each component has at least one vertex. Since $G \neq K_n$, at least one of G_1 and G_2, say G_1, has a pair of vertices x and y that are not adjacent. Let S be all vertices of G except x and y, that is $S = V(G) - \{x, y\}$. Then S is a cut-set of size $n - 2$ and so $\kappa(G) \leq n - 2 < n - 1 \leq \kappa'(G)$.

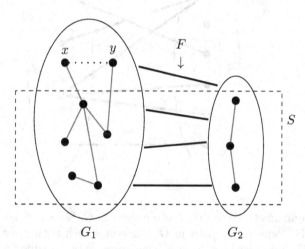

Case 2: There exist nonadjacent vertices x and y with $x \in G_1$ and $y \in G_2$. We will build a cut-set S as follows: Given an edge e from F, if

(i) x is an endpoint of e then pick the other endpoint of e to be in S; that is, if $e = xz$ with $z \in G_2$ then add z to S.

(ii) x is not an endpoint of e then pick the endpoint of e from G_1 to add to S.

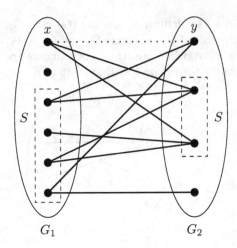

G_1 G_2

Note that $|S| \leq |F|$ since for each vertex in $S \cap G_1$ we have at least one edge incident to it in F and for each vertex in $S \cap G_2$ we have an edge from it to x, as shown in bold below.

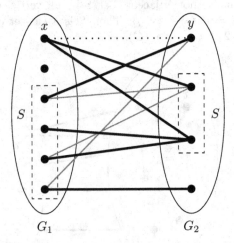

G_1 G_2

It remains to show that S is a cut-set of G. Since G is connected there must be some $x - y$ path in G. Since any such path must use an edge from F, and every edge from F has a representative added to S, any $x - y$ path in G must use a vertex from S. Thus $G - S$ is disconnected and so S is a cut-set for G. Thus $\kappa(G) \leq |S| \leq |F| \leq \kappa'(G)$.

Whitney's Theorem provides an indication that high connectivity (or edge-connectivity) requires a large minimum degree. But is the converse true? Can a graph have high minimum degree but low connectivity? See Exercise 4.7.

4.2   Connectivity and Paths

Now that we have some familiarity with connectivity, we turn to its relationship to paths within a graph. Note that for the remainder of this section, we will assume the graphs are connected, as otherwise the results are trivial. We begin by relating cut-vertices and bridges to paths. You should notice that almost every result for vertices has an edge analog. We begin with the most simple results relating a cut-vertex or a bridge to its presence on a path.

Theorem 4.6 A vertex v is a cut-vertex of a graph G if and only if there exist vertices x and y such that v is on every $x - y$ path.

Proof: First suppose v is a cut-vertex in a graph G. Then $G-v$ must have at least two components. Let x and y be vertices in different components of $G - v$. Since G is connected, we know there must exist an $x - y$ path in G that does not exist in $G - v$. Thus v must lie on this path.

Conversely, let v be a vertex and suppose there exist vertices x and y such that v is on every $x - y$ path. Then none of these paths exist in $G - v$, and so x and y cannot be in the same component of $G - v$. Thus G must have at least two components and so v is a cut-vertex.

Below is the edge version of the result above, the proof of which appears in Exercise 4.17.

Theorem 4.7 An edge e is a bridge of G if and only if there exist vertices x and y such that e is on every $x - y$ path.

The theorem to follow has a similar feel to that about leaves and trees (see Exercise 3.18) and allows us to know we can pick some vertex of a graph not to be a cut-vertex. The second theorem listed below relates bridges to cycles. It should be obvious that any edge along a cycle cannot be a bridge since its removal will only break the cycle, not disconnect the graph; perhaps more surprising is that all edges not on a cycle are in fact bridges. The proof of these results appear in Exercise 4.18 and 4.19.

Theorem 4.8 Every nontrivial connected graph contains at least two vertices that are not cut-vertices.

Theorem 4.9 An edge e is a bridge of G if and only if e lies on no cycle of G.

To arrive at our main result of this section, we need a few additional pieces of terminology. Though their meanings should be obvious from context, we will include formal definitions below. We begin with a way of describing how two paths overlap.

Definition 4.10 Let P_1 and P_2 be two paths within the same graph G. We say these paths are

- *disjoint* if they have no vertices or edges in common.

- *internally disjoint* if the only vertices in common are the starting and ending vertices of the paths.

- *edge-disjoint* if they have no edges in common.

These terms are listed from most restrictive to least restrictive. So two disjoint paths are automatically internally disjoint and edge-disjoint, but two edge-disjoint paths may or may not be internally disjoint.

Our final piece of terminology related to a cut-set or edge-cut is given below, but instead of thinking globally about disconnecting the graph, here we are only concerned with disconnecting two specific vertices from each other.

Definition 4.11 Let x and y be two vertices in a graph G. A set S (of either vertices or edges) *separates* x and y if x and y are in different components of $G - S$. When this happens, we say S is a separating set for x and y.

Note that a cut-set may or may not be a separating set for a specific pair of vertices. Consider the graph G_2 from page 169 (and reproduced on the next page). We have already shown that $\{b, h\}$ is a cut-set and $\{ab, ah\}$ is an edge-cut. But if we want to separate b and c then we cannot use b in the separating set, and using the edges ab and ah will only isolate a, leaving b and c in the same component. However, we can separate b and c using the vertices $\{f, h\}$ and the edges $\{cd, cg, ch\}$. Note that you cannot separate b and c with fewer vertices or edges (try it!).

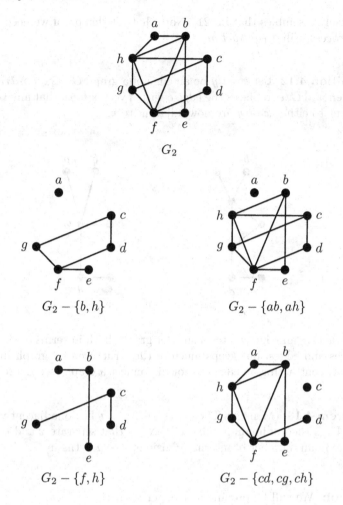

G_2

$G_2 - \{b, h\}$

$G_2 - \{ab, ah\}$

$G_2 - \{f, h\}$

$G_2 - \{cd, cg, ch\}$

4.2.1   Menger's Theorem

The following theorems generalize the results above relating a cut-vertex or bridge to paths in a graph. Menger's Theorem, and the resulting theorems, show the number of internally disjoint (or edge-disjoint) paths directly corresponds to the connectivity (or edge-connectivity) of a graph. For example, in G_2 above we could separate b and c using two vertices and it should be easy to see that $b\,h\,c$ and $b\,e\,f\,d\,c$ are internally disjoint $b-c$ paths. However, if we try to find more than two $b-c$ paths then one of them cannot be internally disjoint from the others (try it!).

Theorem 4.13 (and its edge analog) is named for Karl Menger, the Austrian-American mathematician who first published the result in 1927 [65]. There are many different versions of the proof, and the one presented here most closely resembles that in [21]. Note that for this proof we need an additional process, call a *contraction*.

Definition 4.12 Let $e = xy$ be an edge of a graph G. The **contraction** of e, denoted G/e, replaces the edge e with a vertex v_e so that any vertices adjacent to either x or y are now adjacent to v_e.

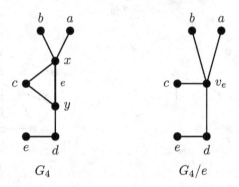

$$G_4 \qquad\qquad G_4/e$$

Contracting an edge creates a smaller graph, both in terms of the number of vertices and edges, but keeps much of the structure of a graph in tact. In particular, contracting an edge cannot disconnect a graph (see Exercise 4.22).

Theorem 4.13 (Menger's Theorem) Let x and y be nonadjacent vertices in G. Then the minimum number of vertices that separate x and y equals the maximum number of internally disjoint $x - y$ paths in G.

Proof: We will be proving a stronger result:

Let $A, B \subseteq V$ and let k equal the minimum number of vertices separating A from B. Then there are k internally disjoint $A - B$ paths.

First note that any path from A to B must travel through any set of vertices separating A and B, and so G cannot have more than k internally disjoint paths. Our goal will be to show that k such paths exist. We will argue by induction on $|E|$.

If G has no edges, then the result is trivially true. Therefore, assume G has at least one edge, say $e = xy$. Suppose that G does not have k internally disjoint $A - B$ paths and consider G/e. If x or y lie in A (or B) then we will consider v_e in A (or B, respectively). Since G does not have k internally disjoint paths, neither can G/e since contracting an edge cannot increase the number of disjoint paths, only possibly decrease them. Thus by the induction hypothesis, G/e contains an $A - B$ separating set Y of less than k vertices since fewer than k internally disjoint $A - B$ paths exist in G/e. Note that v_e must be in Y since otherwise Y would be an $A - B$ separating set of G, contradicting the fact that k is the minimum number of vertices separating A and B. Let $X = Y - \{v_e\} \cup \{x, y\}$. Then X must be an $A - B$ separating set in G of exactly k vertices.

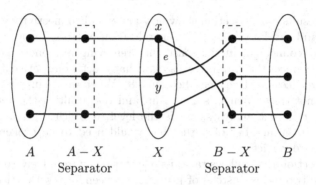

$$A \qquad A - X \qquad X \qquad B - X \qquad B$$
$$\text{Separator} \qquad\qquad \text{Separator}$$

Now consider $G - e$, that is G with the edge e removed. Since $x, y \in X$ and every $A - X$ separating set in $G - e$ is also an $A - B$ separator in G, we know any $A - X$ separating set must also contain at least k vertices. Thus by the induction hypothesis there are k internally disjoint $A - X$ paths in $G - e$. By a similar argument there are k internally disjoint $X - B$ paths in $G - e$. As X separates A and B, these two collections of paths cannot meet outside of X and so they can be combined into $A - B$ paths that are internally disjoint. Thus there are k internally disjoint $A - B$ paths and so Menger's Theorem holds.

An immediate result from Menger's Theorem refers to the global condition of connectivity as opposed to the separation of two specific vertices.

Theorem 4.14 A nontrivial graph G is k-connected if and only if for each pair of distinct vertices x and y there are at least k internally disjoint $x - y$ paths.

By now it shouldn't be too surprising that an edge version exists for the two previous theorems. The proofs of these appear in Exercises 4.32 and 4.33.

Theorem 4.15 Let x and y be distinct vertices in G. Then the minimum number of edges that separate x and y equals the maximum number of edge-disjoint $x - y$ paths in G.

> **Theorem 4.16** A nontrivial graph G is k-edge-connected if and only if for each pair of distinct vertices x and y there are at least k edge disjoint $x - y$ paths.

The results of this section show how paths and connectivity are intricately related, and can be viewed as a generalization of trees. Recall that trees are minimally connected, in the sense that removing any one edge or non-leaf would disconnect the graph. We also saw that exactly one path existed between any two vertices in a tree (see Theorem 3.8). When we are investigating graphs that are not trees, Menger's Theroem (and the resulting theorems) allow us to conduct similar analyses, where the level to which a graph is connected is equal to the number of paths that would need to be broken in order to separate two vertices.

The section that follows takes us one step away from trees to those graphs that have two paths, instead of just one, between any two vertices.

4.3   2-Connected Graphs

As we have already seen, 1-connected graphs are simply those graphs that we more commonly call connected and k-connected graphs can be described in terms of k number of paths between two vertices. So why then do we single out 2-connected graphs? This is in part because they hold a special area in the study of connectivity—they are known to be connected and as we will see cannot contain any cut-vertices. The class of 2-connected graphs provide both some easy results and some more technical and complex areas of study. We begin with a review of our graphs G_2 and G_3 from page 169, reproduced below.

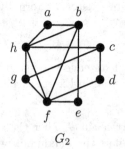
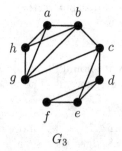

$$G_2 \qquad\qquad\qquad G_3$$

Recall that we showed $\kappa(G_2) = 2 = \kappa'(G_2)$ but that $\kappa(G_3) = 1$ and $\kappa'(G_3) = 2$. So what is the structural difference between G_2 and G_3 that provides the difference in the connectivity measures? Obviously, we can describe it in terms of internally disjoint paths, but there must be some more basic property that separates them. In particular, notice how vertex c seems to be the connecting point between the two halves of G_3; that it, c is a cut-vertex.

Theorem 4.17 A graph G with at least 3 vertices is 2-connected if and only if G is connected and does not have any cut-vertices.

Proof: Assume G is 2-connected. Then any cut-set of G must be of size at least 2. Therefore G must be connected and cannot have a cut-vertex, as in either of these situations we would have a cut-set of size less than 2.

Conversely, suppose G is not 2-connected. Then either G is disconnected or by Menger's Theorem there exist two non-adjacent vertices x and y for which there is exactly one path between them. Removing any vertex along this path will disconnect x and y, and so that vertex serves as a cut-vertex of G.

In Exercise 3.19 we proved that every non-leaf of a tree is a cut-vertex. Therefore no tree is 2-connected and so every 2-connected graph must contain a cycle, or more specifically every vertex lies on a cycle. The proof of the following corollary appears in Exercise 4.21.

Corollary 4.18 A graph G with at least 3 vertices is 2-connected if and only if for every pair of vertices x and y there exists a cycle through x and y.

When first discussing connectivity, we began with what it meant for a graph to be connected, followed by a description of the connected pieces of a graph, which we call *components*. Perhaps not too surprising, we can do the same thing for 2-connected pieces of a graph, called *blocks*.

Definition 4.19 A *block* of a graph G is a maximal 2-connected subgraph of G, that is, a subgraph with as many edges as possible without a cut-vertex.

It is fairly easy to visualize what a component looks like, but what about the blocks of a graph? If G is itself 2-connected then the only block is the entire graph. Otherwise, G is a single vertex (K_1), a single edge (K_2), disconnected, or contains a cut-vertex. From the definition and our results so far, we can see that a block with more than 3 vertices must contain a cycle and all vertices on that cycle must be in the block.

Example 4.3 Determine all blocks for the two graphs below.

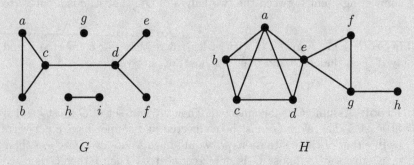

Solution: First note that any isolated vertex will create its own block. Likewise, any component isomorphic to K_2 will also be its own block. The other blocks for graph G are either singular edges (cd, de, and df) or the 3-cycle $abca$, as shown below.

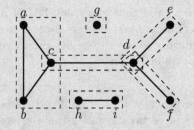

For H, we have one block consisting of the subgraph induced by vertices a, b, c, d, e, the 3-cycle $efgh$, and the edge gh, as shown below.

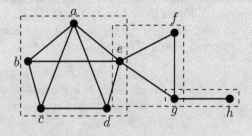

Notice that some blocks within the same component of a graph G will necessarily overlap, and this occurs exactly at any cut-vertex of the larger graph (such as vertex e from graph H in Example 4.3). Moreover, a vertex may be a cut-vertex of G but cannot be a cut-vertex of the blocks to which it belongs. Essentially, every graph can be viewed at being built from its blocks, which are pasted together at the cut-vertices of the graph. Further results on blocks appear in Exercises 4.28 and 4.29.

Blocks will be useful in the proof of Brooks' Theorem from Chapter 6 and Kuratowski's Theorem from Chapter 7, two of the most significant theorems in the areas of graph coloring and planarity, respectively. For the remainder of this section we turn our attention to further results on 2-connected graphs.

Recall that in the proof for Menger's Theorem we use a contraction, where an edge $e = xy$ is replaced by the vertex v_e. A reverse process called a *subdivision* is described below and is useful when discussing 2-connected graphs.

Definition 4.20 Let $e = xy$ be an edge in a graph G. The ***subdivision*** of e adds a vertex v in the edge so as to replace it with the path $x\,v\,y$.

Below is a graph G_5 with two different subdivisions of the edge $e = xy$. The first graph on the left below shows a subdivision where vertex w is added into the edge. Note that we can subdivide the edge e multiple times to create a longer path between x and y to replace edge xy, as shown on the right. This will be used in Chapter 7 when we discuss planar graphs. For now, though, when we describe subdividing an edge we refer to a single vertex addition.

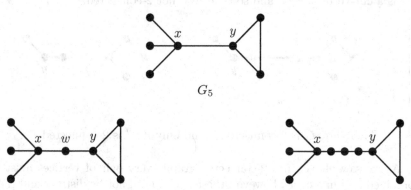

G_5

How does subdividing an edge effect connectivity, in particular with respect to cut-vertices? The result below proves the effect for 2-connected graphs, and a similar result appears in Exercise 4.23 for k-connected graphs.

Theorem 4.21 Let G be a graph and G' the graph obtained by subdividing any edge of G. Then G is 2-connected if and only if G' is 2-connected.

Proof: Let $e = xy$ be an edge of G and let G' be formed by subdividing e into the path $x\,z\,y$.

First suppose G is 2-connected. Then by Corollary 4.18 we know there exists a cycle between any two vertices of G. If a cycle does not use e then it still exists in G'. If a cycle uses e, then by replacing the edge with the path $x\,z\,y$ will maintain the cycle. Thus G' is 2-connected.

Conversely, suppose G is not 2-connected. Clearly if G is disconnected or has fewer than three vertices than G' cannot be 2-connected either. So we may assume G is connected with at least 3 vertices. Thus G must contain a cut-vertex, call it v and consider $G - v$. If e remains in $G - v$ then the subdivision of e would simply occur in one of the components of $G-v$. Thus v is still a cut-vertex in G' and so G' is also not 2-connected.

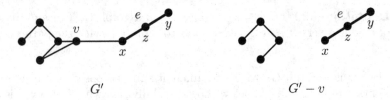

If e does not exist in $G - v$ then either $x = v$ or $y = v$. In either case, z would remain adjacent to the other vertex when v is removed, and so would exist in the same component of $G' - v$. Once again, this implies v is a cut-vertex for G' and so G' is also not 2-connected.

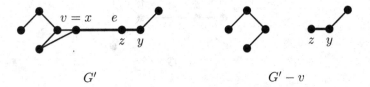

Therefore, G is 2-connected if and only if G' is 2-connected.

As we saw above, in a 2-connected graph every pair of vertices must be contained within a cycle; however, these cycles need not be distinct and these cycles may in fact overlap. One way of describing this is through the use of an ear decomposition. Before we get into the decomposition portion, we need to first understand what is meant by an ear of a graph.

Definition 4.22 An *ear* of a graph is a path P that is contained within a cycle where only the endpoints of P can have degree more than 2 in the graph.

Consider the following graph G_6 on the left. It is clearly 2-connected since it does not have a cut-vertex (check it!). When looking for ears, we need to find vertices of degree 2 and work outward to find a path that is part of a larger cycle. Two ears are shown on the right below, one of which is $g\,a\,b$ and the other $g\,f\,e\,d$.

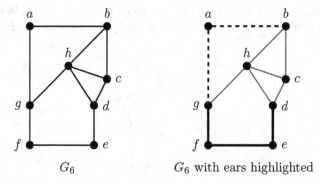

G_6 G_6 with ears highlighted

The two ears shown are the only nontrivial ears of G_6. Since a path can consist of only one edge, any edge of a cycle can technically be an ear, but for any longer path all internal vertices must have degree 2 within the larger graph. While this may seem fairly limiting for practical use, we will make a modification to where we find a succession of paths that are ears on the earlier graph.

> **Definition 4.23** An ***ear decomposition*** of a graph G is a collection $P_0, P_1, \ldots P_k$ so that P_0 is a cycle, P_i is an ear of $P_0 \cup \cdots \cup P_{i-1}$ for all $i \geq 1$, and all edges and vertices are included in the collection.

To find an ear decomposition of a graph, we first must find the starting cycle P_0 and then look for a path whose endpoints are along that cycle. We continue doing this until the entire graph has been added.

> **Example 4.4** Find an ear decomposition of the graph G_6 shown above.
>
> *Solution:* We begin by picking an initial cycle. Here we choose P_0 to be $h\,g\,f\,e\,d\,h$.

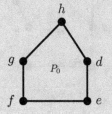

We must now find a path P_1 with endpoints along this cycle. We choose $h\,c\,d$. Note that the internal vertex of this path, c, has degree 2 in the subgraph of G_6 shown below, even though it has degree 3 in G_6. This is acceptable since P_1 is a ear within the graph formed by $P_0 \cup P_1$, namely $G_6[P_0 \cup P_1]$.

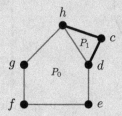

We now find a path P_2 with endpoints from the graph above; we use $h\,b\,c$. As above, the internal vertex of path P_2 is b and has degree 2 in the graph formed by $P_0 \cup P_1 \cup P_2$, namely $G_6[P_0 \cup P_1 \cup P_2]$.

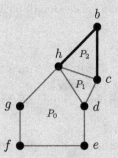

Finally, we conclude our ear decomposition with the path $g\,a\,b$, as shown below.

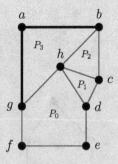

Ear decompositions are not unique; in fact, any cycle in the graph can serve as P_0 and more than one choice for P_1 can even exist with a given starting cycle (see Exercise 4.11). Our final result from this section shows that containing an ear decomposition characterizes 2-connected graphs.

Theorem 4.24 A graph G is 2-connected if and only if it has an ear decomposition.

Proof: Suppose G is 2-connected. We must find an ear decomposition of G. First, since G is 2-connected, we know there exists a cycle in G, call it P_0. If this does not encompass all of G, then we can successively add ears to P_0. Among all possible graphs that can be formed in this way, choose H to be the one with the maximum number of edges. If $H = G$, then we have found an ear decomposition of G. Otherwise, there must exist some edge $e = xy$ that is in G but not H, with one endpoint in H, say x. Note that y cannot also be in H as otherwise e would be an ear of H.

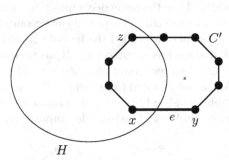

Since G is 2-connected, there must exist some cycle containing e. Moreover, since x is not a cut-vertex of G, there must be a cycle C' containing e that meets H at a vertex other than x. Let z be the first vertex after x on C' that is also in H. Then the portion of C' from x to z that includes e is an ear of H. But this contradicts the choice of H as a graph with the maximum number of edges possible. Thus $H = G$ and we have an ear decomposition of G.

Conversely, suppose G has an ear decomposition $P_0, P_1, \ldots, P_k$. Note that a cycle cannot contain a cut-vertex, so we know P_0 is itself 2-connected. If we can prove that the graph that results from adding an ear to a 2-connected graph remains 2-connected, then this would imply G is itself 2-connected.

Suppose x and y are the endpoints of the ear P to be added to a 2-connected graph G'. Let $e = xy$. Then $G' + e$ is 2-connected since adding an edge cannot decrease the connectivity of a graph. Through successive subdivisions of e we can arrive at the ear P and so $G' \cup P$ is 2-connected by Theorem 4.21. Thus G must be 2-connected.

4.3.1   2-Edge-Connected

Based on previous discussions, it shouldn't be surprising that there are similar notions for graphs that are 2-edge-connected, which can be described as those graphs that are connected but without a bridge. In particular, we extend the ear decomposition idea into its edge analog, called a *closed-ear decomposition*.

> **Definition 4.25** A *closed-ear* in a graph G is a cycle where all vertices have degree 2 in G except for one vertex on the cycle. A *closed-ear decomposition* is a collection $P_0, P_1, \ldots P_k$ so that P_0 is a cycle, P_i is either an ear or closed-ear of $P_0 \cup \cdots \cup P_{i-1}$ for all $i \geq 1$, and all edges and vertices are included in the collection.

The small change in our definition between an ear and closed-ear can most easily be attributed to graphs that are 2-edge-connected, but not 2-connected. Consider the graph from Example 4.4. This graph is 2-edge-connected since it does not have a bridge. Thus the same decomposition we had above still works (since we are allowed to use ears in a closed-ear decomposition). In contrast, the graph G_7 below (sometimes called the bow-tie graph) is 2-edge-connected but not 2-connected since c is a cut-vertex. If we tried to find a regular ear decomposition for G_7 then we would run into a problem in finding P_1 since once the first cycle has been chosen (for example P_0 below) then the remaining portion of the graph would only consist of another 3-cycle. But allowing P_1 to be a closed-ear, we find our closed-ear decomposition.

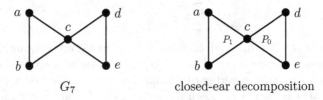

G_7 closed-ear decomposition

The edge analog to Theorem 4.24 is given below and has a very similar proof (see Exercise 4.31).

> **Theorem 4.26** A graph G is 2-edge-connected if and only if it has a closed-ear decomposition.

The majority of this chapter has been devoted to a very theoretical aspect of graph theory. The remainder of this chapter looks at various applications of connectivity, with the largest focus on network flow.

4.4 Network Flow

Digraphs have appeared throughout this text to model asymmetric relationships. For example, at the beginning of Chapter 1, we described game wins as a directed edge in a tournament, and in Chapter 2 we looked at when digraphs, more specifically tournaments, were hamiltonian. This section will focus on a new application for digraphs, one in which items are sent through a network. These networks often model physical systems, such as sending water through pipelines or information through a computer network. The digraphs we investigate will need a starting and ending location, though there is no requirement for the network to be acyclic. In this section, we will need some specialized terminology, in particular, what we mean by a network.

> **Definition 4.27** A *network* is a digraph where each arc e has an associated nonnegative integer $c(e)$, called a *capacity*. In addition, the network has a designated starting vertex s, called the *source*, and a designated ending vertex t, called the *sink*. A *flow* f is a function that assigns a value $f(e)$ to each arc of the network.

Below is an example of a network. Each arc is given a two-part label. The first component is the flow along the arc and the second component is the capacity.

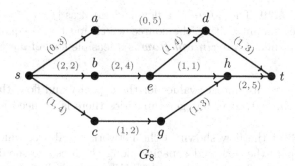

$$G_8$$

The names of the starting and ending vertices are reminiscent of a system of pipes with water coming from the source, traveling through some configuration of the piping to arrive at the sink (ending vertex). Using this analogy further, we can see that some restraints need to be placed on the flow along an arc. For example, flow should travel in the indicated direction of the arcs, no arc can carry more than its capacity, and the amount entering a junction point (a vertex) should equal the amount leaving. These rules are more formally stated in the following definitions.

Definition 4.28 For a vertex v, let $f^-(v)$ represent the total flow entering v and $f^+(v)$ represent the total flow exiting v. A flow is **feasible** if it satisfies the following conditions:

(1) $f(e) \geq 0$ for all edges e

(2) $f(e) \leq c(e)$ for all edges e

(3) $f^+(v) = f^-(v)$ for all vertices other than s and t

(4) $f^-(s) = f^+(t) = 0$

The notation for in-flow and out-flow mirrors that for in-degree and out-degree of a vertex, though here we are adding the flow value for the arcs entering or exiting a vertex. The requirement that a flow is non-negative indicates the flow must travel in the direction of the arc, as a negative flow would indicate items going in the reverse direction. The final condition listed above requires no in-flow to the source and no out-flow from the sink. This is not necessary in theory, but more logical in practice and simplifies our analysis of flow problems.

The network G_8 shown above satisfies the conditions for a feasible flow (check them!). In general, it is fairly easy to verify if a flow is feasible. Our main goal will be to find the best flow possible, called the *maximum flow*.

Definition 4.29 The **value** of a flow is defined as $|f| = f^+(s) = f^-(t)$, that is, the amount exiting the source which must also equal the flow entering the sink. A **maximum flow** is a feasible flow of largest value.

In practice, we use integer values for the capacity and flow, though this is not required. In fact, given integer capacities there is no need for fractional flows.

Look back at the flow shown in the network G_8 above, which has value 3. If we compare the flow and capacity along the arcs, we should see many locations where the flow is below capacity. However, finding a maximum flow is not as simple as putting every arc at capacity—this would likely violate one of the feasibility criteria. For example, if we had a flow of 5 along the arc ad, we would need the flow along dt to also equal 5 to satisfy criteria (3). But in doing so we would violate criteria (2) since the capacity of dt is 3.

The main question in regard to network flow is that of optimization—what is the value of a maximum flow? We could start with a simple feasible flow, as shown above, and use trial and error to keep improving it, though this is not an efficient procedure and does not guarantee the flow we find is indeed maximum. We will discuss an algorithm that not only finds a maximum flow but also provides proof that a larger flow cannot be found. Before we fully discuss the algorithm, we need two additional definitions relating to the flow along a network.

Definition 4.30 Let f be a flow along a network. The *slack* k of an arc is the difference between its capacity and flow; that is, $k(e) = c(e) - f(e)$.

Slack will be useful in identifying locations where the flow can be increased. For example, in the network above $k(sa) = 3, k(sc) = 3$, and $k(sb) = 0$ indicates that we may want to increase flow along the arcs sa and sc but no additional flow can be added to sb. The difficult part is determining where to make these additions. To do this we will build special paths, called *chains*, that indicate where flow can be added.

Definition 4.31 A *chain* K is a path in a digraph where the direction of the arcs are ignored.

In the network shown above, both $s\,a\,d\,t$ and $s\,a\,d\,e\,h\,t$ are chains, though only $s\,a\,d\,e\,h\,t$ is not a directed path since it uses the arc ed in reverse direction.

We now have all the needed elements for finding the maximum flow. The algorithm below is similar to Dijkstra's Algorithm from Section 2.3 which found the shortest path in a graph (or digraph). The format of the Augmenting Flow Algorithm, described below, is an adaption from that given in [79] and based on the Ford-Fulkerson and Edmonds-Karp Algorithms (see [30] and [36]).

Vertices will be assigned two-part labels that aid in the creation of a chain on which the flow can be increased. The first part of the label for vertex y will indicate one of two possibilities: x^- if there is a positive flow along $y \to x$, or x^+ if there is slack along the arc $x \to y$, where the former scenario may allow a reduction along the arc yx in order for additional flow along along another edge out of x, whereas the latter scenario may allow more along the arc xy itself. The second part of the label will indicate the amount of flow that could be adjusted along the arc in question.

Augmenting Flow Algorithm

Input: Network $G = (V, E, c)$, with designated source s and sink t, and each arc is given a capacity c.

Steps:

1. Label s with $(-, \infty)$

2. Choose a labeled vertex x.

    (a) For any arc yx, if $f(yx) > 0$ and y is unlabeled, then label y with $(x^-, \sigma(y))$ where $\sigma(y) = \min\{\sigma(x), f(yx)\}$.

    (b) For any arc xy, if $k(xy) > 0$ and y is unlabeled, then label y with $(x^+, \sigma(y))$ where $\sigma(y) = \min\{\sigma(x), k(xy)\}$.

3. If t has been labeled, go to Step (4). Otherwise, choose a different labeled vertex that has not been scanned and go to Step (2). If all labeled vertices have been scanned, then f is a maximum flow.

4. Find an $s - t$ chain K of slack edges by backtracking from t to s. Along the edges of K, increase the flow by $\sigma(t)$ units if they are in the forward direction and decrease by $\sigma(t)$ units if they are in the backward direction. Remove all vertex labels except that of s and return to Step (2).

Output: Maximum flow f.

    It is important that in Step 2 when we are labeling the neighbors of a vertex x that we first consider arcs into x from unlabeled vertices that have positive flow (part a) and then the arcs out of x to unlabeled vertices with positive slack (part b). These are used to find a chain from s to t onto which flow can be added.

    Before we use the Algorithm on the network G_8 from page 189, we begin with a smaller example to explain the labeling procedure and how it is used to improve the flow.

Example 4.5 Apply the Augmenting Flow Algorithm to the network G_9 shown below.

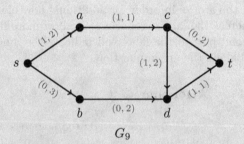

G_9

Solution: We will show edges under consideration in bold.

Step 1: Label s as $(-, \infty)$.

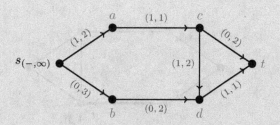

Step 2: Let $x = s$. As there are no arcs to s we will only consider the arcs out of s, of which there are two: sa and sb, which have slack of 1 and 3, respectively. We label a with $(s^+, 1)$, b is labeled $(s^+, 3)$.

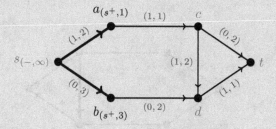

Step 3: As t is not labeled, we will scan either a or b; we choose to start with a, (so $x = a$ in the algorithm). Since the only arc going into a is from a labeled vertex, we need only consider the edges out of a, of which there is only one, ac, which has no slack. Thus c is left unlabeled at this stage.

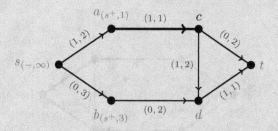

Step 4: As t is not labeled, we will scan b (so $x = b$ in the algorithm). Since the only arc going into b is from a labeled vertex, we need only consider the edges out of b, of which there is only one, bd, with slack of 2. Label d as $(b^+, 2)$ since $o(d) = \min\{\sigma(b), k(bd)\} = \min\{3, 2\} = 2$.

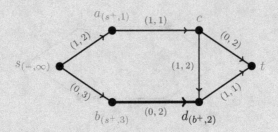

Step 5: As t is not labeled, we will scan d (so $x = b$ in the algorithm). There is one unlabeled vertex with an arc going into d, namely c, which gets a label of $(d^-, 1)$ since $\sigma(c) = \min\{\sigma(d), f(cd)\} = \min\{2, 1\} = 1$. The only arc out of d is dt, which has no slack.

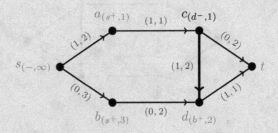

Step 6: As t is not labeled, we will scan c (so $x = c$ in the algorithm). Since the only arc going into c is from a labeled vertex, we need only consider the edges out of c to an unlabeled vertex, of which there is only one, ct, with slack of 2. Label t as $(c^+, 1)$ since $\sigma(t) = \min\{\sigma(c), k(ct)\} = \min\{1, 2\} = 1$.

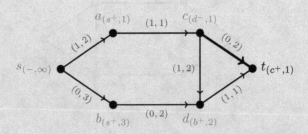

Step 7: Since t is now labeled, we find an $s - t$ chain K of slack edges. Backtracking from t to s gives the chain $s\,b\,d\,c\,t$.

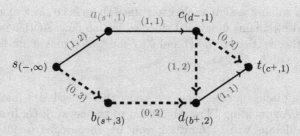

We increase the flow by $\sigma(t) = 1$ units along each of the edges in the forward direction (namely sb, bd, ct) and decrease by 1 along the edges in the backward direction (namely cd). We update the network flow and remove all labels except that for s.

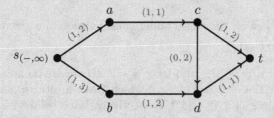

Step 8: As before we will only label the vertices whose arcs from s have slack. We label a with $(s^+, 1)$ and b with $(s^+, 2)$.

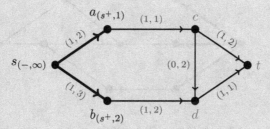

Step 9: We scan either a or b; we begin with a. The only arc from a is to c, but since there is no slack we do not label c. Scanning b we only consider the arc bd, which has slack 1. Label d with $(b^+, 1)$ since $\sigma(d) = \min\{\sigma(b), k(bd)\} = \min\{2, 1\} = 1$.

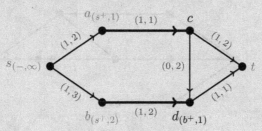

Scanning d will not assign a label to t since there is no slack along the arc dt, and c also remains unlabeled since cd has no flow. At this point there are no further vertices to label and so f must be a maximum flow of G_9, with a value of 2.

Before we discuss further how we know this indeed is a maximum flow, we apply the Augmenting Flow Algorithm to the network G_8 from page 189. Some of the details will be omitted; you are encouraged to fill these in as needed.

Example 4.6 Apply the Augmenting Flow Algorithm to the network G_8 on page 189.

Solution: As before, the edges under consideration in a given step will be shown in bold.

Step 1: Label s as $(-,\infty)$ and let $x = s$. As there are no arcs to s we will only consider the arcs out of s, of which there are three: sa, sb, and sc, which have slack of 3, 0, and 3, respectively. We label a with $(s^+,3)$, b is left unlabeled since there is no slack on sb, and c is labeled $(s^+,3)$.

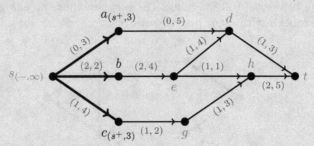

Step 2: As t is not labeled, we will scan either a or c; we choose to start with c. Since the only arc going into c is from a labeled vertex, we need only consider the edges out of c, of which there is only one, cg, with slack of 1. Label g as $(c^+,1)$ since $\sigma(g) = \min\{\sigma(c), k(cg)\} = \min\{3,1\} = 1$.

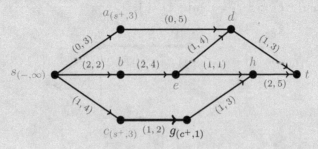

Step 3: We can scan either a or g; we choose g. We need only consider the edges out of g, of which there is only one, gh, with slack of 2. Label h as $(g^+, 1)$ since $\sigma(h) = \min\{\sigma(g), k(gh)\} = \min\{1, 2\} = 1$.

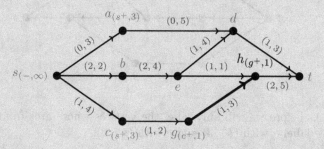

Step 4: We can scan either a or h; we choose h. There is one unlabeled vertex with an arc going into h, namely e, which gets a label of $(h^-, 1)$ since $\sigma(e) = \min\{\sigma(h), f(eh)\} = \min\{1, 1\} = 1$. The only arc out of h is ht, with slack of 3. Label t as $(h^+, 1)$ since $\sigma(t) = \min\{\sigma(h), k(ht)\} = \min\{1, 3\} = 1$.

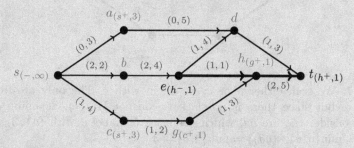

Step 5: Since t is now labeled, we find an $s - t$ chain K of slack edges. Backtracking from t to s gives the chain $s\,c\,g\,h\,t$ as shown below.

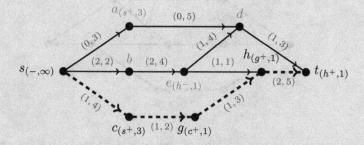

We increase the flow by $\sigma(t) = 1$ units along each of these edges since all are in the forward direction. We update the network flow and remove all labels except that for s.

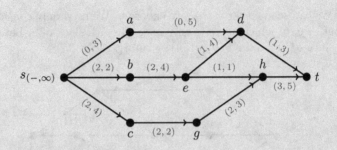

Step 6: As before we will only label the vertices whose arcs from s have slack. We label a with $(s^+, 3)$ and c with $(s^+, 2)$.

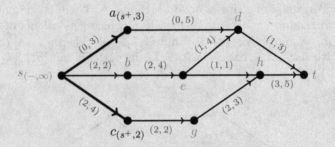

Step 7: We scan either a or c; we begin with c. The only arc from c is to g, but since there is no slack we do not label g. Scanning a we only consider the arc ad, which has slack 5. Label d with $(a^+, 3)$ since $\sigma(d) = \min\{\sigma(a), k(ad)\} = \min\{3, 5\} = 3$.

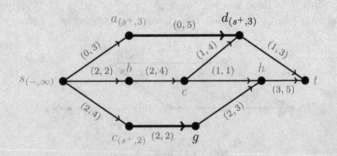

Step 8: Our only unscanned labeled vertex is d. Since e has an arc to d with positive flow, label e as $(d^-, 1)$ since $\sigma(e) = \min\{\sigma(d), f(ed)\} = \min\{3, 1\} = 1$. Also label t with $(d^+, 2)$ since $\sigma(t) = \min\{\sigma(g), k(dt)\} = \min\{3, 2\} = 2$.

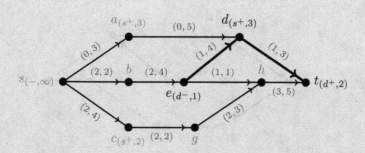

Step 9: Since t is again labeled, we find an $s - t$ chain K' of slack edges. Backtracking from t to s gives the chain $s\,a\,d\,t$, and we increase the flow by $\sigma(t) = 2$ units along each of these edges since all are in the forward direction. We update the network flow and remove all labels except that for s.

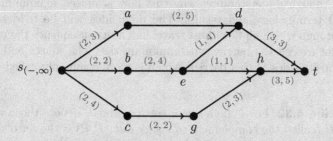

Step 10: The process of again assigning labels is quite similar to the steps above. Some σ values now change along the last chain from which we adjusted the flow. Instead of working through the individual steps, we will pick up after the label for d has been assigned, which is shown below.

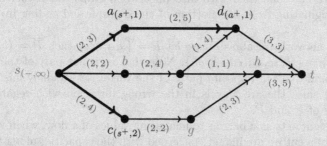

Label e as $(d^-, 1)$ as before since there is flow along the arc ed, but t is not given a label due to no slack along the arc dt. Once this is complete, b will be given a label of $(e^-, 1)$ since $\sigma(b) = \min\{\sigma(e), f(be)\} = \min\{1, 2\} = 1$. No label will be assigned to h since there is no slack along the arc eh. Recall that g also remains unlabeled since cg has no slack.

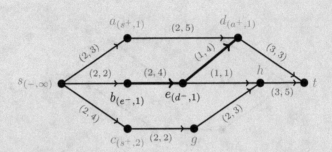

At this point there are no further vertices to label and so f must be a maximum flow, with a value of 6.

When the Augmenting Flow Algorithm halts, a maximum flow has been achieved, though understanding why this flow is indeed maximum requires additional terminology and results. The main idea will be to determine a barrier through which all flow must travel and as a consequence the maximum flow cannot exceed the barrier with minimum size. The source and sink will be on opposite sides of this barrier, which is more commonly called a *cut*.

Definition 4.32 Let P be a set of vertices and $\overline{P}$ denote those vertices not in P (called the complement of P). A *cut* $(P, \overline{P})$ is the set of all arcs xy where x is a vertex from P and y is a vertex from $\overline{P}$. An *s − t cut* is a cut in which the source s is in P and the sink t is in $\overline{P}$.

Note that a cut is a separating set that is reminiscent of an edge-cut, except that we are not concerned with disconnecting a graph but rather describing all arcs originating from one portion of the digraph and ending in the other portion.

In the network G_8 above, if we let $P = \{s, a, e, g\}$ then $\overline{P} = \{b, c, d, h, t\}$ and $(P, \overline{P}) = \{sb, sc, ad, ed, eh, gh\}$. Note that be is not part of the cut even though b and e are in opposite parts of the vertex set (namely b is in $\overline{P}$ and e is in P) since the arc travels in the wrong direction with regards to the definitions of P and $\overline{P}$.

As this cut acts as a barrier to increasing values of a flow, when we discuss the value of a cut we are in fact concerned with the capacities along these arcs rather than their flow. Thus the value of a cut is referred to as its capacity.

Definition 4.33 The *capacity* of a cut, $c(P, \overline{P})$, is defined as the sum of the capacities of the arcs that comprise the cut.

The cut given above has capacity 18 (try it!) and therefore indicates that all feasible flows of G_8 must have value at most 18. Obviously, this is not the best bound for the maximum flow since our work above seems to indicate that the maximum flow of G_8 has value 6. In fact, two easy cuts often provide more useful initial bounds on the value of a flow; the first is where P only consists of the source and the second is where P consists of every vertex except the sink. In the example above, if we let $P = \{s\}$ then the capacity of this cut is $c(P, \overline{P}) = 9$ and if $P' = \{s, a, b, c, d, e, g, h\}$ then it has capacity $c(P', \overline{P'}) = 8$, which are much closer to our conjecture that the maximum flow is 6.

In Section 4.2 we showed that minimum size of a separating set equals the maximum number of internally disjoint paths. A similar result holds for flows and cuts in a network. In fact, Menger's theorem is really just an undirected version of the Max Flow–Min Cut Theorem below, which was published about thirty years after Menger's result [36].

Theorem 4.34 (Max Flow–Min Cut) In any directed network, the value of a maximum $s - t$ flow equals the capacity of a minimum $s - t$ cut.

The difficulty in using this result to prove a flow is maximum is in finding the minimum cut. Luckily, we can use the vertex labeling procedure to obtain our minimum cut.

Min-Cut Method

1. Let $G = (V, A, c)$ be a network with a designated source s and sink t and each arc is given a capacity c.

2. Apply the Augmenting Flow Algorithm.

3. Define an $s - t$ cut $(P, \overline{P})$ where P is the set of labeled vertices from the final implementation of the algorithm.

4. $(P, \overline{P})$ is a minimum $s - t$ cut for G.

By finding a flow and cut with the same value, we now have proof that the flow is indeed maximum.

Example 4.7 Use the Min-Cut Method to find a minimum $s - t$ cut for the network G_8 on page 189 and the network G_9 from Example 4.5.

Solution: The final labeling from the implementation of the Augmenting Flow Algorithm on G_8 produced the following network.

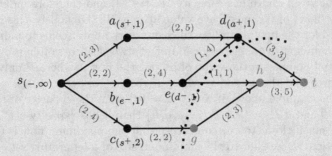

The Min-Cut Method sets $P = \{s, a, b, c, d, e\}$ and $\overline{P} = \{g, h, t\}$. The arcs in the cut are $\{dt, eh, cg\}$, making the capacity of this cut $c(P, \overline{P}) = 3 + 1 + 2 = 6$. Since we have found a flow and cut with the same value, we know the flow is maximum and the cut is minimum.

The final labeling in the network G_9 from Example 4.5 gives $P = \{s, a, b, d\}$ and $\overline{P} = \{c, t\}$. The arcs in the cut set are $\{ac, dt\}$. Note, cd is not in the cut since the arc is in the wrong direction. The capacity of this cut is $c(P, \overline{P}) = 2$, and since we have found a flow and cut with the same value, we know the flow is maximum and the cut is minimum.

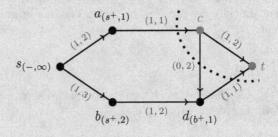

In practice, we can perform the Augmenting Flow Algorithm and the Min-Cut Method simultaneously, thus finding a maximum flow and providing a proof that it is maximum (through the use of a minimum cut) in one complete procedure.

4.5 Centrality Measures

As we have already seen, we measure connectivity of a graph in terms of the number of distinct paths between two vertices or by the number of vertices (or edges) needed to disconnect the graph. This section will investigate another method for evaluating the underlying structure of a graph by identifying central vertices.

In Definition 2.29, we define a vertex to be *central* if its eccentricity equals the radius of the graph. Recall the eccentricity of a vertex x is the maximum distance from x to any other vertex in G; that is $\epsilon(x) = \max_{y \in V(G)} d(x, y)$ and the radius of a graph is the minimum eccentricity among all vertices; that is $rad(G) = \min_{x \in V(G)} \epsilon(x)$. These are useful measures for identifying vertices within short distance of most other vertices, but does that really explain how connected or important a single vertex is to the rest of the graph?

Instead of only relying on path distance, we may want to characterize vertices based on other metrics indicating their relative importance within a graph. These measures are often called *network centralities*, where network here simply means a connected graph.

Consider, for example, your network of friends. If each person were a vertex and an edge represented a friendship, then some of your friends would naturally be friends with each other, and you might find clumps in the vertices, maybe your math major friends, intermural soccer friends, or friends from your dorm; some of these clumps may overlap, so you are not the only connection from these groups of people.

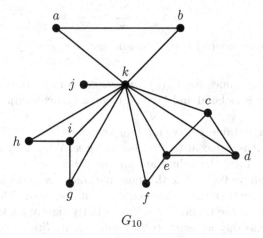

G_{10}

We will use graph G_{10} above as an example for the various centrality measures. However, in most applications we would be working with graphs containing many more vertices (think in the range of a few hundred to many millions) and our analysis would be done using computer software and the adjacency matrix; one such example is shown at the conclusion of this section.

Using a friend network as motivation, how might we identify a central vertex? Perhaps the person with the most friends? This is called the *degree centrality*.

Definition 4.35 Given a graph G, the *degree centrality* of a vertex v is defined as
$$C_d(v) = \deg(v)$$

For graph G_{10} above, it is not too difficult to find the degree centralities, as shown in the table on the next page.

While degree centrality does indicate those vertices with the largest number of direct connections, it seems to oversimplify the analysis. For example, in G_{10} it is quite obvious that vertex k is more influential than the other vertices, but the degree centrality does not allow much distinction between the other vertices.

One attempt to provide further distinction is through a measure called the closeness centrality, which, roughly speaking, looks for how close a vertex is to the other vertices in the graph.

Definition 4.36 Given a graph G, the *closeness centrality* of a vertex v is defined as
$$C_c(v) = \frac{1}{k} \sum_y \frac{1}{d(v,y)}$$

where k is the number of vertices connected to v.

Note that the values for $C_c(v)$ will range from 0 to 1, where 0 would indicate a vertex is isolated and 1 would mean a vertex is adjacent to all other vertices.

As seen in the table below, for G_{10} vertex k has the highest closeness centrality since it is adjacent to all other vertices. What may seem surprising though is how similar the values are for the other ten vertices, with only e having a value above 0.6. While this does not allow us to distinguish between vertices much more than the degree centrality, it does provide some additional information in that the graph as a whole is fairly tightly packed.

Our third centrality measure, *betweennness centrality*, attempts to combat the scenario described above and distinguish between close vertices based on how often a vertex is between other vertices.

Definition 4.37 Given a graph G, the *betweenness centrality* of a vertex v is defined as
$$C_b(v) = \frac{1}{2} \sum_{s \neq t \neq v} \frac{\sigma_{st}(v)}{\sigma_{st}}$$

where the sum is taken over all distinct pairs s and t, σ_{st} is the number of shortest paths from s to t, and $\sigma_{st}(v)$ is the number of these paths that pass through v. Note: we set this ratio to be 0 if there are no paths from s to t.

Betweenness centrality provides a numerical answer for how often a vertex sits on the shortest path between other vertices. The $\frac{1}{2}$ removes the double count of looking at both the path from s to t and the path from t to s. Also the values for $C_b(v)$ will range from 0 to $\frac{(n-1)(n-2)}{2}$, where 0 means a vertex is never in the interior of a shortest path and $\frac{(n-1)(n-2)}{2}$ would be when every path has v as an interior vertex (see Exercise 4.12). The table below lists the various centralities for G_{10}.

	Vertex									
	a	b	c	d	e	f	g	h	i	j	k
C_d	2	2	3	3	4	2	2	2	3	1	10
C_c	0.56	0.56	0.59	0.59	0.63	0.56	0.56	0.56	0.59	0.53	1.00
C_b	0	0	0	0	1	0	0	0	0.5	0	34.5

As expected, vertex k has the highest betweenness centrality and most of the others have value 0. Note how vertices e and i have nonzero betweenness centrality since they lie on some (but not all) of the shortest paths between some vertices. For example, there are two shortest paths from g to h, namely gkh and gih, and so $\sigma_{gh} = 2$ and $\sigma_{gh}(i) = 1$. Since all other vertices in G_{10} that are not adjacent to i do not have shortest paths containing i, we have $C_b(i) = 0.5$.

While vertices e, i, and k having nonzero betweenness centralities does provide some distinction amongst the vertices in G_{10}, we should not ignore the relative sizes of these nonzero values. For this graph, $C_b(k)$ is much larger than $C_b(e)$ and $C_b(i)$, whereas $C_b(e)$ and $C_b(i)$ are almost equal, further emphasizing the central nature of k within the graph.

While each of the measures outlined above can be calculate by hand when working with graphs containing small numbers of vertices, it would be unfeasible to do so with graphs of large size. In fact, most interesting uses of these measures are dealing with graphs containing thousands if not millions of vertices. The example that follows indicates how these measures can be better used to analyze a graph, though all calculations were done using Wolfram Mathmatica, where the measures discussed here are built-in functions.

Example 4.8 The graph below represents a friendship network. Visually, it would appear that vertex 1 seems to play a central role here. We will verify this using the centrality measures listed above.

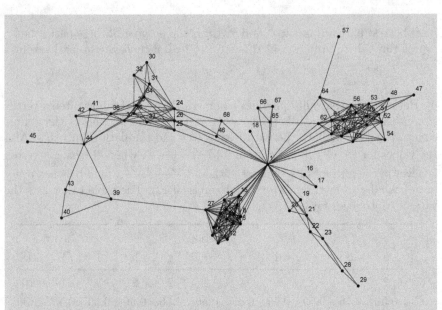

This graph contains 68 vertices, so we will only list the centrality measures for the top 8.

C_d		C_c		C_b	
1	35	1	0.624	1	1670
49	16	27	0.466	24	255
27	16	26	0.463	26	238
*	15	24	0.463	27	217
*	15	58	0.453	39	160
*	15	25	0.450	58	141
*	15	57	0.448	44	125
*	15	**	0.444	25	117

Note that for * above, there are eight vertices all with the same degree centrality of 15 (namely 2, 3, 4, 5, 6, 7, 8, 55, and 58) and for ** above, there are eight vertices all with the same closeness centrality of 0.444 (namely 2, 3, 4, 5, 6, 7, 8, and 61).

These results confirm that vertex 1 plays a central role in this network. Understanding some network dynamics can be useful when trying to combat the spread of information or illness between members of the network.

There are numerous other centrality measures that mathematicians have used to identify key players within a network. In fact, the PageRank Algorithm used by Google to rank webpages is based in part on the Eigenvector Centrality, which uses the adjacency matrix and a unique largest eigenvalue to provide the centrality measure. For further information on network analysis, see [27].

4.6   Exercises

4.1 Find G/e, the graph obtained by contracting edge e, in the graph below with each edge indicated.

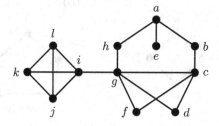

(a) *ae*
(b) *gi*
(c) *ik*
(d) *cg*

4.2 Using the graph from Exercise 4.1, find the graph created by subdividing the edge e indicated below.
(a) *ae*
(b) *cd*
(c) *ik*
(d) *cg*

4.3 For each of the graphs below, determine the blocks of G.

(a)

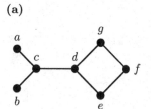

(b)

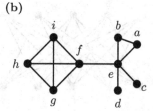

(c)

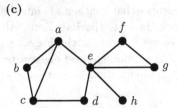

(d)

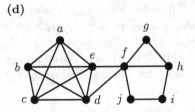

4.4 Given a graph G we can form its *block graph* $B(G)$ where a vertex of $B(G)$ is either a cut-vertex from G or a block of G. Two vertices x, y in $B(G)$ are adjacent if x is a block of G and y is a cut-vertex contained in x. Form the block graph for each of the graphs from Exercise 4.3 above.

4.5 For each of the graphs below, determine $\kappa(G)$ and $\kappa'(G)$.

(a)

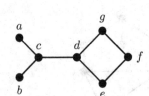

(b)

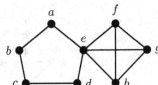

(c)

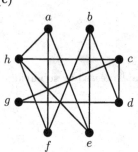

(d)

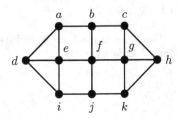

(e)

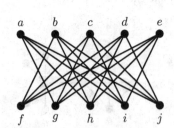

4.6 For each of the problems below, use the Augmenting Flow Algorithm to maximize the flow and the Min-Cut Method to find a minimum cut.

(a)

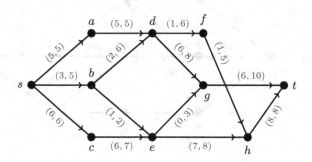

(b)

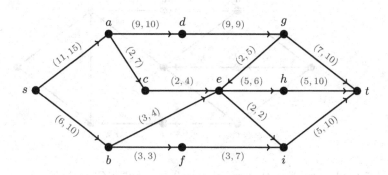

(c)

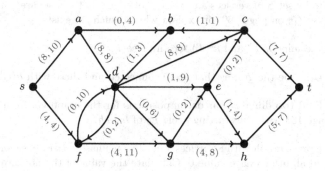

(d)

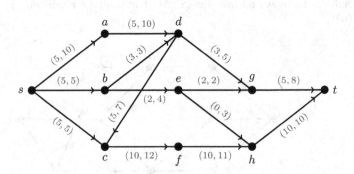

(e)

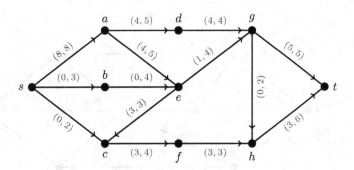

4.7 Find a graph G that satisfies $\delta(G) = 5$ but $\kappa(G) = 1$. Now generalize to $\delta(G) = k$ but $\kappa(G) = 1$.

4.8 Find a set S of vertices and a set T of edges that will separate a and b in G_1, G_2, and G_3 on page 169, or explain why no such set exists.

4.9 Determine $\kappa(T)$ and $\kappa'(T)$ for any tree T.

4.10 Describe the graphs that are 0-connected and those with $\kappa(G) = 0$.

4.11 Find two different ear decompositions for the graph G_6 from Example 4.4 on page 185 with the starting cycle P_0 of $h\,c\,d\,h$.

4.12 Draw a graph with 8 vertices, one of which has nonzero betweenness centrality and all others have value 0. Calculate the value of the nonzero betweenness centrality.

4.13 Explain why K_n does not have a cut-set.

4.14 Determine $\kappa(K_{m,n})$ and $\kappa'(K_{m,n})$, and explain your answer.

4.15 Let G be a connected graph.
   (a) Prove that if a graph G has a bridge then G has a cut-vertex.
   (b) Is the converse true? Prove or give a counterexample.

4.16 Prove that if all the vertices in a graph G have even degree then G does not have any bridges.

4.17 Prove Theorem 4.7: An edge e is a bridge of G if and only if there exist vertices x and y such that e is on every $x - y$ path.

4.18 Prove Theorem 4.8: Every nontrivial connected graph contains at least two vertices that are not cut-vertices.

4.19 Prove Theorem 4.9: An edge e is a bridge of G if and only if e lies on no cycle of G.

4.20 Prove that if G is 2-connected then no vertex can have degree 1.

4.21 Prove Corollary 4.18: A graph G with at least 3 vertices is 2-connected if and only if for every pair of vertices x and y there exists a cycle through x and y.

4.22 Let G be a connected graph with at least 2 vertices. Show that G/e is connected for any edge e of G.

4.23 Let e be any edge of G and G' the graph obtained by subdividing e. Prove that if G is k-connected then so is G'.

4.24 Let G be a connected graph. Prove that if v is a cut-vertex of G then v is not a cut-vertex of $\overline{G}$.

4.25 Prove that if G is a graph with diameter 2 then $\kappa'(G) = \delta(G)$.

4.26 Let G be a 3-regular graph. Prove $\kappa(G) = \kappa'(G)$.

4.27 Let G be a graph with n vertices. Prove that $\kappa(G) = \delta(G)$ if $\delta(G) \geq n - 2$.

4.28 Prove that an edge is a block if and only if it is a bridge.

4.29 Prove two blocks of a graph share at most one vertex. (Hint: use a contradiction argument).

4.30 Prove that the block graph of a connected graph is a tree and all leaves of $B(G)$ are blocks of G.

4.31 Prove Theorem 4.26: A graph G is 2-edge-connected if and only if it has a closed-ear decomposition.

4.32 Prove Theorem 4.15: Let x and y be distinct vertices in G. Then the minimum number of edges that separate x and y equals the maximum number of edge-disjoint $x - y$ paths in G.

4.33 Prove Theorem 4.16: A nontrivial graph G is k-edge-connected if and only if for each pair of distinct vertices x and y there are at least k edge disjoint $x - y$ paths.

5

Matching and Factors

In Chapter 4, we focused on the underlying structure of a graph in terms of its connectivity. In this chapter we again focus on the structure of the graph but from the viewpoint of grouping vertices based on a variety of criteria, mainly in terms of making viable pairings. In the next chapter, we will again group vertices (or edges) together in order to avoid conflict. To begin this chapter, we investigate the optimization of pairings through the use of edge-matchings within a graph, more commonly known as a matching.

Definition 5.1 Given a graph $G = (V, E)$, a ***matching*** M is a subset of the edges of G so that no two edges share an endpoint. The size of a matching, denoted $|M|$, is the number of edges in the matching.

Recall that when two edges do not share an endpoint, we call them *independent* edges, so a matching is just a set of independent edges within a graph.

The most common application of matchings is the pairing of people, usually described in terms of marriages. Other applications of a graph matching are task assignment, distinct representatives, and roommate selection.

Example 5.1 The Vermont Maple Factory just received a rush order for 6-dozen boxes of maple cookies, 3-dozen bags of maple candy, and 10-dozen bottles of maple syrup. Some employees have volunteered to stay late tonight to help finish the orders. In the chart below, each employee is shown along with the jobs for which he or she is qualified. Draw a graph to model this situation and find a matching.

Employee	Task	
Dan	Making Cookies	Bottling Syrup
Jeff	Labeling Packages	Bottling Syrup
Kate	Making Candy	Making Cookies
Lilah	Labeling Packages	
Tori	Labeling Packages	Bottling Syrup

Solution: Model using a bipartite graph where X consists of the employees and Y consists of the tasks. We draw an edge between two vertices a and b if employee a is capable of completing the task b, creating G_1 below.

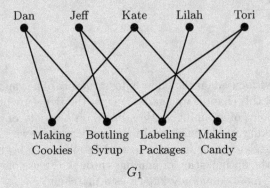

G_1

A matched edge, which is shown in bold below, represents the assignment of a task to an employee. One possible matching is shown below.

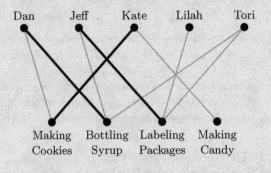

With any matching problem, you should ask yourself what is the important criteria for a solution and how does that translate to a matching. In Example 5.1, is it more important for each employee to have a task or for every task to be completed? We need a way to describe which vertices are the endpoints of a matched edge.

Definition 5.2 A vertex is **saturated** by a matching M if it is incident to an edge of the matching; otherwise, it is called **unsaturated**.

The matching displayed in Example 5.1 has saturated vertices (Dan, Jeff, Kate, Making Cookies, Bottling Syrup, and Labeling Packages) representing the three tasks that will be completed by the three employees. The unsaturated vertices (Making Candy, Lilah, and Tori) represent the tasks that are not assigned or the employees without a task assignment. Is this a good matching? No—some items needed for the order are not assigned and so the order will not be fulfilled. When searching for a matching in a graph, we need to determine what type of matching properly describes the solution.

Definition 5.3 Given a matching M on a graph G, we say M is

- *maximal* if M cannot be enlarged by adding an edge.

- *maximum* if M is of the largest size amongst all possible matchings.

- *perfect* if M saturates every vertex of G.

- an *X-matching* if it saturates every vertex from the collection of vertices X (a similar definition holds for a Y-matching).

Note that a perfect matching is automatically maximum and a maximum matching is automatically maximal, though the reverse need not be true. Consider the two matchings shown in bold below of a graph G_2. The matching on the left is maximal as no other edges in the graph can be included in the matching since the remaining edges require the use of a saturated vertex (either a for edges ac and ae, or d for edge bd). The matching on the right is maximum since there is no way for a matching to contain three edges (since then vertex a would have two matched edges incident to it). In addition, the matching on the right is an X-matching if we define $X = \{a, b\}$. Finally, neither matching is perfect since not every vertex is saturated.

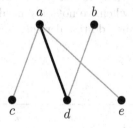

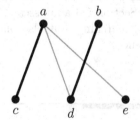

Maximal Matching of G_2 Maximum Matching of G_2

Depending on the application, we are often searching for a perfect matching or an X-matching. However, when neither of these can be found, we need a good explanation as to the size of a maximum matching, which will appear later in Section 5.1.1. For now, we will find a solution to the rush order at the Vermont Maple Factory.

Example 5.2 Determine and find the proper type of matching for the Vermont Maple Factory graph G_1 from Example 5.1.

Solution: Since we need the tasks to be completed but do not need every employee to be assigned a task, we must find an X-matching where X consists of the vertices representing the tasks. An example of such a matching is shown below.

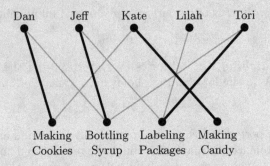

Note that all tasks are assigned to an employee, but not all employees have a task (Lilah is not matched with a task). In addition, this is not the only matching possible. For example, we could have Lilah labeling packages and Tori bottling syrup with Jeff having no task to complete.

Example 5.2 above illustrates that many matchings can have the same size. Thus when finding a maximum matching, it is less important which people get paired with a task than it does that we make as many pairings possible. Section 5.3 addresses the scenario when we not only need to find a maximum matching, but also one that fulfills additional requirements, such as preferences.

5.1 Matching in Bipartite Graphs

As with finding eulerian circuits in Chapter 2, it is often quite clear how to form a matching in a small graph. However, as the size of the graph grows or the complexity increases, finding a maximum matching can become difficult. Moreover, once you believe a maximum matching has been found, how can you convince someone that a better matching does not exist? Consider the graph below with a matching shown in bold.

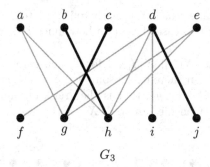

G_3

If you tried to adjust which edges appear in the matching to find one of larger size (try it!), you would find it impossible. But trial and error is a poor strategy for providing a good argument that you indeed have found a maximum matching. The theorem below addresses this through the use of a neighbor set. Recall that given a set of vertices S, the neighbor set $N(S)$ consists of all the vertices incident to at least one vertex from S. Looking back at the graph G_3 above, if we consider $S = \{f, i, j\}$, then $N(S) = \{d\}$ and since at most one of the vertices from S can be paired with d, we know the maximum matching can contain at most 3 edges. Since we found a matching with 3 edges, we know our matching is in fact maximum.

> **Theorem 5.4** (Hall's Marriage Theorem) Given a bipartite graph $G = (X \cup Y, E)$, there exists an X-matching if and only if $|S| \leq |N(S)|$ for any $S \subseteq X$.

This result is named for the British mathematician Philip Hall, who published the original proof in terms of a set theory result in 1935 [45], as well as the many iterations of the result that are often described in terms of pairing boys and girls into marriages. Multiple proofs exist for Hall's Marriage Theorem—the one we reproduce below relies on an induction argument. The end of this section describes the problem Hall was investigating as well as two additional proofs of Hall's Theorem that rely on the other major matching theorems.

Proof: First, suppose G has an X-matching, call it M. Then every vertex in X is saturated by an edge from M. Thus for any $S \subseteq X$ we know there is an M-edge from a vertex $v \in S$ to a vertex $y \in N(S)$ and so $|N(S)| \geq |S|$.

Conversely, call the property "$|S| \leq |N(S)|$ for any $S \subseteq X$" the marriage condition and suppose G satisfies the marriage condition. We will argue by induction on $|X|$ that G has an X-matching. If $|X| = 1$, then since G satisfies the marriage condition we know $|N(X)| \geq 1$ and so there must be an edge to some vertex in Y. Thus G has an X-matching.

Suppose $|X| \geq 2$ and that whenever the marriage condition is met for all X' of size smaller than X that there is a matching that saturates X'. We will consider two cases based on how much larger $N(S)$ is compared to the size of S.

First, suppose $|N(S)| \geq |S| + 1$ for all non-empty proper subsets S of X. Let $x \in X$. Then there exists some $y \in Y$ such that $e = xy$ is an edge of G. Let $G' = G - \{x, y\}$, that is the graph obtained by removing the vertices x and y. Consider any $S \subseteq X - \{x\}$.

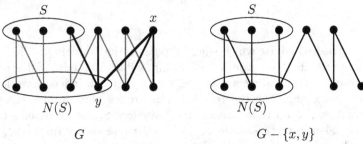

$$G \qquad\qquad\qquad\qquad G - \{x, y\}$$

Then the neighbors of S in G' contains at most 1 fewer than the neighbors of S in G, since the only adjacent vertex possibly missing is y. This gives

$$|N_{G'}(S)| \geq |N_G(S)| - 1 \geq |S|$$

and so G' satisfies the marriage condition. Thus by the induction hypothesis we know G' has a matching M' that saturates $X - \{x\}$. We can obtain an X-matching of G from M' by adding in the edge e.

Finally, suppose $|N(X')| = |X'|$ for some non-empty proper subset X' of X. Let G' be the graph consisting of the vertices in $X' \cup N(X')$ and all edges from X' to $|N(X')|$. Then G' satisfies the marriage condition and since $|X'| < |X|$, we know G' has an X'-matching by the induction hypothesis. It remains to show that $G - G'$ also satisfies the marriage condition.

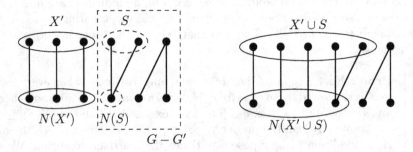

$$G - G'$$

Let S be any subset of $X - X'$. If $|S| > |N_{G-G'}(S)|$ then $S \cup X'$ would have size $|S| + |X'|$ and the neighbors of $S \cup X'$ in G could have size at most $|S| + |N(X')|$. Thus $|S \cup X'| > |N_G(S \cup X')|$, which contradicts that

G satisfies the marriage condition. Thus $G - G'$ also satisfies the marriage condition and so by the induction hypothesis has a matching that saturates $X - X'$. Together with the X'-matching, we have a matching that saturates all of X.

The theorem above is often referred to as Hall's Marriage Theorem since the early examples of matching were often described in terms of marriages between boys and girls within a small town. Note that the marriages considered in this text will be heterosexual marriages simply due to the need for two distinct groups that can only be matched with someone of a different type. Same sex marriages can be viewed as matchings in non-bipartite graphs, which appear in Section 5.2.

Example 5.3 In a small town there are 6 boys and 6 girls whose parents wish to pair into marriages where the only requirement is that a girl must like her future spouse (pretty low standards in my opinion). The table below lists the girls and the boys she likes. Find a pairing with as many marriages occurring as possible.

Girls	Boys She Likes			
Opal	Henry	Jack		
Penny	Gavin	Isa	Henry	Lucas
Quinn	Henry	Jack		
Rose	Kristof	Isa	Jack	
Suzanne	Henry	Jack		
Theresa	Gavin	Lucas	Kristof	

Solution: The information from the table will be displayed using a bipartite graph, where X consists of the girls and Y consists of the boys.

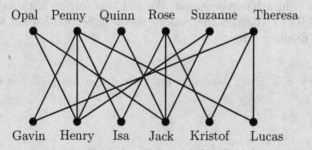

Notice that Opal, Quinn, and Suzanne all only like the same two boys (Henry and Jack), so at most two of these girls can be matched, as shown below in the following graph.

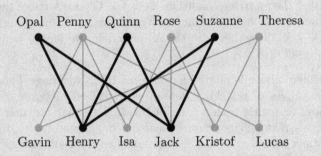

This means at most 5 marriages are possible; one such solution is shown below.

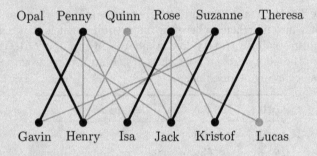

Hall's Marriage Theorem allows us to determine when a graph has a perfect matching or X-matching for bipartite graphs. When the answer to this question is negative, we can still ask for the size of a maximum matching. Hall's Marriage Theorem does not give a definitive answer about the size of a maximum matching but rather gives us the tools to reason why an X-matching does not exist. The next section uses a specific type of path and a collection of vertices as a way to determine not only the size but also produce a maximum matching within a bipartite graph. We end this section with an immediate result from Hall's Theorem that will be useful in Section 5.4 but whose proof appears in Exercise 5.18.

Corollary 5.5 Every k-regular bipartite graph has a perfect matching for all $k > 0$.

5.1.1   Augmenting Paths and Vertex Covers

Consider the graph G_2 on page 215 showing the difference between a maximal and maximum matching, which are reproduced as graphs G_4 and G_5 on the next page. Other than using trial and error to find a better matching, we need a way to determine if a matching is in fact maximum. We do this through the use of alternating and augmenting paths.

Definition 5.6 Given a matching M of a graph G, a path is called

- *M-alternating* if the edges in the path alternate between edges that are part of M and edges that are not part of M.

- *M-augmenting* if it is an M-alternating path and both endpoints of the path are unsaturated by M, implying both the starting and ending edges of the path are not part of M.

Both graphs below have alternating paths; for example, the path $c\,a\,d\,b$ is alternating in both graphs. However, this path is only augmenting in G_4 since both c and b are unsaturated by the matching. If we switch the edges along this path we get a larger matching. This switching procedure removes the matched edges and adds the previously unmatched edges along an augmenting path. Since the path is augmenting, the matching increases in size by one edge. Note that switching along the path $c\,a\,d\,b$ in G_4 produces the matching shown in G_5.

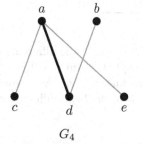 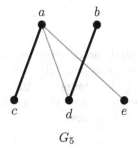

G_4 G_5

We previously discussed that the matching shown in G_5 is the maximum matching for that graph. Based on the discussion above, this should imply that no augmenting path exists (try it!), since otherwise we could switch along such a path to produce a larger matching. This result is stated in the theorem below and was first published by the French mathematician Claude Berge in 1959. Note that unlike Hall's Theorem, Berge's Theorem holds for both bipartite graphs and non-bipartite graphs.

Theorem 5.7 (Berge's Theorem) A matching M of a graph G is maximum if and only if G does not contain any M-augmenting paths.

The proof of Berge's Theorem is interesting in that it uses the contrapositive of each direction of the biconditional, since it is easier to work with graphs that contain M-augmenting paths than to explain why none exist. We will need the following set theory definition and result about two different matchings of a graph in our proof of Berge's Theorem.

Definition 5.8 Let A and B be two sets. Then the **symmetric difference** $A \triangle B$ is all those elements in exactly one of A and B; that is, $A \triangle B = (A - B) \cup (B - A)$.

Below is a Venn diagram representing the symmetric difference. Note that we can also think of the symmetric difference as what remains in the union when we remove the intersection of two sets, that is $A \triangle B = A \cup B - A \cap B$.

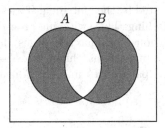

In graph theoretic terms, we can look at the symmetric difference of various subsets of vertices or edges of a larger graph G. Our use here will be in considering the symmetric difference of two matchings, which would consist of the edges in a graph that are in one but not both of two different matchings. An example is shown below.

Example 5.4 Below are two different matchings of a graph G. Find $M_1 \triangle M_2$.

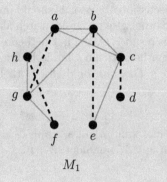

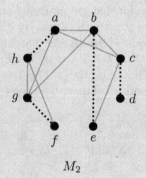

Solution: Note that the symmetric difference will only show those edges involved in exactly one of the two matchings. Thus if an edge is in both matchings or left unmatched in both matchings, it does not appear in $M_1 \triangle M_2$.

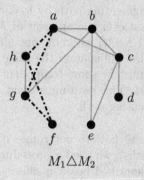

$$M_1 \triangle M_2$$

Notice that the symmetric difference of two matchings may not be itself a matching, as seen in the example above with the cycle created by the vertices a, g, f, h. However, as the lemma below shows, the edges in $M_1 \triangle M_2$ must result in only one of two specific types of structures.

Lemma 5.9 Let M_1 and M_2 be two matchings in a graph G. Then every component of $M_1 \triangle M_2$ is either a path or an even cycle.

Proof: Let $H = M_1 \triangle M_2$. Since M_1 and M_2 are both matchings, we know any vertex has at most one edge incident to it from either matching. Thus each vertex has degree at most 2 in H. Thus every component of H consists of paths and cycles. Moreover, any such path or cycle must alternate between edges in M_1 and M_2, and so every cycle must be of even length.

As the lemma above shows, we can use the symmetric difference of two matchings to gain an understanding of how they relate to each other. In particular, we will use this concept to gain an understanding about their relative sizes, allowing us to prove Berge's Theorem.

Theorem 5.7 (Berge's Theorem, restated) A matching M of a graph G is not maximum if and only if G contains some M-augmenting path.

Proof: First suppose M is a matching of G and G contains an M-augmenting path P. Then we can switch edges along P to produce a larger matching M' and so M is not a maximum matching of G.

Conversely, suppose M is a matching of G and M is not maximum. Then there must exist a different matching M' that is larger than M. We will produce an M-augmenting path by looking at $M \triangle M'$. By Lemma 5.9 above we know that every component of $M \triangle M'$ is either a path or even cycle. Since $|M'| > |M|$ we know not all components can be even cycles, since even cycles will contain the same number of edges from M and M'. Thus there must be some component that is a path with more edges from M' than M. This path must start and end with an M' edge, and so is an M-augmenting path.

Now that we understand how to determine if a matching is maximum (search for augmenting paths), we need a procedure or algorithm to actually find one. The algorithm described below is closely related to the Hungarian Algorithm proposed by Harold Kuhn in 1955 [61]. He named this algorithm in honor of the work of the Hungarian mathematicians Dénes König and Jenö Egerváry whose largest contributions to graph matchings appear later in Theorem 5.11. Also, even though Berge's Theorem holds for graphs that are not bipartite, the algorithm below requires the input of a bipartite graph. A modification for general graphs is discussed in Section 5.2.

Augmenting Path Algorithm

Input: Bipartite graph $G = (X \cup Y, E)$.

Steps:

1. Find an arbitrary matching M.

2. Let U denote the set of unsaturated vertices in X.

3. If U is empty, then M is a maximum matching; otherwise, select a vertex x from U.

4. Consider y in $N(x)$.

5. If y is also unsaturated by M, then add the edge xy to M to obtain a larger matching M'. Return to Step (2) and recompute U. Otherwise, go to Step (6).

6. If y is saturated by M, then find a maximal M-alternating path from x using xy as the first edge.

(a) If this path is M-augmenting, then switch edges along that path to obtain a larger matching M'; that is, remove from M the matched edges along the path and add the unmatched edges to create M'. Return to Step (2) and recompute U.

(b) If the path is not M-augmenting, return to Step (4), choosing a new vertex from $N(x)$.

7. Stop repeating Steps (2)–(4) when all vertices from U have been considered.

Output: Maximum matching for G.

The arbitrary matching in Step (1) could be the empty matching (no edges are initially included), though in practice starting with a quick simple matching allows for fewer iterations of the algorithm. You should not spend much time trying to create an initial maximum matching, but rather choose obvious edges to include.

Example 5.5 Apply the Augmenting Path Algorithm to the bipartite graph G_6 below, where $X = \{a, b, c, d, e, f, g\}$ and $Y = \{h, i, j, k, m, n\}$, with an initial matching shown as dashed lines.

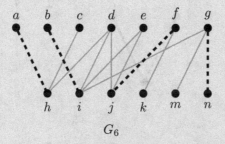

G_6

Solution: During each step shown below, the path under consideration will be in bold, with the matching shown as dashed lines throughout.

Step 1: The unsaturated vertices from X are $U = \{c, d, e\}$.

Step 2: Choose c. The only neighbor of c is h, which is saturated by M. Form an M-alternating path starting with the edge ch. This produces the path $c\,h\,a$, as shown on the next page, which is not augmenting.

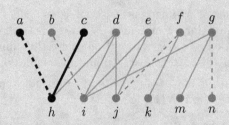

Step 3: Choose a new vertex from U, say d. Then $N(d) = \{h, i, j\}$. Below are the alternating paths originating from d.

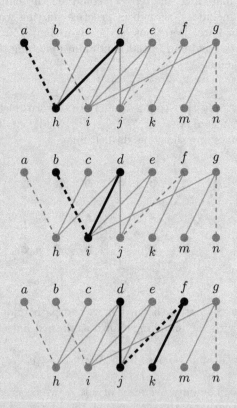

Note that the last path $(djfk)$ is M-augmenting. Form a new matching M' by removing edge fj from M and adding edges dj and fk, as shown in the following graph.

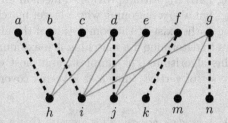

Step 4: Recalculate $U = \{c, e\}$. We must still check c since it is possible for the change in matching to modify possible alternating paths from a previously reviewed vertex; however, the path obtained is $c\,h\,a$, the same as from Step 2.

Step 5: Check the paths from e. The alternating paths are shown below.

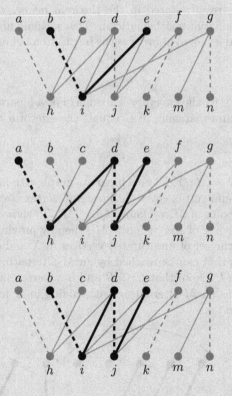

None of these paths are augmenting. Thus no M'-augmenting paths exist in G and so M' must be maximum by Berge's Theorem.

Output: The maximum matching $M' = \{ah, bi, dj, fk, gn\}$ from Step 3.

The Augmenting Path Algorithm provides a method for not only finding a maximum matching, but also a reasoning why a larger matching does not exist since no augmenting paths exist at the completion of the algorithm. However, there are other ways to determine if a matching is maximum without the need to work through this algorithm. The simplest, and most elegant, is through the use of a specific set of vertices known as a vertex cover.

Definition 5.10 A *vertex cover* Q for a graph G is a subset of vertices so that every edge of G has at least one endpoint in Q.

Every graph has a vertex cover (for example Q could contain all the vertices in the graph), yet we want to optimize the vertex cover; that is, find a minimum vertex cover. If every edge has an endpoint to one of the vertices in a vertex cover, then at most one matched edge can be incident to any single vertex in the cover. This result, stated in the theorem below, was first published in 1931 by the Hungarian mathematicians Dénes König and (independently) Jenö Egerváry, and as noted above was the inspiration behind the Augmenting Path Algorithm.

Theorem 5.11 (König-Egerváry Theorem) For a bipartite graph G, the size of a maximum matching of G equals the size of a minimum vertex cover for G.

Proof: Let $G = (X \cup Y, E)$ be a bipartite graph with maximum matching M and minimum vertex cover Q. Since any vertex cover must contain at least one endpoint of M, we know $|Q| \geq |M|$. To show $|Q| = |M|$, we will find a vertex cover of size at most $|M|$, thereby proving $|Q| = |M|$.

Let U be the set of unsaturated vertices in X and define R to be the set of vertices that can be reached by an M-alternating path that begins at a vertex in U. Note that $U \subseteq R$ and any vertex in $X \cap R - U$ must be saturated as an M-alternating path ending in X must conclude with a saturated vertex.

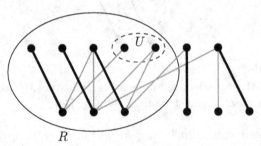

Define $S = X - R$ and $T = Y \cap R$. We will prove $S \cup T$ is a vertex cover of G of size at most $|M|$.

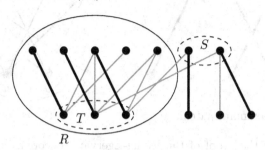

Let $v \in S \cup T$. First note that if $v \in T$ then it must be saturated since otherwise there would be an M-augmenting path in G from a vertex in U ending at v, contradicting the fact that M is maximum. Moreover, if $v \in S$ then v must be saturated as otherwise $v \in U$ and so $v \in R$, which contradicts the choice of S. Since every vertex in $S \cup T$ is saturated, we know $|S \cup T| \le |M|$.

To show $S \cup T$ is a vertex cover we will consider any edge $e = xy$ of G where $x \in X$ and $y \in Y$ and show either $x \in S$ or $y \in T$. If $x \in S$ then we are done. Otherwise $x \in R$.

If $e \in M$ then we know both x and y are saturated. Thus $x \notin U$, but there exists some M-alternating path P from a vertex in U to x. This path must use edge e and so y is also part of P, and so y is also reachable from U by an M-alternating path, making $y \in T$.

Otherwise $e \notin M$. If $x \in U$ then e is itself an M-alternating path and so $y \in T$. Otherwise there exists an M-alternating path P from U to x, and either e is a part of P or the last edge of P must be an edge from M and $e = xy$ can be added to the end of this path. In either case $y \in T$.

Thus every edge of G has an endpoint in $S \cup T$ and so this is a vertex cover of size $|M|$, proving a minimum vertex cover must be of the same size of a maximum matching.

Consider the graph G_3 from page 217 (reproduced on the next page) in the discussion leading to Hall's Marriage Theorem. Based on the König-Egerváry Theorem, to show the matching in bold is maximum we need to find a vertex cover of size 3. One such cover is shown; you should check that every edge has an endpoint that is either d, g or h.

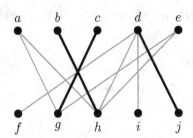

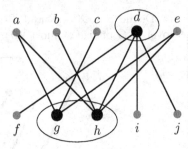

G_3 maximum matching G_3 vertex cover

Note that the proof of the König-Egerváry Theorem provides a method for finding a vertex cover, which we describe below. But first, we should take a moment to appreciate the form of the König-Egerváry Theorem. The statement has a similar feel to the work we did on network flow in Section 4.4, and in particular Theorem 4.34 relating the size of a maximum flow to that of a minimum cut. These types of results in mathematics are special, in that by maximizing one quantity we are also minimizing another and the optimal answer for both is found when they are equal. These types of problems are often given in terms of a linear programming question, where a quantity is being optimized given a set of constraints.

Returning to the vertex covers, in most cases a minimum vertex cover for a bipartite graph will require some vertices from both pieces of the vertex partition. It should not come to much surprise that the Augmenting Path Algorithm can be used to find a vertex cover. In fact, many texts include the procedure below as part of the algorithm itself.

Vertex Cover Method

1. Let $G = (X \cup Y, E)$ be a bipartite graph.

2. Apply the Augmenting Path Algorithm and mark the vertices considered throughout its final implementation.

3. Define a vertex cover Q as the unmarked vertices from X and the marked vertices from Y.

4. Q is a minimum vertex cover for G.

In Step 2, a vertex is marked if it was considered during the final step in the implementation of the Augmenting Path Algorithm. Note that this is not just the vertices in U, the unsaturated vertices from X, but also any vertex that was reached through an alternating path that originated at a vertex from U. Thus the unmarked vertices will be those that are never mentioned during the final step of the implementation of the Augmenting Path Algorithm.

Example 5.6 Apply the Vertex Cover Method to the output graph from Example 5.5.

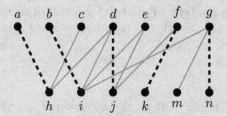

Solution: Recording the vertices considered throughout the last step of the Augmenting Path Algorithm, the marked vertices from X are a, b, c, d, and e, and the marked vertices from Y are h, i, and j. This produces the vertex cover $Q = \{f, g, h, i, j\}$ of size 5, shown below. Recall that the maximum matching contained 5 edges.

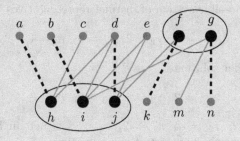

Note that more than one minimum vertex cover may exist for the same graph, just as more than one maximum matching may exist. In the example above, we found one such vertex cover through the matching found using the Augmenting Path Algorithm, though the set $Q' = \{g, h, i, j, k\}$ is also a valid minimum vertex cover (as shown below).

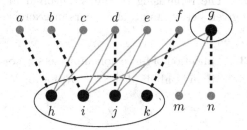

By finding a minimum vertex cover, we are able to definitively answer how large of a matching can be formed. Vertex covers provide another avenue for proving a maximum matching has been found.

Before we return to Hall's Theorem, we will discuss an additional application of graph matching, called a *System of Distinct Representatives*.

Definition 5.12 Given a collection of finite nonempty sets $S_1, S_2, \ldots, S_n$ (where $n \geq 1$), a **system of distinct representatives**, or **SDR**, is a collection $r_1, r_2, \ldots, r_n$ so that r_i is a member of set S_i and $r_i \neq r_j$ for all $i \neq j$ (for all $i, j = 1, 2, \ldots, n$).

Example 5.7 Find an SDR for the collection of sets given below:

$$S_1 = \{1, 2, 3, 5\}$$
$$S_2 = \{2, 4, 8\}$$
$$S_3 = \{2, 6\}$$
$$S_4 = \{4, 8\}$$

Solution: One possible system of distinct representatives is

$$r_1 = 1, \quad r_2 = 2, \quad r_3 = 6, \quad r_4 = 4$$

Note that another solution is

$$r_1 = 2, \quad r_2 = 4, \quad r_3 = 6, \quad r_4 = 8$$

whereas $r_1 = 2$, $r_2 = 4$, $r_3 = 2$, $r_4 = 8$ is not an SDR since 2 is representing more than one set at the same time.

In less technical terms, the idea of an SDR is that a collection of groups each need their own representative and no two groups can have the same representative. Hall's original paper was not concerned with a matching in a bipartite graph, but rather with determining the necessary conditions for obtaining an SDR. The rephrasing of Hall's Theorem in terms of distinct representatives is given below; its proof appears in Exercise 5.22.

Theorem 5.13 A collection of finite nonempty sets $S_1, S_2, \ldots, S_n$ (where $n \geq 1$) has an SDR if and only if $\left| \bigcup_{i \in R} S_i \right| \geq |R|$ for all $R \subseteq \{1, 2, \ldots, n\}$.

The next example shows how to model an SDR problem in terms of a graph matching.

Example 5.8 During faculty meetings at a small liberal arts college, multiple committees provide a report to the faculty at large. These committees often overlap in membership, so it is important that, for any given year, a person is not providing the report for more than one committee. Find a system of distinct representatives for the groups listed below.

Committee	Members		
Admissions Council	Ivan	Leah	Sarah
Curriculum Committee	Kyle	Leah	
Development and Grants	Ivan	Kyle	Norah
Honors Program Council	Norah	Sarah	Victor
Personnel Committee	Sarah	Victor	

Solution: Begin by forming a bipartite graph where X consists of the committees and Y the faculty members, and draw an edge between two vertices if a person is a member of that committee.

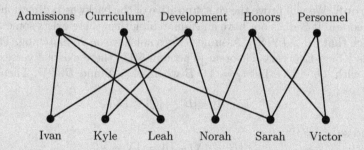

A system of distinct representatives is modeled as a matching. Note that we are interested in each committee having a representative, not every person being a representative. Thus we want an X-matching. One such matching is shown below.

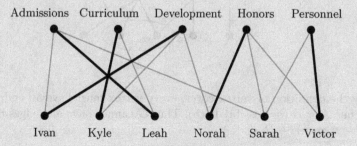

We conclude this section by revisiting Hall's Theorem and providing two alternative proofs.

5.1.2 Hall's Theorem Revisited

As we noted above, there are multiple proofs of Hall's Theorem, and the one we presented relied on induction and a basic structural argument. Below we include two additional proofs of Hall's Theorem. The first relies on the König-Egerváry Theorem and vertex covers; the second uses Berge's Theorem and the notion of augmenting paths. Note that for both we are omitting the forward direction of the proof as it remains the same as in the proof shown on page 217.

> **Theorem 5.4** (Hall's Marriage Theorem) Given a bipartite graph $G = (X \cup Y, E)$, there exists an X-matching if and only if $|S| \leq |N(S)|$ for any $S \subseteq X$.

Proof: We will prove the contrapositive of the backwards direction of the theorem: If G does not have an X-matching then there exists some $S \subseteq X$ such that $|S| > |N(S)|$. Assume G does not have an X-matching. Then by the König-Egerváry Theorem (Theorem 5.11) there exists a vertex cover Q with $|Q| < |X|$. Let $Q = A \cup B$ where $A \subseteq X$ and $B \subseteq Y$. Then

$$|A| + |B| = |Q| < |X|$$

and so

$$|B| < |X| - |A| = |X - A|$$

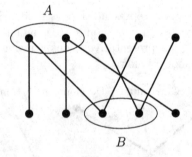

By the definition of vertex cover, every vertex must have an endpoint in either A or B (or possibly both). Thus G cannot have any edges between $X - A$ and $Y - B$.

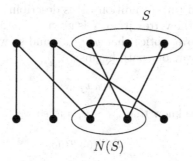

Let $S = X - A$. Then any neighbor for a vertex in S must be in B. Thus

$$|N(S)| \leq |B|$$
$$< |X - A| = |S|$$

Thus we have found at least one set S in X that satisfies $|S| > |N(S)|$. Therefore G has an X-matching if and only if $|S| \leq |N(S)|$ for any $S \subseteq X$.

For our final proof of Hall's Theorem, we will be building an augmenting path, thereby proving the given matching is not maximum. To do this, we will need to be careful in the way that we pick the vertices. Before we work through the proof, we begin with a small example to provide clarity in some of the fairly cumbersome notation to follow.

Consider the graph below with a matching M shown in bold.

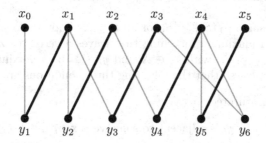

We will construct a sequence of distinct vertices $x_0, y_1, x_1, y_2, \ldots$, with $x_i \in X$ and $y_i \in Y$ so that the following three conditions are met:

(i) x_0 is unsaturated.

(ii) for all $i \geq 1$, y_i is adjacent to some vertex $x_{f(i)} \in \{x_0, \ldots, x_{i-1}\}$

(iii) $x_i y_i \in M$ for all $i \geq 1$

This sequence is building an alternating path from x_0 where the vertices from X and Y with the same index are a part of the matching M, as noted in conditions (i) and (iii). Condition (ii) is describing another neighbor of y_i, where the index on the vertex from X is given in terms of some function $f(i)$.

In the graph above, notice that $y_6 \sim x_4$, and so we could write $f(6) = 4$, or written another way our path begins

$$y_6 \, x_{f(6)}$$

By condition (iii) we know that $x_4 \sim y_4$ and so we can extend the path to

$$y_6 \, x_{f(6)} \, y_{f(6)}$$

By condition (ii) we know there must be some $x_i \sim y_{f(6)}$ where $i < f(6)$. In the graph above, we see that $x_2 \sim y_4$, and moreover $y_2 \sim x_2$. Then our path $y_6 \, x_4 \, y_4 \, x_2 \, y_2$ would be given by

$$y_6 \, x_{f(6)} \, y_{f(6)} \, x_{f(f(6))} \, y_{f(f(6))}$$

Note that $y_2 \sim x_1$, and $x_1 \sim y_1$, and by condition (ii) we know that $y_1 \sim x_0$. Thus our path ends as $y_6 \, x_4 \, y_4 \, x_2 \, y_2 \, x_1 \, y_1 \, x_0$, which is written as

$$y_6 \, x_{f(6)} \, y_{f(6)} \, x_{f(f(6))} \, y_{f(f(6))} \, x_{f(f(f(6)))} \, y_{f(f(f(6)))} \, x_{f(f(f(f(6))))}$$

or more simply as

$$y_6 \, x_{f(6)} \, y_{f(6)} \, x_{f^2(6)} \, y_{f^2(6)} \, x_{f^3(6)} \, y_{f^3(6)} \, x_{f^4(6)}$$

where $f^4(6) = 0$. In the proof below we will use the sequence of this form of maximal length and show that it must be augmenting.

Proof: Assume $|N(S)| \geq |S|$ for all $S \subseteq X$. Suppose for a contradiction that M is a maximum matching that leaves a vertex of X unsaturated. Let $x_0, y_1, x_1, y_2, \ldots$ with $x_i \in X$ and $y_i \in Y$ be a maximal sequence of distinct vertices so that the following three conditions are met:

(i) x_0 is unsaturated.

(ii) for all $i \geq 1$, y_i is adjacent to some vertex $x_{f(i)} \in \{x_0, \ldots, x_{i-1}\}$

(iii) $x_i y_i \in M$ for all $i \geq 1$

First note that by condition (ii) we know $y_1 \sim x_0$, and since $x_1 y_1 \in M$, the edge between x_0 and y_1 must be unmatched.

Let z be the final vertex in our sequence. If $z = x_k$ then $S = \{x_0, x_1, \ldots, x_k\}$ has size $k+1$, and so by our initial assumption we know $|N(S)| \geq k+1$. Thus there must be some $y \notin \{y_1, \ldots, y_k\}$ that is adjacent to some x_i from S, satisfying condition (ii) above. Thus y could be added to the sequence as y_{k+1}, contradicting the maximality of the length of the sequence. Thus the final vertex z of the sequence must be from Y, call it y_k.

Form the path

$$P = y_k \, x_{f(k)} \, y_{f(k)} \, x_{f^2(k)} \cdots y_{f^{r-1}(k)} \, x_{f^r(k)}$$

This path must end at x_0, so $f^r(k) = 0$. Also the path P is alternating where the last edge is unmatched, since x_0 is unsaturated, and the first edge $y_k x_{f(k)}$ is unmatched since the second edge $x_{f(k)} y_{f(k)} \in M$ by condition (iii). Thus to prove P is an augmenting path, it remains to show that y_k is unsaturated.

If y_k were saturated, then there exists some $x \in X$ such that $y_k x \in M$. Since y_k is the last vertex of the sequence, we know that $x \in \{x_0, x_1, \ldots, x_{k-1}\}$, and so $x = x_i$ for some $0 \le i \le k - 1$. But then $y_k x_i \in M$, implying $y_k = y_i$, a contradiction since the sequence must contain distinct vertices.

Thus y_k and x_0 are both unsaturated vertices, making P an augmenting path. This contradicts Berge's Theorem since M was chosen to be a maximum matching. Thus every vertex in X must be saturated and so M is an X-matching.

5.2 Matching in General Graphs

Up to this point we have only discussed matchings inside of bipartite graphs, although Berge's Theorem was stated to hold for non-bipartite graphs as well. Although bipartite graphs model many problems that are solved using a matching, there are some problems that are best modeled with a graph that is not bipartite. Consider the following scenario:

Bruce, Evan, Garry, Hank, Manny, Nick, Peter, and Rami decide to go on a week-long canoe trip in Guatemala. They must divide themselves into pairs, one pair for each of four canoes, where everyone is only willing to share a canoe with a few of the other travelers.

Modeling this as a graph cannot result in a bipartite graph since there are not two distinct groups that need to be paired, but rather one large group that must be split into pairs.

Example 5.9 The group of eight men from above have listed who they are willing to share a canoe with. This information is shown in the following table, where a Y indicates a possible pair. Note that these relationships are symmetric, so if Bruce will share a canoe with Manny, then Manny is also willing to share a canoe with Bruce. Model this information as a graph. Find a perfect matching or explain why no such matching exists.

	Bruce	Evan	Garry	Hank	Manny	Nick	Peter	Rami
Bruce	·	·	·	·	Y	Y	·	Y
Evan	·	·	·	Y	Y	Y	·	·
Garry	·	·	·	Y	·	·	Y	Y
Hank	·	Y	Y	·	·	·	Y	·
Manny	Y	Y	·	·	·	·	·	Y
Nick	Y	Y	·	·	·	·	·	Y
Peter	·	·	Y	Y	·	·	·	·
Rami	Y	·	Y	·	Y	Y	·	·

Solution: The graph is shown below where an edge represents a potential pairing into a canoe.

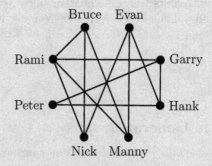

Note that Peter can only be paired with either Garry or Hank. If we choose to pair Peter and Garry, then Hank must be paired with Evan, leaving Nick and Manny to each be paired with one of Rami and Bruce. One possible matching is shown below. Since all people have been paired, we have a perfect matching.

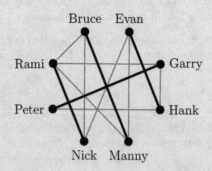

Finding matchings in general graphs is often more complex than in bipartite graphs, in part due to the fact that only some of the results from Section 5.1 still apply. In particular, Berge's Theorem holds (M is maximum if and only if G has no M-augmenting paths) yet the Augmenting Path Algorithm only works on bipartite graphs. The main sticking point is that in finding alternating paths from an unsaturated vertex, there could be more than one path to x and only investigating one would miss an augmenting path. For example, if we are searching for alternating paths from u to x in the graph G_7 below, we might choose $u\,a\,b\,x$. If instead we chose $u\,a\,b\,c\,x$, we could continue the alternating path to find an augmenting path to y (namely, $u\,a\,b\,c\,x\,y$).

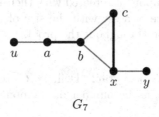

G_7

Jack Edmonds devised a modification to the Augmenting Path Algorithm that works on general graphs, accounting for these odd cycles [28]. This algorithm is more powerful than we need for small examples, as these can usually be determined through inspection and some reasoning. We include a discussion of Edmonds' Blossom Algorithm at the end of this section.

Example 5.10 Halfway through the canoe trip from Example 5.9, Rami will no longer share a canoe with Garry, and Hank angered Evan so they cannot share a canoe. Update the graph model and determine if it is now possible to pair the eight men into four canoes.

Solution: The updated graph only removes two edges, but in doing so the graph is now disconnected. A disconnected graph could have a perfect matching so long as each component itself has a perfect matching.

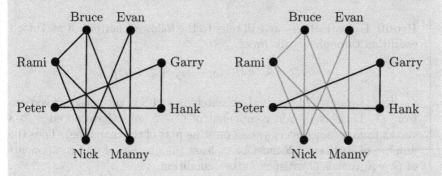

In this case, a perfect matching is impossible since Peter, Garry, and Hank form one component and so at most two of them can be paired. Likewise, the other component contains five people and so at most four can be paired. Thus there is no way to pair the eight men into four canoes.

The theorem below is one of the classic matching theorems for general graphs, due to the British mathematician William Tutte, who made significant contributions to graph theory after World War II, where he served as a code breaker [80]. The result relates the number of vertices removed from a graph to the number of odd components created with their removal, similar to Hall's Theorem relating the size of a set with the size of its neighbor set. We will use the following notation throughout the proof:

Definition 5.14 Let G be a graph. Define $o(G)$ to be the number of odd components of G, that is the number of components containing an odd number of vertices.

In Example 5.10 above, $o(G) = 2$ since it contains two components, one with 5 vertices (with Bruce, Evan, Manny, Nick, and Rami) and the other with 3 (Gary, Hank, and Peter).

One direction of the proof to follow is fairly straightforward, whereas the other will make use of an induction argument on the number of vertices in G. Also, recall that a proper subset of vertices is a subset that does not include all of the vertices of the graph; we will use the notation $S \subsetneq V(G)$ to indicate that S is indeed a proper subset.

Theorem 5.15 (Tutte's Theorem) A graph G has a perfect matching if and only if for every proper subset of vertices S the number of odd components of $G - S$ is at most $|S|$, that is $o(G - S) \leq |S|$.

Proof: For ease of use, we will refer to the following statement as Tutte's condition throughout this proof:

$$o(G - S) \leq |S| \text{ for every } S \subsetneq V(G)$$

First, suppose G has a perfect matching and S is any subset of vertices from G. Then every odd component of $G - S$ must have an edge to a vertex from S (since every vertex must be part of the matching). Thus the number of vertices in S must be at least the number of odd components of $G - S$, that is G satisfies Tutte's condition.

Conversely, assume $|G| = n$. We will show that G has a perfect matching whenever it satisfies Tutte's condition. First note that if $S = \emptyset$, then $|S| = 0$ and $G - S = G$. Thus

$$o(G) = o(G - S) \le |S| = 0$$

and so G must only have even components. Thus n must be even. Moreover, for every vertex we add to S, we will affect the size of exactly one component from $G - S$, and so $|S|$ and $o(G - S)$ must have the same parity.

We will argue by induction on n (where n must always be an even integer) that if G satisfies Tutte's condition then G will have a perfect matching.

First suppose $n = 2$. Since G cannot have any odd components, we know $G = K_2$. Thus G has a perfect matching, namely the single edge in G.

Now suppose for some $n \ge 4$ that all graphs H of even order $n' < n$ satisfying Tutte's condition have a perfect matching. Let G be a graph of order n satisfying Tutte's condition. We will consider whether $o(G - S) < |S|$ for all proper subsets S or if there exists some $S \subsetneq V(G)$ where $o(G - S) = |S|$. Note we only need to consider sets S with $2 \le S \le n$.

Case 1: Suppose $o(G - S) < |S|$ for all $S \subsetneq V(G)$. Since $o(G - S)$ and $|S|$ are of the same parity, we know $o(G - S) \le |S| - 2$ for all $S \subsetneq V(G)$ with $2 \le |S| \le n$. Let $e = xy$ be some edge of G and let $G' = G - \{x, y\}$. Let $T \subsetneq V(G')$ and let $T' = T \cup \{x, y\}$. Then $|T'| = |T| + 2$.

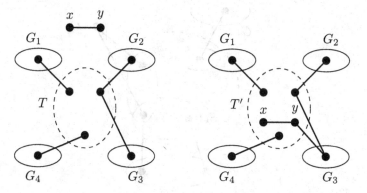

If $o(G' - T) > |T|$ then $o(G' - T) > |T'| - 2$. Note that $o(G' - T) = o(G - T')$ and so $o(G - T') \ge |T'|$, which is a contradiction to the fact that G satisfies Tutte's condition. Thus $o(G' - T) \le |T|$, and so by the induction hypothesis G' has a perfect matching M. Together with the edge $e = xy$, $M \cup e$ is a perfect matching of G.

Case 2: Suppose there exists some subset $S' \subsetneq V(G)$ satisfying

$$(\ast)\quad o(G - S') = |S'|$$

Among all possible sets S', pick S to be the largest possible one satisfying condition $(*)$, say $|S| = k$.

Let $G_1, G_2, \ldots, G_k$ be the odd components of $G - S$. Suppose G_0 were some even component of $G - S$, and let $v_0 \in V(G_0)$. Then $G_0 - \{v_0\}$ would consist of at least one odd component. Let $S' = S \cup \{v_0\}$.

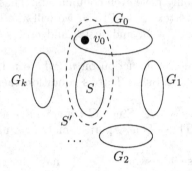

Thus $o(G - S') \geq k + 1 = |S'|$. But then $o(G - S') = |S'|$ since G satisfies Tutte's condition, and so S' is a larger set than S satisfying $(*)$, a contradiction to our choice of S. Thus $G - S$ can only consist of odd components.

For $i = 1, \ldots, k$ let S_i denote the vertices of S adjacent to at least one vertex of G_i. Since G does not contain any odd components, we know that $S_i \neq \emptyset$ for all i.

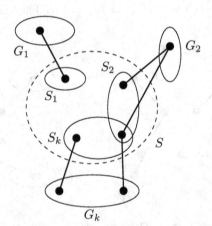

Moreover, let R be the union of j of the S_i's. Then $R \subseteq S$ and in $G - R$ the only odd components that remain are the G_i for which $S_i \subseteq R$. Thus $o(G-R) = j$ and since $o(G-R) \leq |R|$, we know $|R| \geq j$. Then by Theorem 5.13 we can find a system of distinct representatives for $S_1, \ldots, S_k$; that is, there exists distinct vertices $v_1, \ldots, v_k$ so that $v_i \in S_i$ and for some $u_i \in G_i$ we have $v_i \sim u_i$.

We will show each $G_i - u_i$ has a perfect matching, which we can combine together with $v_i u_i$ to get a perfect matching of G. To do this, we simply must show that $G_i - u_i$ satisfies Tutte's condition. Note that since each G_i is an odd component, we know that $G_i - u_i$ has an even number of vertices.

For any i, consider $W \subsetneq V(G_i - u_i)$. If $o(G_i - u_i - W) > |W|$ then we know $o(G_i - u_i - W) \geq |W| + 2$. But then the odd components in $G - (S \cup W \cup \{u_i\})$ are exactly the odd components of G, except G_i, along with the odd components of $G_i - u_i - W$.

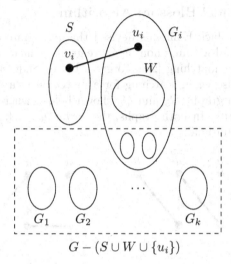

$$G - (S \cup W \cup \{u_i\})$$

This implies

$$
\begin{aligned}
o(G - (S \cup W \cup \{u_i\})) &= o(G_i - u_i - W) + o(G - S) - 1 \\
&\geq |S| + |W| + 1 \\
&= |S \cup W \cup \{u_i\}|
\end{aligned}
$$

But this contradicts the maximality of S. Thus for each i we know $o(G_i - u_i - W) \leq |W|$, that is $G_i - u_i$ satisfies Tutte's condition. Thus by the induction hypothesis applied to every $G_i - u_i$ we know that $G_i - u_i$ has a perfect matching. Since $|S| = k$, $v_1 u_1, v_2 u_2, \ldots, v_k u_k$ forms a matching that saturates S and together with each of the perfect matchings of the $G_i - u_i$ we have a prefect matching of G.

The following result can be seen as a corollary to Tutte's Theorem, as a quick proof uses a counting argument between the degrees of each vertex and the number of odd components (see Exercise 5.23). What is more interesting is this result was proven by Julius Petersen in 1891, more than fifty years before Tutte's Theorem was first published. His original proof was much more complex (as Tutte's theorem had not been discovered) and appeared in the same paper as Theorem 5.20, which appears in Section 5.4 about factors. Petersen is best known for the Petersen graph, which is seen in Exercise 5.3(f) and will also appear in Chapter 6 when we discuss graph colorings.

Corollary 5.16 Every cubic graph without any bridges has a perfect matching.

5.2.1 Edmonds' Blossom Algorithm

As noted above, Jack Edmonds devised this procedure so that augmenting paths can be found within a non-bipartite graph, which can then be modified to create a larger matching [28]. To get a better understanding of the complexities that arise when searching for a matching in a non-bipartite graph, consider the two graphs G_8 and G_9 shown below, where only the graph on the left is bipartite. In both graphs, the vertex u is left unsaturated by the matching shown in bold.

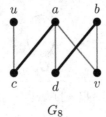
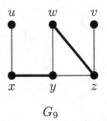

$$G_8 \qquad\qquad\qquad G_9$$

Notice that in G_8 above, when looking for an alternating path out of u, if the end of the path is in the same partite set as u, then this path must have even length and the last edge of the path would be from the matching. This implies if an alternating path from u to a saturated vertex s has even length, then all paths from u to s have even length, and they must all use as their last edge the only matched edge out of s. However, in the non-bipartite graph on the right, this is not the case. For example, we can find an alternating path of length 4 from u to z, namely $u\,x\,y\,w\,z$, but another alternating path from u to z exists of length 3 where the last edge is not matched, namely $u\,x\,y\,z$. This is caused by the odd cycle occurring between y, w, and z. Note that the vertex from which these two paths diverge (namely y) is entered by a matched edge (xy) and has two possible unmatched edges out (yw and yz). This configuration is the basis behind the blossom.

Definition 5.17 Given a graph G and a matching M, a *flower* is the union of two M-alternating paths from an unsaturated vertex u to another vertex v where one path has odd length and the other has even length.

The ***stem*** of the flower is the maximal common initial path out of u, that ends at a vertex b, called the ***base***. The ***blossom*** is the odd cycle that is obtained by removing the stem from the flower.

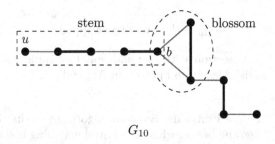

G_{10}

A few additional details can be discerned from these definitions. First, the stem must be of even length since the last edge must be from the matching M. Next, the blossom is an odd cycle C_{2k+1} where exactly k edges are from M, since the two edges from the base on the cycle must both be unmatched edges. Moreover, every vertex of the blossom must be saturated by M since traveling some direction along the cycle will end with an edge from M. Finally, the only edges that come off the blossom that are also from M must be from the stem. Thus any non-stem edges coming off the blossom must be unmatched.

Edmonds' Blossom Algorithm starts in the same way as the Augmenting Path Algorithm, where we examine alternating paths originating from an unsaturated vertex. Where these algorithms differ is when a blossom is encountered. When all alternating paths from u are being explored, if two different paths to a vertex are found to end in different types of edges (namely matched or unmatched), then a blossom has been discovered. We can then contract the blossom, much in the same way we contract an edge from Chapter 4 in the proof of Menger's Theorem. The contraction of the blossom from G_{10} above is shown below.

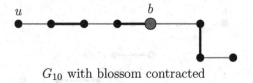

G_{10} with blossom contracted

If we find an augmenting path in the contracted graph, then we can find an augmenting path in the original graph by choosing the correct direction along the blossom. Now just like in the Augmenting Path Algorithm we can swap edges along this path, creating a larger matching while still maintaining the properties of a matching.

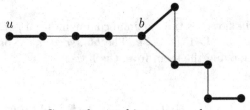

G_{10} with matching swapped

We conclude this section with one additional example below. The pseudocode for Edmonds' Algorithm appears in Appendix E.

Example 5.11 Use Edmonds' Blossom Algorithm to find a maximum matching on the graph below, where the initial matching is shown in bold.

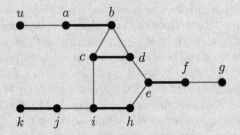

Solution: We begin by looking for alternating paths out of u. Note that we will explore all paths simultaneously (essentially building a Breadth-First Tree). At b we have two options for finding an alternating path, namely $uabc$ and $uabd$. If we take either path out to their next vertex, we get the ending vertex of the other previous path (such as $uabdc$ and vertex c from the path $uabc$). This implies that we have found a blossom, with odd cycle $bcdb$ and base vertex b. We contract this blossom to a vertex B as shown below.

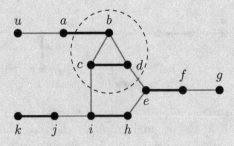

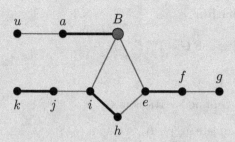

Again we build out alternating paths from u. In doing so, we find the path $u\,a\,B\,e\,f\,g$, which is augmenting since it is alternating with both endpoints being unsaturated. Thus we retrace this path in the original graph as $u\,a\,b\,c\,d\,e\,f\,g$ and swap all edges along this path to obtain a larger matching.

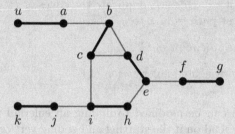

The matching shown above is maximum since all vertices are saturated (and so is also a perfect matching).

5.2.2   Chinese Postman Problem

In Section 2.1.5 we introduced eulerizations and weighted-eulerizations as a way to modify a graph to ensure an eulerian circuit exists. Recall this was useful for many applications where an exhaustive route is needed, such as snowplows, mail delivery, and 3D printing. In small examples we did not need any fancy techniques as it was fairly easy to see the solution, or to use a few instances of trial and error. The weighted-eulerization problem, commonly called the *Chinese Postman Problem*, originates from the Chinese mathematician Guan Meigu who proposed the problem in 1960 and was solved about a decade later by Jack Edmonds and Ellis Johnson [29][42]. Their algorithm uses both Dijkstra's Algorithm for finding a shortest path (see Section 2.3.1) and a matching in a complete graph.

Postman Algorithm

Input: Weighted graph $G = (V, E, w)$.

Steps:

1. Find the set S of odd vertices in G.

2. Form the complete graph K_n where $n = |S|$.

3. For each distinct pair $x, y \in S$, find the shortest path P_{xy} and its total weight $w(P_{xy})$.

4. Define the weight of the edge xy in K_n to be $w(xy) = w(P_{xy})$)

5. Find a perfect matching M of K_n of least total weight.

6. For each edge $e = xy \in M$, duplicate the edges of $P_{x,y}$ corresponding to the shortest path from x to y, creating G^*.

7. Find an eulerian circuit of G^*.

Output: Optimal weighted-eulerization of G.

This algorithm can be modified to finding an eulerian trail, but the formulation would depend on if the starting and ending vertices are fixed. While we have not explicitly described how to find a perfect matching in a complete graph, it should be clear that this can be accomplished in small examples through a brute force attack; more sophisticated techniques exist for large graphs (see [29]).

Example 5.12 Apply the Postman Algorithm to the graph below.

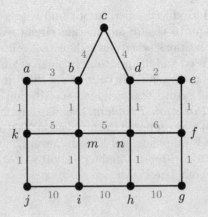

Solution: To begin, we see the odd vertices are b, d, f, h, i, and k. We form the weighted graph K_6 as shown on the left below, where the weight of each edge corresponds to the weight of the shortest path between the endpoints of that edge. For example, $w(bd) = 7$ since the shortest path from b to d in the graph above is $b\,m\,n$ of total weight 7. Next, we find a minimum weight perfect matching of K_6, as shown on the right below; the total weight is 14.

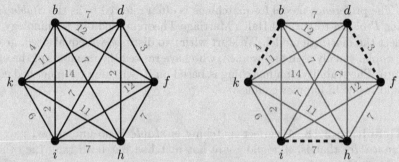

The matched edges are converted back into the shortest paths as shown below. Now that all the vertices have even degree, we can apply Fleury's Algorithm or Hierholzer's Algorithm as in Section 2.1.4.

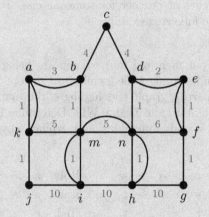

5.3 Stable Matching

In our discussion of matchings so far, we have only been concerned with finding the largest matching possible; but in many circumstances, one pairing may be preferable over another. When taking preferences into account, we no longer focus on whether two items can be paired but rather which pairing is best for the system. Initially, we will only consider those situations that can be modeled by a bipartite graph with an equal number of items in X and Y. In addition, previous models demonstrated undesirable pairings by leaving off the edge in the graph; but with preferences added, we include every edge possible in the bipartite graph. This means the underlying graph will be a complete bipartite graph.

The preference model for matchings is often referred to as the *Stable Marriage Problem* to parallel Hall's Marriage Theorem, and our terminology will reflect the marriage model. We start with two distinct, yet equal sized, groups of people, usually men and women, who have ranked the members of the other group. The stability of a matching is based on if switching two matched edges would result in happier couples.

Definition 5.18 A perfect matching is *stable* if no unmatched pair is *unstable*; that is, if x and y are not matched but both rank the other higher than their current partner, then x and y form an unstable pair.

In essence, when pairing couples into marriages we want to ensure no one will leave their current partner for someone else. To better understand stability, consider the following example.

Example 5.13 Four men and four women are being paired into marriages. Each person has ranked the members of the opposite sex as shown below. Draw a bipartite graph and highlight the matching Anne–Rob, Brenda–Ted, Carol–Stan, and Diana–Will. Determine if this matching is stable. If not, find a stable matching and explain why no unstable pair exists.

Anne:	t > r > s > w	Rob: a > b > c > d
Brenda:	s > w > r > t	Stan: a > c > b > d
Carol:	w > r > s > t	Ted: c > d > a > b
Diana:	r > s > t > w	Will: c > b > a > d

Solution: The complete bipartite graph appears on the next page with the matching in bold. This matching is not stable since Will and Brenda form an unstable pair, as they prefer each other to their current mate.

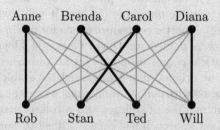

Switching the unstable pairs produces the matching below (Anne ↔ Rob, Brenda ↔ Will, Carol ↔ Stan, and Diana ↔ Ted).

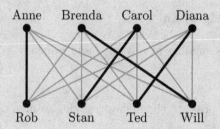

Note that this matching is not stable either since Will and Carol prefer each other to their current mate. Switching the pairs produces a new matching (Anne ↔ Rob, Brenda ↔ Stan, Carol ↔ Will, Diana ↔ Ted) shown below.

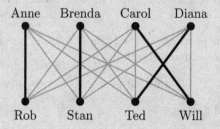

This final matching is in fact stable due to the following: Will and Carol are paired, and each other's first choice; Anne only prefers Ted over Rob, but Ted does not prefer Anne over Diana; Brenda does not prefer anyone over Stan; Diana prefers either Rob or Stan over Ted, but neither of them prefers Diana to their current mate.

While searching for an unstable pair and switching the matching may eventually result in a stable matching (in fact, it will always eventually land on a stable matching), a better procedure exists for finding a stable matching. The algorithm below is named for David Gale and Lloyd Shapley, the two American mathematicians and economists who published this algorithm in 1962 [38]. Though the original paper discussed the stable marriage problem, it was mainly concerned with college admissions where the size of the applicant pool differs from the number of colleges and a college accepts more than one student. Their brilliant approach has been modified to work on problems that cannot be modeled by a complete bipartite graph, a few of which appear in the next section. In addition, their work led to further studies on economic markets, one of which awarded Shapley (along with his collaborator Alvin Roth) the 2012 Nobel Prize in Economics.

Gale-Shapley Algorithm

Input: Preference rankings of n women and n men.

Steps:

1. Each man proposes to the highest ranking woman on his list.

2. If every woman receives only one proposal, this matching is stable. Otherwise move to Step (3).

3. If a woman receives more than one proposal, she

   (a) accepts if it is from the man she prefers above all other currently available men and rejects the rest; or,

   (b) delays with a maybe to the highest ranked proposal and rejects the rest.

4. Each man now proposes to the highest ranking unmatched woman on his list who has not rejected him.

5. Repeat Steps (2)–(4) until all people have been paired.

Output: Stable Matching.

The Gale-Shapley Algorithm is written so that the men are always proposing, giving the women the choice to accept, reject, or delay. This produces an asymmetric algorithm, meaning a different outcome could occur if the women were proposing, which is demonstrated in the next two examples.

Example 5.14 Apply the Gale-Shapley Algorithm to the rankings from Example 5.13, which are reproduced below.

Anne:	$t > r > s > w$	Rob: $a > b > c > d$
Brenda:	$s > w > r > t$	Stan: $a > c > b > d$
Carol:	$w > r > s > t$	Ted: $c > d > a > b$
Diana:	$r > s > t > w$	Will: $c > b > a > d$

Solution: For record keeping, in each round of proposals we indicate if a proposal is accepted by a check ($\checkmark$), a rejection by an X and a delay as ?.

Step 1: The initial proposals are Rob $\to$ Anne, Stan $\to$ Anne, Ted $\to$ Carol, and Will $\to$ Carol. Note that we think of these proposals as simultaneous and a woman does not make her decision until all proposals are made for a given round.

Step 2: Since not all the proposals are different, Anne and Carol need to make some decisions. First, neither Rob nor Stan are Anne's top choice so she rejects the lower ranked one (Stan) and says maybe to the other (Rob). Next, Will is Carol's top choice so she accepts his proposal and rejects Ted.

Rob	Anne	?
Stan	Anne	X
Ted	Carol	X
Will	Carol	$\checkmark$

Step 3: The remaining men (Rob, Stan, and Ted) propose to the next available woman on their preference list. Rob proposes again to Anne since she delayed in the last round. Stan proposes to Brenda; he cannot propose to Anne since she rejected him previously and cannot propose to Carol since she is already paired. Ted proposes to Diana.

Step 4: Since all the proposals are different (no woman received more than one proposal), the women must all accept.

Rob	Anne	$\checkmark$
Stan	Brenda	$\checkmark$
Ted	Diana	$\checkmark$
Will	Carol	$\checkmark$

Output: The matching shown above is stable.

As noted above, the Gale-Shapley Algorithm is asymmetric and in fact favors the group making the proposals. In the previous example, two men are paired with their top choice, one with his second, and one with his third. Even though the same holds for the women (two first choices, one second, and one third), if the women were the ones proposing we would likely see an improvement in their overall happiness. This is outlined in the example below.

Example 5.15 Apply the Gale-Shapley Algorithm to the rankings from Example 5.13 with the women proposing.

Solution:

Step 1: The women all propose to their top choice. The initial proposals are Anne → Ted, Brenda → Stan, Carol → Will, and Diana → Rob.

Step 2: Since all the proposals are different (no man received more than one proposal), the men must all accept.

Anne	Ted	✓
Brenda	Stan	✓
Carol	Will	✓
Diana	Rob	✓

Output: The matching shown above is stable.

Examples 5.14 and 5.15 demonstrate two important properties of stable matching. First, the group proposing in the Gale-Shapley Algorithm is more likely to be happy. This is especially true if the top choices are all different for the proposers, as happened above. In Example 5.15, the men were required to accept the proposal they received since all the proposal were different. This ensured the women were all paired with their first choice, whereas only one man was paired with his first choice, two with their third choice, and one with his fourth choice. Though this may seem more imbalanced than the matching in Example 5.14, it is still a stable matching, which demonstrates the second item about stable matchings. There is no guarantee that a unique stable matching exists. In fact, many examples have more than one stable matching possible. The important concept to remember is that for a complete bipartite graph with rankings, a stable matching will *always* exist. If we generalize this to other types of graphs, the same may not hold.

One last note on the history of this procedure. Though the algorithm is correctly attributed to Gale and Shapley, it had been implemented about a decade earlier in the pairing of hospitals and residents. The residency selection process was poorly managed prior to the foundation of the National

Resident Matching Program in 1952 by medical students. In 1984 Alvin Roth proved that the algorithm used by the NRMP was a modification of the Gale-Shapley Algorithm. Its implementation had the hospitals "proposing" to the medical students, which meant that the residency programs were favored over the applicants. The algorithm was readjusted in 1995 to have the applicants proposing to the residency programs, ensuring the applicants' preferences are favored. The NRMP today encompasses more than 40,000 applicants and 30,000 positions [76].

There has been extensive study into variations on the Stable Marriage Problem described above. We will discuss two of these: Unacceptable Partners and Stable Roommates. We will begin with the Unacceptable Partners problem since it still has a bipartite graph as its underlying structure, whereas the Stable Roommates problem is based on an underlying graph that is not bipartite. Further generalizations to the Stable Marriage Problem can be found in Exercise 5.14, [43] and [46].

5.3.1 Unacceptable Partners

Look back at the preferences in Example 5.13. By having each person rank all others of the opposite sex, we assume that all of these potential matches are acceptable. This is very clearly not accurate to a real world scenario— some people should never be married if even they are the only pair left. To adjust for this, we introduce the notion of an *unacceptable partner*. Consider the rankings below, where if a person is missing from the ranking, then they are deemed unacceptable (so Will is unacceptable to Diana and only Anne and Brenda are acceptable to Stan).

Anne: $t > r > s > w$	Rob: $a > b > c > d$
Brenda: $w > r > t$	Stan: $a > b$
Carol: $w > r > s > t$	Ted: $c > d > a > b$
Diana: $s > r > t$	Will: $c > b > a$

We are still looking for a matching in a bipartite graph, only now the graph is not complete. We must adjust our notion of a stable matching, since it is possible that not all people could be matched (think of a confirmed bachelor; he would label all women as unacceptable). Under these new conditions, a matching (with unacceptable partners) is *stable* so long as no unmatched pair x and y exist such that x and y are both acceptable to each other, and each is either single or prefers the other to their current partner. To account for this new definition of stable, we must make two minor adjustments to the Gale-Shapley Algorithm.

Gale-Shapley Algorithm (with Unacceptable Partners)

Input: Preference rankings of n women and n men.

Steps:

1. Each man proposes to the highest ranking woman on his list.

2. If every woman receives only one proposal from someone they deem acceptable, they all accept and this matching is stable. Otherwise move to Step (3).

3. If the proposals are not all different, then each woman:

   (a) rejects a proposal if it is from an unacceptable man;

   (b) accepts if the proposal is from the man she prefers above all other currently available men and rejects the rest; or

   (c) delays with a maybe to the highest ranked proposal and rejects the rest.

4. Each man now proposes to the highest ranking unmatched woman on their list who has not rejected him.

5. Repeat Steps (2)–(4) until all people have been paired or until no unmatched man has any acceptable partners remaining.

Output: Stable Matching.

The major change here is in dealing with unacceptable partners. As with the original form of the Gale-Shapley Algorithm, this new version always produces a stable matching. In addition, the algorithm is written so that the men are proposing, but as before, this can be modified so that the women are proposing.

Example 5.16 Apply the Gale-Shapley Algorithm to the rankings on page 255 to find a stable matching.

Solution:

Step 1: The initial proposals are Rob $\to$ Anne, Stan $\to$ Anne, Ted $\to$ Carol, and Will $\to$ Carol.

Step 2: Since not all the proposals are different, Anne and Carol need to make some decisions. First, since Stan is unacceptable to Anne she rejects him and since Rob is not her top choice she says maybe. Next, Will is Carol's top choice so she accepts his proposal and rejects Ted.

Rob	Anne	?
Stan	Anne	X
Ted	Carol	X
Will	Carol	✓

Step 3: The remaining men propose to the next available woman on their preference list. Rob proposes again to Anne since she delayed in the last round. Stan proposes to Brenda; he cannot propose to Anne since she rejected him previously. Ted proposes to Diana.

Step 4: Even though all proposals are different, Brenda rejects Stan since he is an unacceptable partner. The other two women say maybe since their proposals are not from their top choice.

Rob	Anne	?
Stan	Brenda	X
Ted	Diana	?
Will	Carol	✓

Step 5: Rob proposes again to Anne and Ted proposes to Diana since the women said maybe in the last round. Stan does not have any acceptable partners left, and so must remain single.

Step 6: Anne and Diana accept their proposals since all proposals are different.

Rob	Anne	✓
Stan		
Ted	Diana	✓
Will	Carol	✓

Output: The matching shown above is stable. Note that Stan and Brenda remain unmatched.

As the example above illustrates, in a scenario with unacceptable partners a stable matching can exist with not all people paired. The version of this

example where the women propose appears in Exercise 5.12. A modification
of this procedure to allow for situations with an unequal number of men and
women appears in Exercise 5.14.

5.3.2   Stable Roommates

The Stable Roommate Problem is a modification of the Stable Marriage problem, only now the underlying graph is not bipartite. Each person ranks the
others and we want a stable matching; that is, a matching so that two unpaired people do not both prefer each other to their current partner. Similar to
the Augmenting Path Algorithm being unavailable for use on general graphs,
the Gale-Shapley Algorithm does not work on the Stable Roommate Problem
since it is based on two distinct groups ranking members of the other group.
This first example finds all possible pairings and examines if they are stable.

Example 5.17 Four women are to be paired as roommates. Each woman
has ranked the other three as shown below. Find all possible pairings and
determine if any are stable.

$$\text{Emma:} \quad l \; > \; m \; > \; z$$
$$\text{Leena:} \quad m \; > \; e \; > \; z$$
$$\text{Maggie:} \quad e \; > \; z \; > \; l$$
$$\text{Zara:} \quad e \; > \; l \; > \; m$$

Solution: There are three possible pairings, only one of which is stable.

- Emma ↔ Leena and Maggie ↔ Zara
  This is stable since Emma is with her first choice and the only person
  Leena prefers over Emma is Maggie, but Maggie prefers Zara over
  Leena.

- Emma ↔ Maggie and Leena ↔ Zara
  This is not stable since Emma prefers Leena over Maggie and Leena
  prefers Emma over Zara.

- Emma ↔ Zara and Leena ↔ Maggie
  This is not stable since Emma prefers Maggie over Zara and Maggie
  prefers Emma over Leena.

Recall the Gale-Shapley Algorithm showed that a stable matching on a
bipartite graph (where the partition sets have equal size) will always exist,
yet it is possible in the Stable Roommate Problem that a stable matching will
fail to exist.

Example 5.18 Before the four women from Example 5.17 are paired as roommates, Maggie and Zara get into an argument, causing them to adjust their preference lists. Determine if a stable matching exists.

$$
\begin{array}{rccccc}
\text{Emma:} & l & > & m & > & z \\
\text{Leena:} & m & > & e & > & z \\
\text{Maggie:} & e & > & l & > & z \\
\text{Zara:} & e & > & l & > & m
\end{array}
$$

Solution: There are three possible pairings, none of which are stable.

- Emma ↔ Leena and Maggie ↔ Zara
  This is not stable since Leena prefers Maggie over Emma and Maggie prefers Leena over Zara.

- Emma ↔ Maggie and Leena ↔ Zara
  This is not stable since Emma prefers Leena over Maggie and Leena prefers Emma over Zara.

- Emma ↔ Zara and Leena ↔ Maggie
  This is not stable since Emma prefers Maggie over Zara and Maggie prefers Emma over Leena.

The previous two examples used an exhaustive method to find a stable matching or to determine if no such matching exists. While this is not too difficult with a small number of vertices, it becomes computationally impractical as the number of vertices grows. In fact, the number of possible ways to pair n people (where n is even) is $(n-1)!!$, called n *double factorial*. For a given integer k, $k!!$ is defined as the product of all even integers less than or equal to k if k is even and the product of all odd integers less than or equal to k if k is odd. In Examples 5.17 and 5.18 above, four people required $3!! = 3 \cdot 1 = 3$ possible groups of pairings. Thus for a Stable Roommate Problem, we would need to check 15 pairings when there are 6 people $((6-1)!! = 5*3*1)$ and check 105 for 8 people (such as in Example 5.9 when pairing people into canoes). An efficient algorithm for the Stable Roommate Problem was first published in 1985 by the computer scientist Robert Irving; for further information, see [43].

5.4   Factors

In the search for a prefect matching in a graph G, we are in essence looking for a spanning subgraph H that consists of independent edges. But what if we do not want to restrict ourselves to just edges? As we have seen before, mathematicians will take a result and generalize it to see what additional results can be produced. To that end, we finish this chapter on the generalization of matchings, called factors. We will revisit factors next chapter when discussing edge-coloring (see Section 6.3).

Definition 5.19 Let G be a graph with spanning subgraph H and let k be a positive integer. Then H is a **k-factor** of G if H is a k-regular.

Note that a perfect matching is a 1-factor since the requirement that the edges in a matching are independent results in each vertex to have degree 1. So what would a 2-factor entail? Each vertex must have degree 2, so the spanning subgraph H would consist of some collection of cycles.

Example 5.19 Find a 2-factor for the graph shown below.

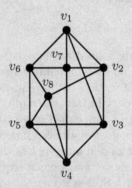

Solution: More than one 2-factor exists for the graph above, as we simply need to find a spanning subgraph that is 2-regular. One such solution is given below. Note that this solution consists of two components, one of which is isomorphic to C_3 and the other to C_5.

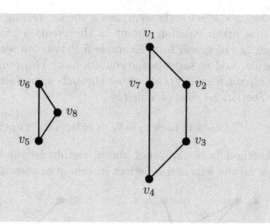

Recall that Tutte's Theorem above (see Theorem 5.15) determined exactly when a graph contains a 1-factor but what if we want a 2-factor? The theorem below is due to the Danish mathematician Julius Petersen, and is one of his main contributions to graph theory, the others being Corollary 5.16 about matchings in a cubic graph and a specific graph, widely called the Petersen graph in his honor, that is used as a counterexample to a famous graph coloring conjecture. The Petersen graph appears in Exercise 5.3(f).

Theorem 5.20 If G is a $2k$-regular graph, then G has a 2-factor.

Proof: First, we will assume G is connected as otherwise we could apply the argument below to each component of G. Next, since every vertex of G has even degree we know G must contain an eulerian circuit C by Theorem 2.7. We will create a bipartite graph G' based upon this eulerian circuit and use a matching on G' to produce our 2-factor of G.

Let $V(G) = \{v_1, v_2, \ldots, v_n\}$ and define the vertices of G' as $x_1, x_2, \ldots, x_n$ and $y_1, y_2, \ldots y_n$ so that $x_i y_j$ is an edge of G' when v_i immediately precedes v_j on the eulerian circuit C. Since G is $2k$-regular, we know C enters and exits each vertex of G exactly k times, and so G' is a k-regular bipartite graph. Thus by Corollary 5.5, G' contains a perfect matching M.

Using the matching M as our guide, we can construct the 2-factor H of G. Note that the edge incident to x_i in M will represent an edge exiting v_i, whereas an edge incident to y_i will represent an edge entering v_i. Thus to find H, we start at v_1 and take the edge to v_i that arises from the matched edge $x_1 y_i$. The next edge in the 2-factor will be from v_i to v_j arising from the matched edge $x_i y_j$. Thus will continue until all vertices are listed, creating a 2-regular spanning subgraph of G.

While not the most complex proof we have encountered in this text, the proof of Petersen's 2-factor theorem uses some interesting techniques, in particular the use of an eulerian circuit in the creation of a bipartite graph. Looking back at the graph from Example 5.19, we can see that this graph is in fact 4-regular and so satisfies the conditions of Theorem 5.20. We can find an eulerian circuit for the graph, either through inspection or the techniques outlined in Section 2.1, one of which is

$$C : v_1 \, v_3 \, v_5 \, v_4 \, v_8 \, v_2 \, v_7 \, v_6 \, v_5 \, v_8 \, v_6 \, v_1 \, v_2 \, v_3 \, v_4 \, v_7 \, v_1$$

Using the method from the proof above, we obtain the bipartite graph G' shown below on the left, and a perfect matching as shown on the right.

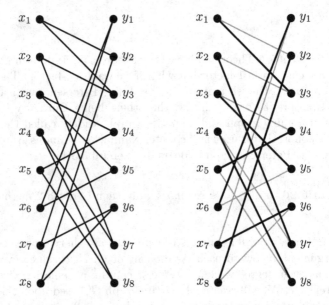

Translating this matching into a 2-factor gives the graph below. Note that this is not the same 2-factor we gave in Example 5.19. In fact, if you combined these two 2-factors we would obtain our original graph.

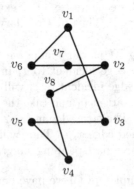

The previous example demonstrates a different way to look at factors, where we are not just concerned with obtaining a k-factor, but now want to find a collection of k-factors that encompass the entire graph. These are called *factorizations*.

Definition 5.21 A *k-factorization* of G is a partition of the edges into disjoint k-factors.

As to be expected, having a k-factorization is more difficult than simply containing a k-factor. For example, as we have previously noted a 1-factor is equivalent to having a perfect matching in a graph. But what would a 1-factorization be? It would mean that a graph can be partitioned into distinct perfect matchings $M_1, M_2, \ldots M_k$ so that each matching spans the graph, no two matchings have an edge in common, and all edges appear exactly once in one of these matchings. Finding a perfect matching was difficult enough!

However, it should be clear that some conditions can be placed on a graph to have a 1-factorization, in particular the graph must have an even number of vertices and must be regular! Once a perfect matching has been found, we could remove all those edges and examine the graph that remains. If we are to find another 1-factor (perfect matching), then no vertex can be isolated, and so as we continue this process until all edges are saturated by exactly one perfect matching, we must have all vertices to be of the same degree. But is it enough for a graph to be regular to have a 1-factorization?

Example 5.20 Determine if the graph below has a 1-factorization.

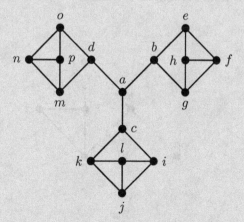

Solution: This graph is 3-regular, implying any 1-factorization would be made from three disjoint 1-factors (or perfect matchings). Suppose we have such a 1-factorization, say M_1, M_2, and M_3, which will be indicated by various line styles. Then each edge out of a would be from a different 1-factor. Without loss of generality, we will assume ab is from M_1. Thus the other two edges out of b would be from M_2 and M_3, and by symmetry we can assume be is from M_2 and bg is from M_3.

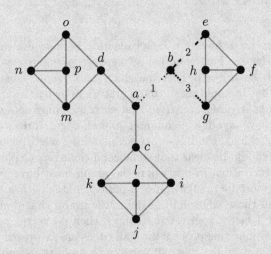

Since the other edges out of e cannot also be from M_2, and each vertex must be incident to an edge from each 1-factor, by symmetry we can assume ef is from M_1 and eh is from M_3.

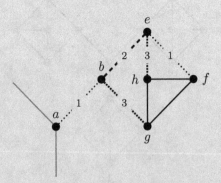

This forces fh to be from M_2 and gh from M_1.

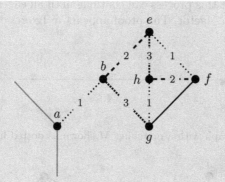

But then fg cannot be in any of M_1, M_2, or M_3 since either f or g is incident to a edge from each of these 1-factors. Thus the graph above does not have a 1-factorization.

Although a graph being regular is necessary, it is not a sufficient condition. We will see in Section 6.3.1 that having a 1-factorization is in fact equivalent to asking if a k-regular graph has an edge-coloring using exactly k colors. An edge-coloring assigns labels to each of the edges so that no two edges with the same label share an endpoint. Thus which 1-factor an edge belongs to (say M_1) is equivalent to the color for an edge (say color 1).

Determining if a graph has a 1-factorization is not an easy question to answer in the general case; however, there is one special case where we can get a definitive answer without a complicated proof.

Proposition 5.22 Every k-regular bipartite graph has a 1-factorization for all $k \geq 1$.

The proof of the Proposition above appears in Exercise 5.24 and relies on a fairly straightforward induction argument. In Section 6.3.1 we will show that K_{2n} also has a 1-factorization and discuss other results on graphs with 1-factorizations.

Surprisingly, 2-factorizations are not much more difficult to prove than 2-factors. In fact, Theorem 5.20 above is just a simplified version of Petersen's original result, stated below.

Theorem 5.23 A graph G has a k-factorization if and only if G is $2k$-regular.

We have already shown that a $2k$-regular graph has a 2-factor, what remains to show is when we remove the 2-factor the resulting graph also contains a 2-factor, and that this process can continue until all edges in the graph have been included in a 2-factor. The proof appears in Exercise 5.25.

5.5   Exercises

5.1 Below is a graph with a matching M shown as dotted lines.

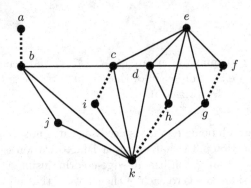

(a) Find an alternating path starting at a. Is this path augmenting?

(b) Find an augmenting path in the graph or explain why none exists.

(c) Is M a maximum matching? maximal matching? perfect matching? Explain your answer. If M is not maximum, find a matching that is maximum.

5.2 Each of the graphs below has a matching shown as dotted lines. Complete the following steps for both:

(i) Find an alternating path starting at vertex a.

(ii) Is this path augmenting? Explain your answer.

(iii) Use the Augmenting Path Algorithm to find a maximum matching.

(iv) Use the Vertex Cover Method to find a minimum vertex cover.

(a)

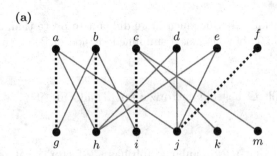

(b)

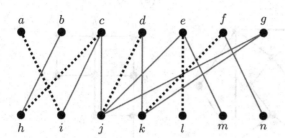

5.3 Find a maximum matching for each of the graphs below. Include an explanation as to why the matching is maximum.

(a)

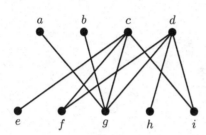

(b)

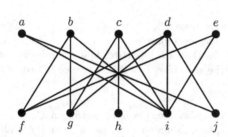

(c)

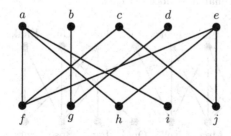

(d)   **(e)**

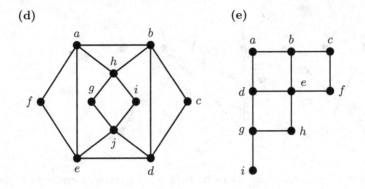

(f)

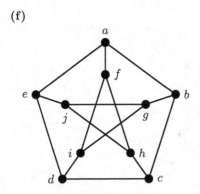

5.4 Using the graphs from Exercise 5.3,
   (a) Determine which graphs are bipartite.
   (b) For each of the graphs that are bipartite, find a minimum vertex cover. Verify that the size of the matching found in Exercise 5.3 equals the size of your vertex cover.

5.5 The Roanoke Ultimate Frisbee League is organizing a Contra Dance. The fifteen members must be split into male-female pairs, though not all people are willing to dance with each other. The graph below models those who can be paired (as both people find the other acceptable). Find a maximum matching and explain why a larger matching does not exist.

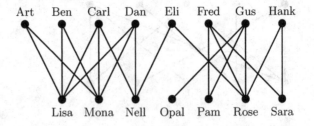

5.6 Seven committees must elect a chairperson to represent them at the end-of-year board meeting; however, some people serve on more than one committee and so cannot be elected chairperson for more than one committee. Based on the membership lists below, determine a system of distinct representatives for the board meeting.

Committee	Members			
Benefits	Agatha	Dinah	Evan	Vlad
Computing	Evan	Nancy	Leah	Omar
Purchasing	George	Vlad	Leah	
Recruitment	Dinah	Omar	Agatha	
Refreshments	Nancy	George		
Social Media	Evan	Leah	Vlad	Omar
Travel Expenses	Agatha	Vlad	George	

5.7 Each year, the chair of the mathematics department must determine course assignments for the faculty. Each professor has submitted a list of the courses he or she wants to teach. Find a system of assignments where each professor will teach exactly one of the remaining courses or explain why none exists.

Professor	Preferred Courses			
Dave	Abstract Algebra	Real Analysis	Statistics	Calculus
Roland	Statistics	Geometry	Calculus	
Chris	Calculus	Geometry		
Adam	Statistics	Calculus		
Hannah	Abstract Algebra	Real Analysis	Topology	
Maggie	Abstract Algebra	Real Analysis	Geometry	Topology

5.8 Instead of pairing a professor with only one course of their preference from Exercise 5.7, now the mathematics department chair must pair each professor with two of the courses from their (expanded) list.

   (a) Describe how to turn this into a matching problem where a solution is given in terms of a perfect matching.

   (b) Find a perfect matching for the professors and their preferred course list shown below or explain why none exists.

Professor	Preferred Courses		
Dave	Abstract Algebra	Real Analysis	Number Theory
	Calculus II	Calculus I	Statistics
Roland	Vector Calculus	Discrete Math	Statistics
	Calculus II	Geometry	Calculus I
Chris	Vector Calculus	Real Analysis	Discrete Math
	Statistics	Geometry	Calculus I
Adam	Statistics	Calculus I	Differential Equations
	Geometry	Number Theory	
Hannah	Abstract Algebra	Real Analysis	Number Theory
	Linear Algebra	Topology	
Maggie	Abstract Algebra	Real Analysis	Linear Algebra
	Geometry	Topology	Calculus II

5.9 The students in a geometry course are paired each week to present homework solutions to the class. In the table below, a possible pair is indicated by a Y. Find a way to pair the students or explain why none exists.

	Al	Brie	Cam	Fred	Hans	Megan	Nina	Rami	Sal	Tina
Al	.	Y	.	.	Y	.	.	Y	.	Y
Brie	Y	.	Y	Y	Y	Y	.	Y	Y	.
Cam	.	Y	.	.	Y	.	Y	Y	.	.
Fred	.	Y	.	.	.	Y	Y	.	.	Y
Hans	Y	Y	Y	.	.	.	.	Y	Y	Y
Megan	.	Y	.	Y	.	.	Y	Y	Y	.
Nina	.	.	Y	Y	.	Y	.	.	Y	.
Rami	Y	Y	Y	.	Y	Y	.	.	Y	.
Sal	.	Y	.	.	Y	Y	Y	Y	.	.
Tina	Y	.	.	Y	Y	.	.	.	.	.

5.10 Apply the Gale-Shapley Algorithm to the set of preferences below with
(a) the men proposing
(b) the women proposing

Alice: $r > s > t > v$		Rich: $a > d > b > c$	
Beth: $s > r > v > t$		Stefan: $a > c > d > b$	
Cindy: $v > t > r > s$		Tom: $c > b > d > a$	
Dahlia: $t > v > s > r$		Victor: $c > d > b > a$	

5.11 Apply the Gale-Shapley Algorithm to the set of preferences below
with
   (a) the men proposing
   (b) the women proposing

Edith:	l	>	n	>	o	>	m	>	p	Liam:	f	> e	> h	> g	> i
Faye:	n	>	l	>	m	>	o	>	p	Malik:	e	> i	> g	> f	> h
Grace:	p	>	m	>	o	>	n	>	l	Nate:	f	> g	> i	> h	> e
Hanna:	p	>	n	>	o	>	l	>	m	Olaf:	i	> e	> f	> g	> h
Iris:	p	>	o	>	m	>	n	>	l	Pablo:	f	> h	> g	> e	> i

5.12 Apply the Gale-Shapley Algorithm (with Unacceptable Partners) to the
preferences from Example 5.16 with the women proposing.

5.13 Apply the Gale-Shapley Algorithm (with Unacceptable Partners) to the
preferences below with
   (a) the men proposing
   (b) the women proposing

Edith:	l	>	n	>	m					Liam:	f	> e	> h	> g	
Faye:	n	>	l	>	m	>	o	>	p	Malik:	e	> h	> i	> f	
Grace:	m	>	o	>	n	>	l			Nate:	g	> f	> i		
Hanna:	p	>	o	>	l	>	m			Olaf:	i	> e	> f		
Iris:	p	>	m	>	n	>	l			Pablo:	f	> h	> g	> i	

5.14 In each of the examples where the Gale-Shapley Algorithm was utilized, it
was required that the number of men equals the number of women. Just as we
were able to modify the algorithm for instances where some people are deemed
unacceptable, we can modify the algorithm to account for unequal numbers. To
do this, we introduce ghost participants in order to equalize the gender groups.
These ghosts are deemed unacceptable by those of the opposite sex, and in turn
find no person of the opposite sex acceptable. Using this modification, find a
stable set of marriages for the preferences listed below.

Alice:	p	>	r	>	s	>	t		Peter:	b	> a	> c	> d	> e
Beth:	r	>	p	>	s	>	t		Rich:	c	> b	> e	> d	> a
Carol:	t	>	p	>	s	>	r		Saul:	a	> b	> c	> d	> e
Diana:	t	>	s	>	r	>	p		Teddy:	e	> c	> d	> a	> b
Edith:	r	>	s	>	t	>	p							

5.15 Using the graph from Example 5.19 on page 260, find a different eulerian
circuit and follow the construction from Theorem 5.20 to find a different 2-factor
for the graph.

5.16 Determine the size of a maximum matching of C_n, P_n, K_n, and $K_{m,n}$ for
all possible values of m and n.

5.17 Explain why a hamiltonian graph has a 2-factor. Must every graph with a 2-factor be hamiltonian? Prove or give a counterexample.

5.18 Prove Corollary 5.5: Every k-regular bipartite graph has a perfect matching for all $k > 0$.

5.19 Prove that every tree has at most one perfect matching.

5.20 Recall that $\Delta(G)$ denotes the maximum degree of G. Use the König-Egervary Theorem to prove that every bipartite graph G has a matching of size at least $\dfrac{|E|}{\Delta}$.

5.21 Recall that $\delta(G)$ denotes the minimum degree of G. Prove that if G is a graph with $2n$ vertices and $\delta \geq n$ then G has a perfect matching.

5.22 Prove Theorem 5.13: A collection of finite nonempty sets $S_1, S_2, \ldots, S_n$ (where $n \geq 1$) has an SDR if and only if $\left| \bigcup_{i \in R} S_i \right| \geq |R|$ for all $R \subseteq \{1, 2, \ldots, n\}$.

5.23 Prove Corollary 5.16: Every cubic graph without any bridges has a perfect matching. (Hint: Use Tutte's Theorem)

5.24 Prove Proposition 5.22: Every k-regular bipartite graph has a 1-factorization for all $k \geq 1$. (Hint: Argue by induction on k).

5.25 Prove Theorem 5.23: A graph G has a k-factorization if and only if G is $2k$-regular.

5.26 In Section 3.4.1, we discussed an approximation algorithm for the metric Traveling Salesman Problem that made use of a minimum spanning tree. Christofides' Algorithm, outlined below, uses the same basic backbone of using a minimum spanning tree, but also makes use of a matching within the graph. Christofides' Algorithm, first introduced in 1976, had the best performance ratio of any approximation algorithm until a modified version was published in 2015.

Apply Christofides' Algorithm to the graph from Example 3.10 and Exercise 3.9.

Christofides' Algorithm

Input: Weighted complete graph K_n, where the weight function w satisfies the triangle inequality.

Steps:

1. Find a minimum spanning tree T for K_n.
2. Let X be the set of all vertices of odd degree in T.

3. Create a weighted complete graph H whose vertex set is X and where the weight of the edges is taken from the weights given in K_n.

4. Find a perfect matching M on H of minimum weight.

5. Create a graph G by adding the matched edges from M to the tree T obtained in Step (1).

6. Find an eulerian circuit for G.

7. Convert the eulerian circuit into a hamiltonian cycle by skipping any previously visited vertex (except for the starting and ending vertex).

8. Calculate the total weight.

Output: Hamiltonian cycle for K_n.

6

Graph Coloring

In the previous chapter we discussed the application of graph matching to a problem where items from two distinct groups must be paired. An important aspect of this pairing is that no item could be paired more than once. Compare that with the following scenario:

> Five student groups are meeting on Saturday, with varying time requirements. The staff at the Campus Center need to determine how to place the groups into rooms while using the fewest rooms possible.

Although we can think of this problem as pairing groups with rooms, there is no restriction that a room can only be used once. In fact, to minimize the number of rooms used, we would hope to use a room as often as possible. This chapter explores graph coloring, a strategy often used to model resource restrictions. We will explore graph coloring in terms of both vertices and edges, though most of our time will be spent on coloring vertices. But before we get into the heart of coloring, we begin with a historically significant problem, known as the Four Color Theorem.

6.1 Four Color Theorem

In 1852 Augustus De Morgan sent a letter to his colleague Sir William Hamilton (the same mathematician who introduced what we now call hamiltonian cycles) regarding a puzzle presented by one of his students, Frederick Gutherie (though Gutherie later clarified that the question originated from his brother, Francis). This question was known for over a century as the *Four Color Conjecture*, and can be stated as

> Any map split into contiguous regions can be colored using at most four colors so that no two bordering regions are given the same color.

An important aspect of this conjecture is that a region, such as a country or state, cannot be split into two disconnected pieces. For example, the state of Michigan is split into the Lower Peninsula and the Upper Peninsula and so is not a contiguous region; thus the contiguous United States does not satisfy the hypothesis of the Four Color Conjecture. However, it is still possible to color the lower 48 states using 4 colors (try it!).

The Four Color Conjecture started as a map coloring problem, yet migrated into a graph coloring problem. In the late 19th century, Alfred Kempe studied the dual problem where each region on a map was represented by a vertex and an edge exists between two vertices if their corresponding regions share a border. This approach was extensively used in the mid-20th century as the study of graph theory exploded with the advent of the computer. The search for a proper map coloring is now reduced to a proper vertex coloring (more commonly referred to as just a coloring) for a *planar graph*. A graph is planar if it can be drawn so that no edges cross. We will study planar graphs extensively in Chapter 7.

Definition 6.1 A proper *k-coloring* of a graph G is an assignment of colors to the vertices of G so that no two adjacent vertices are given the same color and exactly k colors are used.

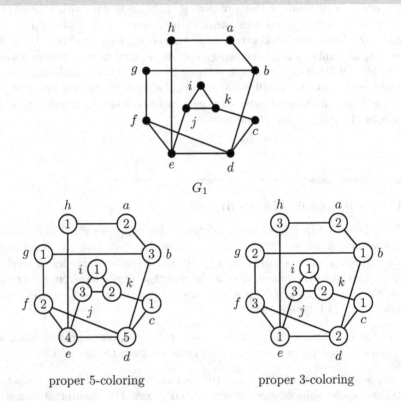

proper 5-coloring   proper 3-coloring

Above is a graph with two different proper colorings of the vertices. Note that beyond small examples, we rarely use color names (red, blue, green, etc.) but rather refer to color numbers (color 1, 2, 3, etc.) since names of colors get more complicated as we move beyond the standard 6 to 10 colors. The two colorings given above use a different number of colors, but are both proper since no two vertices of the same color are adjacent.

Definition 6.2 Given a proper k-coloring of G, the ***color classes*** are sets $S_1, \ldots S_k$ where S_i consists of all vertices of color i.

The first coloring of G_1 has color classes $S_1 = \{c, g, h, i\}$, $S_2 = \{a, f, k\}$, $S_3 = \{b, j\}$, $S_4 = \{e\}$, $S_5 = \{d\}$ and the second coloring has color classes $S_1 = \{b, c, e, i\}$, $S_2 = \{a, d, g, k\}$, $S_3 = \{f, h, j\}$. Recall that two vertices are independent if they have no edges between them. Thus a color class must consist of independent vertices. We will see there is a relationship between independent sets and coloring a graph.

Definition 6.3 The ***independence number*** of a graph G is $\alpha(G) = n$ if there exists a set of n vertices with no edges between them but every set of $n + 1$ vertices contains at least one edge.

Most problems on graph coloring are optimization problems since we want to minimize the number of colors used; that is, find the lowest value of k so that G has a proper k-coloring. The example below demonstrates how a map coloring relates to a vertex coloring of a graph.

Example 6.1 Find a coloring of the map of the counties of Vermont and explain why three colors will not suffice.

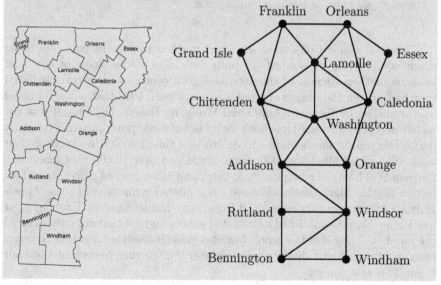

Solution: First note that each county is given a vertex and two vertices are adjacent in the graph when their respective counties share a border. One possible coloring is shown below.

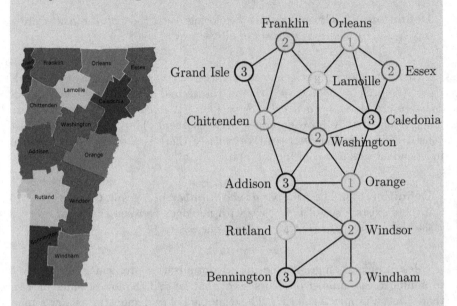

Note that Lamoille County is surrounded by five other counties. If we try to alternate colors amongst these five counties, for example Orleans – 1, Franklin – 2, Chittenden – 1, Washington – 2, we still need a third color for the fifth county (Caledonia – 3). Since Lamoille touches each of these counties, we know it needs a fourth color.

The Four Color Conjecture intrigued mathematicians in part due to its simplicity but also because of the numerous false proofs. Some of the most brilliant mathematicians of the 19th and 20th centuries incorrectly believed they had proven the conjecture and it was not until 1976 that a correct proof was published by Kenneth Appel and Wolfgang Haken. Their proof was not widely accepted for some time both for its use of a computer program (the first major theorem in mathematics to do so) and the difficulty in checking their work. The proof filled over 900 pages, contained over 10,000 diagrams, took thousands of hours of computational time, and when printed stood about four feet in height. Mathematicians have since offered refinements to the Appel-Haken argument; in particular, Niel Robertson, Daniel Sanders, Paul Seymour and Robin Thomas published an updated version in 1994 that not only reduced the number of pages in the proof, but also made it feasible for others to verify their results on a standard home computer. For further history of the Four Color Theorem, see [85].

6.2   Vertex Coloring

For the remainder of this chapter, we will explore graph colorings for graphs that may or may not be planar, mainly since we already know that planar graphs need at most 4 colors and so there is not much room for further exploration. Any graph we consider can be simple or have multi-edges but cannot have loops, since a vertex with a loop could never be assigned a color. In any graph coloring problem, we want to determine the smallest value for k for which a graph has a k-coloring. This value for k is called the *chromatic number* of a graph.

> **Definition 6.4** The *chromatic number* $\chi(G)$ of a graph is the smallest value k for which G has a proper k-coloring.

In order to determine the chromatic number of a graph, we often need to complete the following two steps:

(1) Find a vertex coloring of G using k colors.

(2) Show why fewer colors will not suffice.

At times it can be quite complex to show a graph cannot be colored with fewer colors. There are a few properties of graphs and the existence of certain subgraphs that can immediately provide a basis for these arguments.

Look back at Example 6.1 about coloring the counties in Vermont and the discussion of alternating colors around a central vertex. In doing so, we were using one of the most basic properties in graph coloring: the number of colors needed to color a cycle. Recall that a cycle on n vertices is denoted C_n. The examples below show optimal colorings of C_3, C_4, C_5, and C_6.

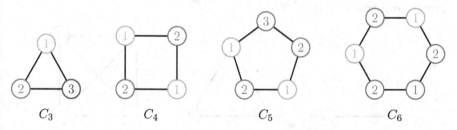

C_3 C_4 C_5 C_6

Notice that in all the graphs we try to alternate colors around the cycle. When n is even, we can color C_n in two colors since this alternating pattern can be completed around the cycle. When n is odd, we need three colors for C_n since the final vertex visited when traveling around the cycle will be adjacent to a vertex of color 1 and of color 2. This was demonstrated in the coloring of the five counties surrounding Lamoille County in Example 6.1.

The next structure that provides additional reasoning for the lower bound of the chromatic number is based upon an odd cycle. Again, referencing the coloring of the counties in Vermont, we used the odd cycle around Lamoille County to explain why at least 3 colors were needed. However, we showed that in fact 4 colors were required since the Lamoille vertex is adjacent to each of vertices in the surrounding odd cycle. This structure is often referred to as a wheel.

Definition 6.5 A *wheel* W_n is a graph in which n vertices form a cycle around a central vertex that is adjacent to each of the vertices in the cycle.

The first few wheels are shown below. Note that when n is odd, we get a scenario similar to that of Lamoille county from Example 6.1, and thus requiring 4 colors. In general, we can use odd wheels to explain why 3 colors will not suffice.

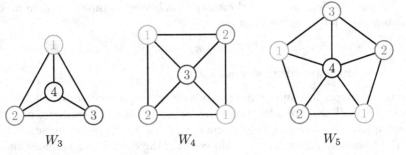

W_3 $\qquad\qquad$ W_4 $\qquad\qquad$ W_5

The final structure we search for within a graph is based on the notion of a complete graph. Recall that in a complete graph each vertex is adjacent to every other vertex in the graph. Thus if we assign colors to the vertices, we cannot use a color more than once. Possible colorings of a few complete graphs are shown below.

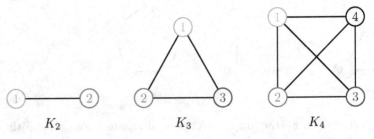

K_2 $\qquad\qquad$ K_3 $\qquad\qquad$ K_4

When a complete graph appears as a subgraph within a larger graph, we call it a *clique*.

Definition 6.6 A *clique* in a graph is a subgraph that is itself a complete graph. The *clique size* of a graph G, denoted $\omega(G)$, is the largest value of n for which G contains K_n as a subgraph.

Knowing the clique size of a graph automatically provides a nice lower bound for the chromatic number. For example, if G contains K_5 as a subgraph, then we know this portion of the graph needs at least 5 colors. Thus $\chi(G) \geq 5$. Thus when trying to argue that fewer colors will not suffice, we look for odd cycles (which require 3 colors), odd wheels (which require 4 colors), and cliques (which require as many colors as the number of vertices in the clique). Below is a summary of our discussion so far regarding lower bounds for the chromatic number of a graph.

Special Classes of Graphs with known $\chi(G)$

- $\chi(C_n) = 2$ if n is even $(n \geq 2)$

- $\chi(C_n) = 3$ if n is odd $(n \geq 3)$

- $\chi(K_n) = n$

- $\chi(W_n) = 4$ if n is odd $(n \geq 3)$

One note of caution: a graph can have a chromatic number that is much larger than its clique size. In fact, Jan Mycielski showed that there exist graphs with an arbitrarily large chromatic number yet have a clique size of 2. We often refer to graphs with $\omega(G) = 2$ as *triangle-free*. Mycielski's proof provided a method for finding a triangle-free graph that requires the desired number of colors[67].

Example 6.2 Mycielski's Construction is a well-known procedure in graph theory that produces triangle-free graphs with increasing chromatic numbers. The idea is to begin with a triangle-free graph G where $V(G) = \{v_1, v_2, \ldots, v_n\}$ and add new vertices $U = \{u_1, u_2, \ldots, u_n\}$ so that $N(u_i) = N(v_i)$ for every i; that is, add an edge from u_i to v_j whenever v_i is adjacent to v_j. In addition, we add a new vertex w so that $N(w) = U$; that is, add an edge from w to every vertex in U. The resulting graph will remain triangle-free but need one more color than G. If you perform enough iterations of this procedure, you can obtain a graph with $\omega(G) = 2$ and $\chi(G) = k$ for any desired value of k.

Consider G to be the complete graph on two vertices, K_2, which is clearly triangle free and has chromatic number 2, as shown in the following graph.

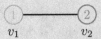

After the first iteration of Mycielski's Construction, we get the graph shown below on the left. Notice that u_1 has an edge to v_2 since v_1 is adjacent to v_2. Similarly, u_2 has an edge to v_1. In addition, w is adjacent to both u_1 and u_2. The graph on the right below is an unraveling of the graph on the left. Thus we have obtained C_5, which we know needs 3 colors.

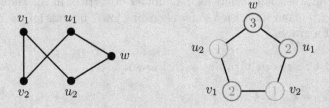

After the second iteration, we obtain the graph shown below. The outer cycle on 5 vertices represents the graph obtained above in the first iteration. The inner vertices are the new additions to the graph, with u_1 adjacent to v_2 and v_5 since v_1 is adjacent to v_2 and v_5. Similar arguments hold for the remaining u-vertices and the center vertex w is adjacent to all of the u-vertices. A coloring of the graph is shown below on the right. Note that the outer cycle needs 3 colors, as does the group of u-vertices. This forces w to use a fourth color. In addition, no matter which three vertices you choose, you cannot find a triangle among them, and so the graph remains triangle-free.

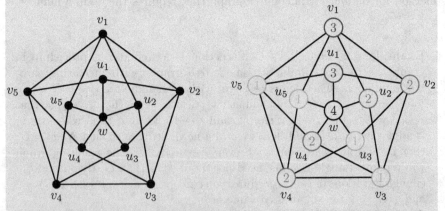

If we continue this procedure through one more step, we obtain a graph needing 5 colors with a clique size of 2.

Although Mycielski's Construction should warn you not to rely too heavily on the clique size of a graph, most real world applications have a chromatic number close to their clique size.

The discussion so far has focused on lower bounds for the chromatic number of a graph. However, when searching for an optimal coloring, it is often useful to know upper bounds as well. Combined with the lower bounds we found above, we get a nice range from which to narrow our search for an optimal coloring. Perhaps the most useful of results for upper bounds is the following theorem due to the English mathematician Rowland Leonard Brooks and published in 1941 [10].

Theorem 6.7 (Brooks' Theorem) Let G be a connected graph and Δ denote the maximum degree among all vertices in G. Then $\chi(G) \leq \Delta$ as long as G is not a complete graph or an odd cycle. If G is a complete graph or an odd cycle then $\chi(G) = \Delta + 1$.

The reasoning behind Brooks' Theorem is that if all the neighbors of x have been given different colors, then one additional color is needed for x. If x has the maximum degree over all vertices in G, then we have used $\Delta + 1$ colors for x and its neighbors. Perhaps more surprising is that unless a graph equals K_n or C_m (for an odd m), the neighbors of the vertex of maximum degree cannot all be given different colors and so the bound tightens to Δ. For example, the third graph from Mycielski's construction in Example 6.2 has a maximum degree of 5. Since this graph is neither a complete graph nor an odd cycle (although it does contain an odd cycle), we know the chromatic number is at most 5. In addition, since the clique size is 2, we know the chromatic number must be at least 2. As we showed above, the correct value was 4. The formal proof of Brooks Theorem is enlightening in how it combines connectivity with graph colorings, but is fairly complex and relies on a specific algorithm for coloring, First-Fit, which is a type of on-line coloring. These will be studied further in Section 6.4.1, and therefore we will delay for now the formal proof of Brooks' Theorem. However, we will revisit the proof after our discussion of First-Fit, and it can be found on page 323.

To summarize our discussion so far, we have:

$$\omega(G) \leq \chi(G) \leq \Delta(G) + 1$$

One final note of caution: although combining Brooks' Theorem with the known chromatic numbers of specific subgraph structures (such as complete graphs and odd wheels) narrows the range of possible values of the chromatic number of a graph, in practice the range between the maximum degree and the clique size of a graph can be quite large. The results we have discussed so far are simply tools to aid in further examination of a graph's structure.

6.2.1 Coloring Strategies

The bounds above provide starting points for determining the range in which to search for a proper k-coloring of a graph. The process for finding a minimum coloring is not trivial, though we will discuss some strategies for determining the chromatic number of a graph. We will see variations of graph coloring in Section 6.4.

In our discussion of Brooks' Theorem, we noted that if every neighbor of a vertex has a different color, then one additional color would be needed for that vertex. This implies that large degree vertices are more likely to increase the value for the chromatic number of a graph and thus should be assigned a color earlier rather than later in the process. In addition, it is better to look for locations in which colors are forced rather than chosen; that is, once an initial vertex is given color 1, look for cliques within the graph containing that vertex as these have very clear restrictions on assigning future colors.

Example 6.3 Every year on Christmas Eve, the Petrie family compete in a friendly game of Trivial Pursuit. Unfortunately, due to longstanding disagreements and the outcome of previous years' games, some family members are not allowed on the same team. The table below lists the ten family members competing in this year's Trivial Pursuit game, where an entry of N in the table indicates people who are incompatible. Model the information as a graph and find the minimum number of teams needed to keep the peace this Christmas.

	Betty	Carl	Dan	Edith	Frank	Henry	Judy	Marie	Nell	Pete
Betty	.	.	N	.	.	.	.	N	N	N
Carl	.	.	N	N	.	.	.	N	.	.
Dan	N	N	.	N	.	.	.	N	N	N
Edith	.	N	N	.	.	.	.	N	.	.
Frank	.	.	.	.	.	N	N	.	.	N
Henry	.	.	.	.	N	.	.	.	.	N
Judy	.	.	.	.	.	N	.	.	N	N
Marie	N	N	N	N	.	.	.	.	N	N
Nell	N	.	N	.	.	.	N	N	.	N
Pete	N	.	N	.	N	N	N	N	N	.

Solution: Each person will be represented by a vertex in the graph and an edge indicates two people who are incompatible. Colors will be assigned to the vertices, where each color represents a Trivial Pursuit team.

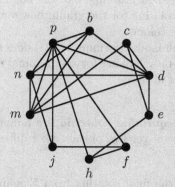

At our initial step, we want to find a vertex of highest degree (p) and give it color 1. Once p has been assigned a color, we look at its neighbors with high degree as well, namely d (degree 6), m and n (both of degree 5). These four vertices are also all adjacent to each other (forming a K_4 shown in bold below on the left) and so must use three additional colors.

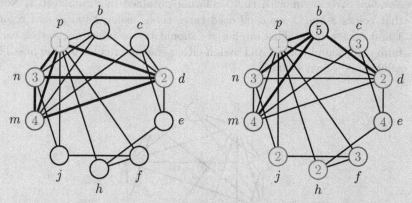

Finally, b has the next highest degree (4) and is also adjacent to all the previously colored vertices (forming a K_5) and so a fifth color is needed. The remaining vertices all have degree 3 and can be colored without introducing any new colors. One possible solution is shown above on the right. This solution translates into the following teams:

Team	Members		
1	Pete		
2	Dan	Henry	Judy
3	Carl	Frank	Nell
4	Edith	Marie	
5	Betty		

The coloring obtained in Example 6.3 was not unique. There are many ways to find a proper coloring for the graph; however, every proper coloring would need at least five colors.

In terms of the graph model (forming teams) does the solution above seem fair? Often we are not only looking for the minimal k-coloring, but also one that adds in a notion of fairness.

Definition 6.8 An *equitable coloring* is a minimal proper coloring of G so that the number of vertices of each color differs by at most one.

By this definition, the final coloring from Example 6.3 is not equitable. Note that not all graphs have equitable coloring using exactly $\chi(G)$ colors.

Example 6.4 Find an equitable coloring for the graph from Example 6.3.

Solution: We begin with the 5-coloring obtained in Example 6.3. Note that colors 2 and 3 are each used three times, color 4 twice, and colors 1 and 5 each once. This implies we should try to move one vertex each from color 2 and color 3 and assign either color 1 or color 5. One possible solution is shown below.

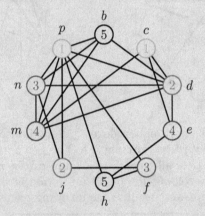

A common strategy for coloring is to begin by finding all vertices that can be given color 1, and once that is done find all the vertices that can be given color 2, and so on. The problem with this strategy is that you may choose to give vertex x color 1 which can necessitate the addition of a new color for vertex y when if x was given color 2, then y could be colored using one of the previously used colors. For example, the first coloring of graph G_1 on page 276, reproduced below, used independent sets to find a proper coloring, with vertices c, g, h, i given color 1, followed by a, f, k given color 2, etc. However, as we saw there was a proper coloring using only 3 colors.

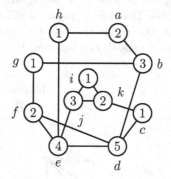

A better strategy is to focus on locations that *force* specific colors to be used rather than *choose* which color to use. Coloring G_1 in this way, we would start with i, j, k getting colors 1, 2, and 3 since these vertices form K_3. Moving outward we can give e color 1, requiring d and f to use colors 2 and 3. We continue with the rest of the vertices to obtain an optimal coloring using 3 colors.

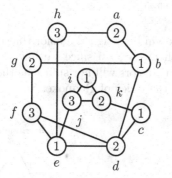

Below is a summary of the coloring strategies we have discussed so far.

Basic Coloring Strategies

- Begin with vertices of high degree.

- Look for locations where colors are forced (cliques, wheels, odd cycles) rather than chosen.

- When these strategies have been exhausted, color the remaining vertices while trying to avoid using any additional colors.

Example 6.5 Due to the nature of radio signals, two stations can use the same frequency if they are at least 70 miles apart. An edge in the graph below indicates two cities that are at most 70 miles apart, necessitating different radio stations. Determine the fewest number of frequencies need for each city shown below (not drawn to scale) to have its own municipal radio station.

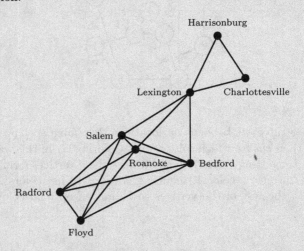

Solution: Each vertex will be assigned a color that corresponds to a radio frequency. This graph has $\chi = 5$ since we have a 5-coloring, as shown below, and fewer than 5 colors will not suffice as there is a K_5 among the vertices representing the cities of Roanoke, Salem, Bedford, Floyd, and Radford.

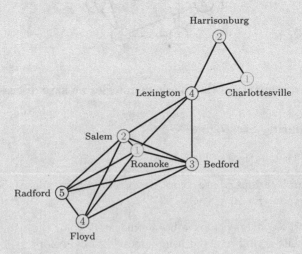

6.2.2 General Results

Up to this point, we have mainly viewed graph coloring from an applied perspective, that is we have focused on using graph coloring to model a specific problem and how to find an optimal vertex coloring. However, coloring provides a wealth of information about the underlying structure of the graph. The remainder of this section will introduce some more theoretical results on graph coloring.

First we begin with a basic counting argument relating the number of edges of a graph with its chromatic number. In essence we can create an upper bound based on the number of edges in a graph rather than the maximum degree.

Proposition 6.9 Let G be a graph with m edges. Then

$$\chi(G) \le \frac{1}{2} + \sqrt{2m + \frac{1}{4}}$$

Proof: Assume $\chi(G) = k$. First note that there must be at least one edge between color classes since otherwise two color classes without any edges between them could have been given the same color. Now, if we viewed each color class as a vertex and represented any edge between color classes as a singular edge in this new graph, we would obtain the complete graph K_k.

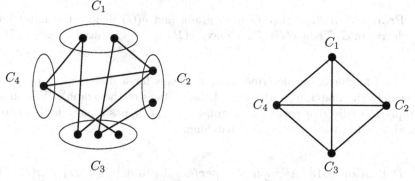

Thus G must have at least as many edges as the complete graph K_k, that is $m \ge \dfrac{k(k-1)}{2}$. We can rewrite this as $2m \ge k(k-1) = k^2 - k$. Completing the square gives us $2m + \frac{1}{4} \ge k^2 - k - \frac{1}{4} = (k - \frac{1}{2})^2$. Thus $k \le \frac{1}{2} + \sqrt{2m + \frac{1}{4}}$.

The next two results find upper bounds on the chromatic number based on structures within the graph, the first based on the length of the longest path and the second using induced subgraphs. The proof for both of these appear in the Exercises.

Proposition 6.10 Let G be a graph and $l(G)$ be the length of the longest path in G. Then $\chi(G) \leq 1 + l(G)$.

The next result relies on the notion of an induced subgraph, which was defined on page 6, but we restate it below.

Definition 6.11 Given a graph $G = (V, E)$, an *induced subgraph* is a subgraph $G[V']$ where $V' \subseteq V$ and every available edge from G between the vertices in V' is included.

Another way of thinking of an induced subgraph is that we remove all edges from the graph that do not have both endpoints in the vertex set V'.

The main reason we need induced subgraphs for coloring problems is that if we took any subgraph and colored it, we may be missing edges that would indicate two vertices need different colors in the larger graph. In this regard, if we think of starting from a smaller portion of a graph and moving outward to the entire graph, we would want to start with the subgraph induced from a small subset of the vertices.

Proposition 6.12 Let G be a graph and $\delta(G)$ denote the minimum degree of G. Then $\chi(G) \leq 1 + \max_H \delta(H)$ for any induced subgraph H.

In Chapter 5, we describe what it means for a matching to be perfect (namely the matching saturates all the vertices of the graph). We can also apply the adjective perfect to a graph, though it has almost nothing to do with the definition of a perfect matching.

Definition 6.13 A graph G is *perfect* if and only if $\chi(H) = \omega(H)$ for all induced subgraphs H.

It may be tempting to simplify this definition to only consider how the chromatic number and clique size of the entire graph are related. But we must resist this temptation and continually think about how this equality must hold for *all* induced subgraphs.

Before we delve too much into the theory of perfect graphs, we note that even cycles are perfect but not odd cycles of length greater than 3. First note that $\omega(C_n) = 2$ for all values of $n \geq 4$. However, since G is an induced subgraph of itself, and whenever n is odd that $\chi(C_n) = 3$, we know that any odd cycle of length at least 5 cannot be perfect. For the even cycles, whenever we consider a proper induced subgraph, we will either have an edge or a set of independent vertices. In both cases, the induced subgraph will satisfy $\chi(H) = \omega(H)$. Some sample induced subgraphs are shown for C_4 below.

Example 6.6 Determine if either of the two graphs below are perfect.

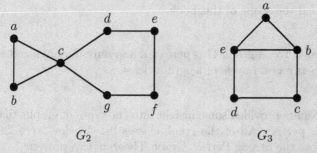

G_2 G_3

Solution: Without much work we can see that both graphs above satisfy $\chi(G) = \omega(G)$. However, if we look at the subgraph H induced by $\{c, d, e, f, g\}$ in G_2 we see that H is just C_5 and so $\chi(H) = 3$ even though $\omega(H) = 2$. Thus G_2 is not perfect.

However, G_3 is in fact perfect. If an induced subgraph H contains $\{a, b, e\}$, then $\omega(H) = 3 = \chi(H)$; otherwise one of $\{a, b, e\}$ will not be in H and so $\omega(H) \leq 2$ and without much difficulty we can show $\omega(H) = \chi(H)$. A few illustrative induced subgraphs are shown below.

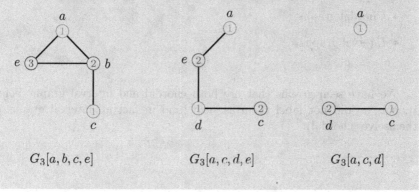

$G_3[a, b, c, e]$ $\qquad$ $G_3[a, c, d, e]$ $\qquad$ $G_3[a, c, d]$

Perfect graphs were introduced in a 1961 paper by Claude Berge (whose name should be familiar from Chapter 5) [5]. In this paper, Berge proposed two conjectures about perfect graphs, the first of which is given below and was proven in 1972 by Lovász, and is commonly called the Perfect Graph Theorem [62].

Theorem 6.14 A graph G is perfect if and only if $\overline{G}$ is perfect.

As we have already seen, having an induced graph isomorphic to C_5 would disqualify a graph from being perfect. In fact, this is the basis for the other major conjecture from Berge's paper given below. It remained unsolved for an additional thirty years beyond the Perfect Graph Theorem and is commonly called the Strong Perfect Graph Theorem. The proof was published in 2006 by Maria Chudnovsky, Neil Robertson, Paul Seymour and Robin Thomas [14] and is beyond the scope of this book.

Theorem 6.15 A graph G is perfect if and only if no induced subgraph of G or $\overline{G}$ is an odd cycle of length at least 5.

These results provided some insight into the types of graphs that could be classified as perfect. All of the graph classes listed below were known to be perfect before the Strong Perfect Graph Theorem was proven.

Perfect Graphs

The following classes of graphs are known to be perfect:

- Trees

- Bipartite graphs

- Chordal graphs

- Interval graphs

We have seen graphs that are both chordal and interval graphs before, though we did not label them as such (and in fact all interval graphs are themselves chordal).

Definition 6.16 A graph G is **chordal** if any cycle of length four or larger has an edge (called a chord) between two nonconsecutive vertices of the cycle.

Looking back at some of the graphs we have already seen in this chapter, we can identify the graph from Example 6.1 as chordal whereas G_1 from page 276 is not since there is a 5-cycle without a chord (namely the cycle created by $a\,b\,d\,e\,h\,a$).

Definition 6.17 A graph G is an **interval graph** if every vertex can be represented as a finite interval and two vertices are adjacent whenever the corresponding intervals overlap; that is, for every vertex x there exists an interval I_x and xy is an edge in G if $I_x \cap I_y \neq \emptyset$.

The adjective of chordal has more of a theoretical importance, whereas interval graphs arise from many applications.

Example 6.7 Five student groups are meeting on Saturday, with varying time requirements. The staff at the Campus Center need to determine how to place the groups into rooms while using the fewest rooms possible. The times required for these groups is shown in the table below. Model this as a graph and determine the minimum number of rooms needed.

Student Group	Meeting Time
Agora	13:00–15:30
Counterpoint	14:00–16:30
Spectrum	9:30–14:30
Tupelos	11:00–12:00
Upstage	11:15–15:00

Solution: First we display the information in terms of the intervals. Although this step is not necessary, sometimes the visual aids in determining which vertices are adjacent.

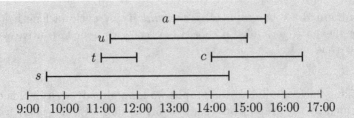

Below is the graph where each vertex represents a student group and two vertices are adjacent if their corresponding intervals overlap.

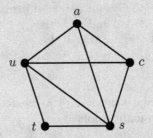

A proper coloring of this graph is shown below. Note that four colors are required since there is a K_4 subgraph with $a, c, s,$ and u.

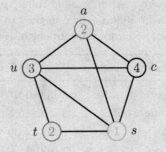

It should be noted that in most applications of interval graphs, you are given the intervals and must form the graph. A much harder problem is determining if an interval representation of a graph exists and then finding one. But once an interval representation is known, coloring interval graphs is quite easy by simply coloring the vertices based on when the corresponding interval is first seen as we sweep from left to right.

Example 6.8 Eight meetings must occur during a conference this upcoming weekend, as noted below. Determine the minimum number of rooms that must be reserved.

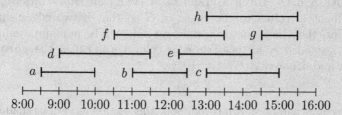

Solution: Each meeting is represented by a vertex, with an edge between meetings that overlap and colors indicating the room in which a meeting will occur. If we color the vertices according to their start time (so in the order a, d, f, b, e, c, h, g), we get the coloring below.

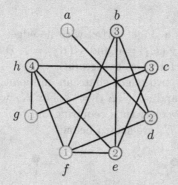

Note that four meeting rooms are needed since there is a point at which four meetings are all in session, which is demonstrated by the K_4 among the vertices $c, e, f,$ and h.

6.3 Edge Coloring

Up to this point we have discussed coloring the vertices of a graph. This section focuses on a different aspect of graph coloring where instead of assigning colors to the vertices of a graph, we will instead assign colors to the edges of a graph. Such colorings are called *edge-colorings* and have their own set of definitions and notations, many of which are analogous to those for vertex colorings from above. Similar to our previous study of vertex colorings, we will only consider simple graphs, that is graphs without multi-edges. (Note that a graph with a loop cannot be edge-colored).

Definition 6.18 Given a graph $G = (V, E)$ an ***edge-coloring*** is an assignment of colors to the edges of G so that if two edges share an endpoint, then they are given different colors. The minimum number of colors needed over all possible edge-colorings is called the ***chromatic index*** and denoted $\chi'(G)$.

Edge colors will be shown throughout this section as various shades of gray and line styles.

Example 6.9 Recall that the chromatic number for any complete graph is equal to the number of vertices. Find the chromatic index for K_n for all n up to 6.

Solution: Since K_1 is a single vertex with no edges and K_2 consists of a single edge, we have $\chi'(K_1) = 0$ and $\chi'(K_2) = 1$. Due to their simplicity, a drawing is omitted for these two graphs.

For K_3 since any two edges share an endpoint, we know each edge needs its own color and so $\chi'(K_3) = 3$. For K_4 we can color opposite edges with the same color, thus requiring only 3 colors. Optimal edge-colorings for K_3 and K_4 are shown below.

Edge-coloring K_5 and K_6 is not quite so obvious as those from above. Since no two adjacent edges can be given the same color, we know every edge out of a vertex must be given different colors. Since every vertex in K_5 has degree 4, we know at least 4 colors will be required. Start by using 4 colors out of one of the vertices of the K_5. As shown in the next graph on left, we started at a. Moving to the edges incident to b, we attempt to use our pool of 4 previously used colors; however, one of these is unavailable since it has already been used on the ab edge. A possible coloring of the edges incident to b is given in the next graph on the right.

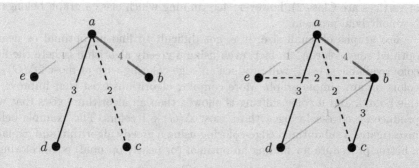

At this point c is left with two incident edges to color but only one color available, namely the one used on edge ab, thus requiring a fifth color to be used. A proper edge-coloring of K_5 is shown below on the left. A similar procedure can be used to find a coloring of K_6 using only 5 colors, as shown below on the right.

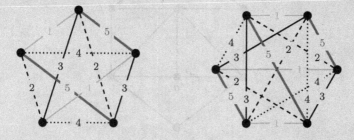

In general, $\chi'(K_n) = n-1$ when n is even and $\chi'(K_n) = n$ when n is odd.

From the example above we conclude that if a vertex x has degree k, then the entire graph will need at least k colors since each of the edges incident to x will need a different color. Thus the chromatic index must be at least the maximum degree, $\Delta(G)$, of the graph. But notice that for odd values of n, that K_n required one more color than the maximum degree (for example, $\chi'(K_5) = 5$ and $\Delta(K_5) = 4$). In fact, any graph will either require $\Delta(G)$ or $\Delta(G)+1$ colors to color its edges. This is a much tighter bound than we were able to find for the chromatic number of a graph and was proven in 1964 by the Ukrainian mathematician, Vadim Vizing.

Theorem 6.19 (Vizing's Theorem) $\Delta(G) \leq \chi'(G) \leq \Delta(G)+1$ for all simple graphs G.

Graphs are referred to as Class 1 if $\chi'(G) = \Delta(G)$ and Class 2 if $\chi'(G) = \Delta(G) + 1$. Some graph types are known to be in each class (for example, bipartite graphs are Class 1 and regular graphs with an odd number of vertices are Class 2); however, determining which class a graph belongs to is a nontrivial problem.

For graphs of small size, it is not difficult to find an optimal or nearly optimal edge-coloring. In fact, even using a greedy algorithm (where the first color available is used) will produce an edge-coloring with at most $2\Delta(G) - 1$ colors on any simple graph. More complex algorithms exist that improve on this bound, and if color shifting is allowed then an algorithm exists that will produce an edge-coloring with at most $\Delta(G) + 1$ colors. The example below investigates a suboptimal edge-coloring using a greedy algorithm and explains a better procedure for finding an optimal (or nearly optimal) edge-coloring.

Example 6.10 Consider the graph G_4 below and color the edges in the order $ac, fg, de, ef, bc, cd, dg, af, bd, bg, bf, ab$ using a greedy algorithm.

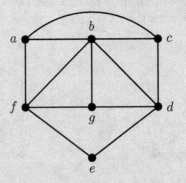

Step 1: Since the first three edges ac, fg, and de are not adjacent, we give each of them the first color.

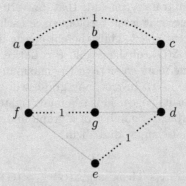

Step 2: Now since ef is adjacent to a previously colored edge, we need a second color for it. Moreover, bc must also use a second color and since these are not adjacent they can have the same color.

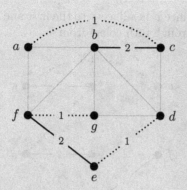

Step 3: Edge *cd* is adjacent to edges using the first two colors, so a third color is needed. Edge *dg* can use the second color ,and edge *af* must use the third color.

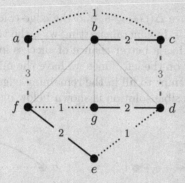

Step 4: Edge *bd* needs a fourth color since it is adjacent to edges using each of the previous three colors. However, *bg* can be given the third color.

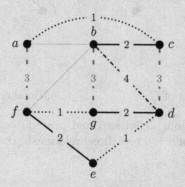

Step 5: Edge *bf* needs a fifth color since it is adjacent to the first three colors through *f* and adjacent to the fourth color through *b*. Moreover, *ab*

needs a sixth color since it is adjacent to the first and third colors through *a* and the second through fifth colors from *b*.

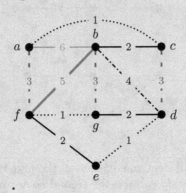

In the example above $\Delta(G_4) = 5$, and the edge-coloring above uses 6 colors; however $\chi'(G_4) = 5$. In general, starting with the vertex of highest degree and coloring its edges has a better chance of success in avoiding unnecessary colors, as shown below on the left. Once we have the minimum number of colors established, we attempt to fill in the remaining edges without introducing an extra color; one possible solution is shown below on the right.

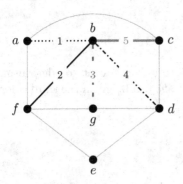   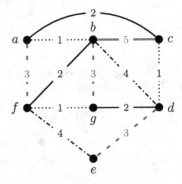

initial coloring of G_4 optimal edge-coloring of G_4

Beyond the analogous definitions and procedures between edge-coloring and vertex-coloring, there is a very direct relationship between the two through the use of a *line graph*.

Definition 6.20 Given a graph $G = (V, E)$, the **line graph** $L(G) = (V', E')$ is the graph formed from G where each vertex x' in $L(G)$ represents the edge x' from G and $x'y'$ is an edge of $L(G)$ if the edges x' and y' share an endpoint in G.

Below is the graph G_4 from Example 6.10 and its line graph. Notice that the vertex e_1 in $L(G_4)$ is adjacent to e_2 and e_4 through the vertex a in G_4 and e_1 is adjacent to e_3 and e_8 through the vertex c in G_4.

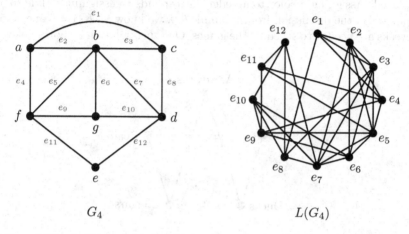

$$G_4 \qquad\qquad\qquad L(G_4)$$

From this definition, it should be clear how edge-coloring and vertex-coloring are related. In particular, if edge e_1 is given the color blue in G, this would correspond to coloring vertex e_1 in $L(G)$ with blue. This correspondence provides the following result.

Theorem 6.21 Given a graph G with line graph $L(G)$, we have $\chi'(G) = \chi(L(G))$.

The result above shows we can find an edge-coloring of any graph by simply vertex-coloring its line graph. However, as we saw above, vertex-coloring is in itself not an easy problem. However, other results on line graphs provide some interest, namely if G is eulerian then $L(G)$ is hamiltonian!

Applications of edge-coloring abound, in particular scheduling independent tasks onto machines and communicating data through a fiber-optic network. The example below relates edge-coloring to Section 1.1 and how to schedule games between teams in a round-robin tournament.

Example 6.11 The five teams from Section 1.1 (Aardvarks, Bears, Cougars, Ducks, and Eagles) need to determine the game schedule for the next year. If each team plays each of the other teams exactly once, determine a schedule where no team plays more than one game on a given weekend.

Solution: Represent each team by a vertex and a game to be played as an edge between the two teams. Then exactly one edge exists between every pair of vertices and so K_5 models the system of games that must be played. Assigning a color to an edge corresponds to assigning a time to a game. By the discussion from Example 6.9, we know $\chi'(K_5) = 5$ and so 5 weeks are needed to schedule the games. One such solution is shown below.

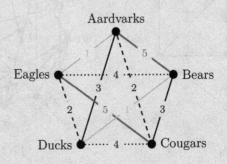

Week	Games	
1	Aardvarks vs. Bears	Cougars vs. Eagles
2	Aardvarks vs. Cougars	Ducks vs. Eagles
3	Aardvarks vs. Ducks	Bears vs. Cougars
4	Aardvarks vs. Eagles	Bears vs. Ducks
5	Bears vs. Eagles	Cougars vs. Ducks

Although the example above is fairly easy to solve as the graph model is a complete graph, the same procedure can be extended to a larger number of teams where each team only plays a subset of all teams in the league. In particular, edge-coloring can be used to determine team schedules in the National Football League!

6.3.1 1-Factorizations Revisited

Recall that in Section 5.4 we discussed a variation on matching in which the edges of a graph would be partitioned into disjoint 1-factors, called a 1-factorization. If we give each edge of a 1-factor the same color, then a 1-

factorization could be represented as an edge-coloring of a graph. However, not every edge-coloring can represent a 1-factorization, since a 1-factor must also span the graph, requiring every vertex to be incident to an edge of every color. Thus we need a graph to be k-regular and have chromatic index k to have a 1-factorization. The graph G_5 below is 4-regular with chromatic index 4 and contains a 1-factorization.

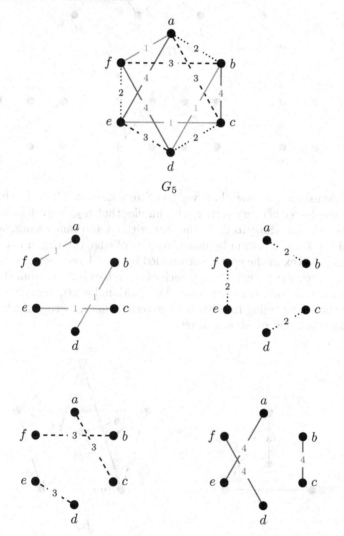

G_5

Proposition 5.22 states that every k-regular bipartite graph has a 1-factorization. The proof is not constructive, but does illustrate that we could find a 1-factorization of a bipartite by simply recursively removing perfect matchings from the graph, as shown in graph G_6.

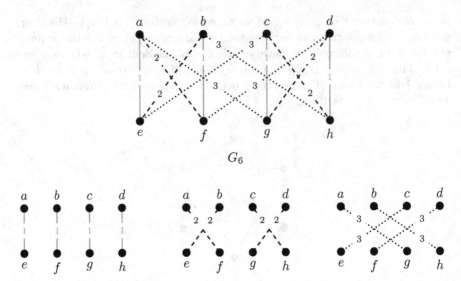

G_6

In Example 6.9, we saw that K_{2n} has chromatic index $2n-1$, which is the same as the degree of every vertex. This implies that K_{2n} has a 1-factorization, but does not show how to find one. But without too much work, we can in fact find the 1-factorization by using the idea of edge coloring. First, we draw K_{2n} with a vertex in the center surrounded by $2n-1$ vertices arranged as the corners of a regular polygon. Then each of the edges coming from that center vertex will be given a different color. All additional edges are given the same color as the edge coming from the center vertex that is perpendicular to it. A few small examples are shown below.

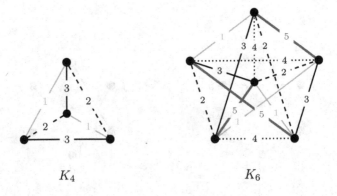

K_4 K_6

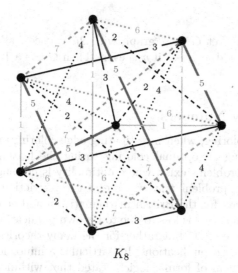

K_8

Recall that K_{2n-1} also has chromatic index $2n-1$ even though the graph is $2n-2$ regular (so K_5 is 4 regular with $\chi' = 5$). This means that complete graphs on an odd number of vertices do not have 1-factorizations! Additionally, there are many regular graphs that do not have 1-factorizations, the most famous of which is the Petersen graph shown below. Note that this graph is cubic (3-regular) but has chromatic index 4, meaning it cannot have a 1-factorization!

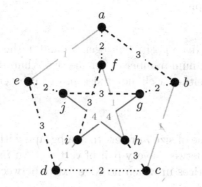

edge-coloring of Petersen graph

We have now some knowledge about complete graphs and bipartite graphs that have 1-factorizations, but many more types of graphs remain that we have yet to identify as having 1-factorizations. In fact, Béla Csaba, Daniela Kühn, Allan Lo, Deryk Osthus and Andrew Treglown recently proved the following theorem relating the degree to the number of vertices in a regular graph that guarantees a 1-factorization, the proof of which is beyond the scope of this text [18].

Theorem 6.22 Let G be a graph with $2n$ vertices, all of which have degree n if n is odd or $n - 1$ if n is even. Then G has a 1-factorization.

6.3.2   Ramsey Numbers

As with vertex-coloring, when investigating edge-colorings we are often concerned with finding an optimal coloring (using the least number of colors possible). Other problems exist, where optimality is no longer the goal. One such edge-coloring problem relaxes some of the restrictions on coloring the edges, and is named for the British mathematician and economist, Frank P. Ramsey. Ramsey's legacy is less so due to his own publications, mainly due to his death at the age of 26, but rather for the many theories and results that arose from his limited publications. In particular a minor lemma in his 1928 paper "On a problem of formal logic" stated that within any system there exists some underlying order [71]. Although a simple concept, it birthed an area of mathematics now known as Ramsey Theory.

Ramsey Theory can be described in different forms, so we will naturally use the graph theoretic version. In particular, we will discuss Ramsey numbers as they relate to coloring the edges of a graph. Unlike our edge-colorings above in which no two edges can be given the same color if they have a common endpoint, here we will be concerned with specific monochromatic structures within the larger graph.

Definition 6.23 Given positive integers m and n, the **Ramsey number** $R(m, n)$ is the minimum number of vertices r so that all simple graphs on r vertices contain either a clique of size m or an independent set of size n.

Recall that a clique of size m refers to a subgraph with m vertices in which there exists an edge between every pair of vertices. An *independent set* of size n is a group of n vertices in which no edge exists between any two of these n vertices.

To get a better handle on this technical definition, Ramsey numbers are often described in terms of guests at a party. For example, if you wanted to find $R(3, 2)$, then you would be asking how many guests must be invited so that at least 3 people all know each other or at least 2 people do not know each other. Try it!

Ramsey numbers can be viewed as coloring the edges of a complete graph using two colors, say red and blue, so that either a red clique of size m exists or a blue clique of size n exists. You can view the blue clique as the edges that would not exist in the graph, thus making their endpoints an independent set of vertices. Proving $R(m,n) = r$ requires two steps: first, we find an edge-coloring of K_{r-1} without a red m-clique and without a blue n-clique; second, we must show that any edge-coloring of K_r will have either a red m-clique or a blue n-clique. These steps should feel familiar, as they mirror our discussion of the chromatic number of a graph. We will indicate the two types of lines as bold or dashed.

Example 6.12 Determine $R(3,3)$.

Solution: First note that we are searching for a 3-clique of either color, also known as a monochromatic triangle. However, $R(3,3) > 5$ as the edge-coloring of K_5 below does not contain a monochromatic triangle.

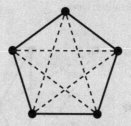

Next, consider K_6. Since each vertex has degree 5 and we are coloring the edges using only two colors, we know every vertex must have at least 3 adjacent edges of the same color. Suppose a has three adjacent dashed edges, as shown below. All other edges are shown in gray to indicate we do not care (yet) which color they have.

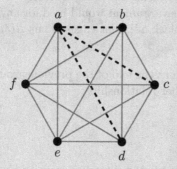

If any one of the edges between b, c, and d is dashed, then a dashed triangle exists using the edges back to a. One possibility is shown in the next graph.

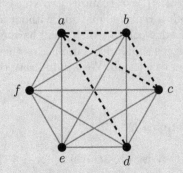

Otherwise, none of the edges between b, c and d are dashed, creating a bold triangle among these vertices.

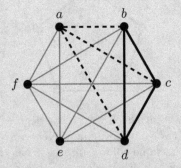

Although the argument above used vertex a with three adjacent dashed edges, a similar argument would hold for any vertex of K_6 and for a vertex with three black adjacent edges. Thus $R(3,3) = 6$.

Using the discussion at the end of the previous example for inspiration, we note two Ramsey number relationships:

- $R(m,n) = R(n,m)$

- $R(2,n) = n$

The first shows that we can interchange m and n without impacting the Ramsey number, as can be seen by simply switching the color on every edge from blue to red and vice versa. The explanation for the second relationship appears in Exercise 6.17.

Although the solution for $R(3,3)$ was not too difficult to determine, increasing m and n both by one value greatly increases the complexity of a solution. In fact, $R(4,4) = 18$ was proven in 1955 and yet $R(5,5)$ is still unknown. The following table lists some known values and bounds for small values of m and n at the time of publication. A single value in a column indicates the exact value is known; otherwise, the two values given are the upper and lower bounds as stated. Further discussion and results can be found at [74].

m	3	4	5		6		7	
n			lower	upper	lower	upper	lower	upper
3	6	9	14		18		23	
4		18	25		36	41	49	61
5			43	48	58	87	80	143
6					102	165	115	298
7							205	540

6.4 Coloring Variations

The previous sections should show just how varied and deep of a field is graph coloring. This section will highlight additional variations of graph coloring, specifically vertex colorings. Within each of these we will see applications of graph coloring that can inform why this new version of coloring is worthy of study.

6.4.1 On-line Coloring

On-line coloring differs from a general coloring in that the vertices are examined one at a time (hence they are seen in a linear manner, or "on a line"). Often, we are restricted to situations where portions of the graph are visible at different times and so a vertex must be assigned a color without all the information available. The notion of an on-line coloring relies on a specific type of subgraph, called an *induced subgraph*, which was defined on page 6 and discussed in our section on graph coloring results (see page 289). Recall that we need induced subgraphs for coloring problems since if we only took any subgraph and colored it, we may be missing edges that would indicate two vertices need different colors in the larger graph. On-line coloring algorithms require a vertex to be colored based only upon the induced subgraph containing that vertex and the previously colored vertices.

Definition 6.24 Consider a graph G with the vertices ordered as $x_1 \prec x_2 \prec \cdots \prec x_n$. An *on-line algorithm* colors the vertices one at a time where the color for x_i depends on the induced subgraph $G[x_1, \ldots, x_i]$

which consists of the vertices up to and including x_i. The maximum number of colors a specific algorithm $\mathcal{A}$ uses on any possible ordering of the vertices is denoted $\chi_{\mathcal{A}}(G)$.

Many different on-line algorithms exist, some of which can be quite complex. Mathematicians are often interested in finding an on-line algorithm that works well on a specific type of graph, or in showing how the underlying structure of specific types of graphs limits the performance of any on-line algorithm. We will focus on a greedy algorithm called *First-Fit* that uses the first available color for a new vertex.

First-Fit Coloring Algorithm

Input: Graph G with vertices ordered as $x_1 \prec x_2 \prec \cdots \prec x_n$.

Steps:

1. Assign x_1 color 1.

2. Assign x_2 color 1 if x_1 and x_2 are not adjacent; otherwise, assign x_2 color 2.

3. For all future vertices, assign x_i the least number color available to x_i in $G[x_1, \ldots, x_i]$; that is, give x_i the first color not used by any neighbor of x_i that has already been colored.

Output: Coloring of G.

One of the benefits of First-Fit is the ease with which it is applied. When a new vertex is encountered, we simply need to examine its neighbors that have already been colored and give the new vertex the least color available.

Example 6.13 Apply the First-Fit Algorithm to the graph from Example 6.3 if the vertices are ordered alphabetically.

Solution: To emphasize that only some of the graph is available at each step of the algorithm, only the edges to previously considered vertices will be drawn.

$$b$$
$$①$$

Step 1: Color b with 1.

Step 2: Color c with 1 since b and c are not adjacent.

Step 3: Color d with 2 since d is adjacent to a vertex of color 1.

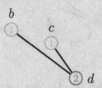

Step 4: Color e with 3 since e is adjacent to a vertex of color 1 (c) and a vertex of color 2 (d).

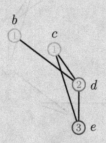

Step 5: Color f with 1 since f is not adjacent to any previous vertices.

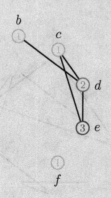

Step 6: Color h with 2 since h is adjacent to a vertex of color 1 (f).

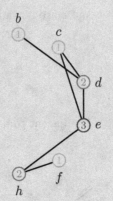

Step 7: Color j with 2 since j is adjacent to a vertex of color 1 (f).

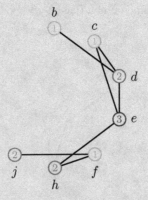

Step 8: Color m with 3 since m is adjacent to vertices of color 1 (b, c) and a vertex of color 2 (d).

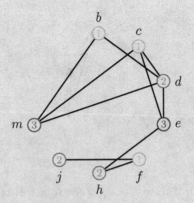

Step 9: Color n with 4 since n is adjacent to vertices of color 1, 2, and 3.

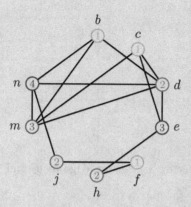

Step 10: Color p with 5 since p is adjacent to vertices of color 1, 2, 3, and 4.

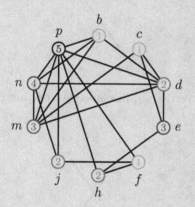

As the example above shows, First-Fit can perform quite well given the proper ordering. Unfortunately, given the right graph and wrong order of the vertices, First-Fit can perform remarkably poorly, as seen below.

Example 6.14 A vertex will be revealed one at a time, along with any edges to previously seen vertices. The First-Fit Algorithm will be applied in the order the vertices are seen.

Step 1: The first vertex is v_1. It is given color 1.

v_1

Step 2: The second vertex is v_2. Since there is no edge to v_1, it is also given color 1.

v_1

v_2

Step 3: The third vertex, v_3, has an edge to v_2 and so must be assigned color 2.

Step 4: The fourth vertex, v_4, has an edge to v_1 but not v_3 and so is also given color 2.

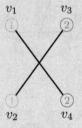

Step 5: The fifth vertex, v_5, is adjacent to a vertex of color 1 (v_2) and a vertex of color 2 (v_4). It must be assigned color 3.

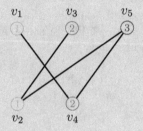

Step 6: The sixth vertex, v_6, is also adjacent to a vertex of color 1 (v_1) and a vertex of color 2 (v_3) but not a vertex of color 3 so it can also be given color 3.

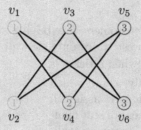

If we continue in this fashion, after $2t$ steps we will have used t colors.

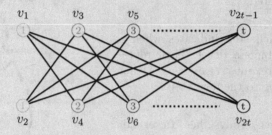

However, this graph is bipartite and every bipartite graph can be colored using 2 colors, as shown below.

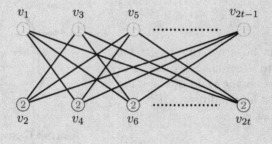

First-Fit is in fact one of the worst performers for on-line algorithms since it uses very little knowledge of how the previous vertices were treated. When evaluating a coloring algorithm's performance, we look at the worst case scenario, that is we are asking what is the most number of colors the algorithm could use on any possible ordering of the vertices, denoted $\chi_{FF}(G)$. It is possible for an on-line algorithm (even First-Fit) to provide an optimal coloring, but in many cases this does not occur.

On-line coloring algorithms perform nicely on interval graphs. In particular, as we saw above, when the interval representation is known and the vertices are ordered by starting time of their intervals, then First-Fit will produce a coloring using exactly $\chi(G)$. However, if the intervals are not ordered by their starting time or a different on-line algorithm is used, the optimal coloring may not be found.

Example 6.15 Ten customers are buying tickets for various trips along the Pacific Northwest train route shown on the right. Each person must be assigned a seat when a ticket is purchased and you only know which seats have been previously assigned. Using a random assignment of seat numbers as the information becomes available, find a way to minimize the number of seats required.

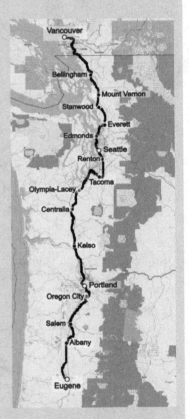

Solution: Each step indicates when a new person buys a ticket. Their representative vertex must be assigned a color before moving to the next step.

Step 1: Cathy buys her ticket for Bellingham to Edmonds. She is assigned seat 1, as shown in the graph below.

$①\ c$

Step 2: Fiona buys her ticket next for Bellingham to Renton. Since she and Cathy will be on the train at the same time, she must have a different seat. She is assigned seat 2.

Amtrak Cascades Route

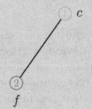

Step 3: Next, Greg buys a ticket for Vancouver to Mount Vernon. His

trip overlaps with those of both Cathy and Fiona, necessitating the use of another seat. He is assigned seat 3.

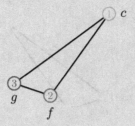

Step 4: Ben buys the next ticket for Portland to Oregon City. Since his trip does not overlap with any of the earlier purchased tickets, he can be assigned any seat. We choose seat 1.

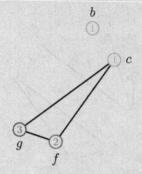

Step 5: Next Ingrid buys a ticket for Oregon City to Albany. Similar to Ben, she can also be assigned any seat. We choose seat 3.

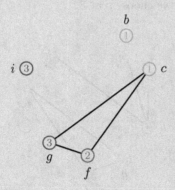

Step 6: Jessica buys the next ticket for Albany to Eugene. As in the previous two steps, she can be assigned any seat. We choose seat 2.

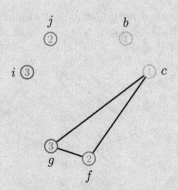

Step 7: Howard buys a ticket for Centralia to Eugene. Since his route overlaps with those of Ben, Jessica, and Ingrid, he must be assigned to a new seat, number 4.

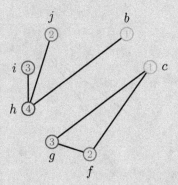

Step 8: Dana buys the next ticket for Renton to Tacoma. He can be assigned any seat, so we choose seat 3.

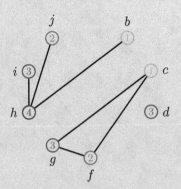

Step 9: Aiden buys a ticket for Seattle to Oregon City. Since his trip

overlaps with someone in seats 1 through 4, we need a new seat for him, number 5.

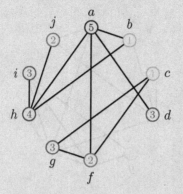

Step 10: Emily is the last person to buy a ticket for a trip from Everett to Kelso. Since her trip overlaps with someone in each of the previously assigned seats, we need a sixth seat for Emily.

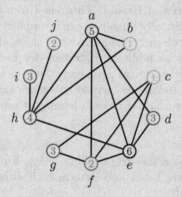

In the example above, we did not use First-Fit but rather a more random choice for an on-line algorithm. In Exercise 6.1 First-Fit is applied to this graph using the same order of the vertices, resulting in an improvement on the number of colors (or seats). However, neither of these algorithms produces an optimal coloring as the chromatic number for this graph is 3 since the clique size is 3 (for example, there is a K_3 with vertices a, b, and h), and since this is an interval graph we know $\chi(G) = \omega(G)$. One optimal coloring is shown in the next graph.

In general, First-Fit can be shown to be no worse than roughly $5\chi(G)$ on interval graphs (which means First-Fit will never use more than five times the optimal) and there exists an on-line algorithm that uses at most $3\chi(G)$ colors on interval graphs (see [57],[58]). Though this may seem quite large, recall that in Example 6.14 we had a graph with $2t$ vertices using t colors when only 2 were needed, implying a performance of $\frac{n}{4}\chi(G)$ for bipartite graphs with n vertices.

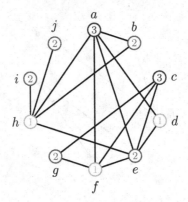

optimal coloring for Example 6.15

Even with all its limitations, First-Fit is a useful strategy when attempting to prove bounds on the chromatic number of a graph. In particular, we will use it below in our proof of Brooks' Theorem (from Section 6.2.2). On-line coloring, and more specifically First-Fit, can be applied to other graph coloring variations (even edge-coloring). One can view these algorithms from a more applied perspective, as an additional strategy for tackling a graph coloring problem, or through a theoretical perspective, where we want to think about the limiting behavior. Below we include one more example of each, though these just scratch the surface of the abundance of problems surrounding on-line coloring.

Example 6.16 Color the graph below using First-Fit using the two different vertex orders listed and determine if either finds the optimal coloring for the graph.

(a) $a \prec b \prec c \prec d \prec e \prec f \prec g \prec h$

(b) $e \prec h \prec b \prec d \prec c \prec g \prec f \prec a$

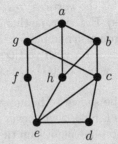

Solution: Below are the two First-Fit colorings of the graph based on the order given. Only the vertex order from (b) gives the optimal coloring since $\omega(G) = 3$.

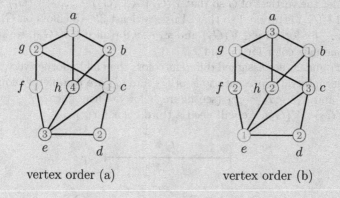

vertex order (a)   vertex order (b)

The example below demonstrates the power of on-line coloring from a theoretical viewpoint and was published by H.A. Kierstead and William Trotter in 1981. The proof will make use of an interval representation of a graph and a specific ordering of the vertices.

Example 6.17 Prove there exists an interval graph G using at least $3\omega(G) - 2$ colors.

Solution: We will show that for all online algorithms $\mathcal{A}$ there exists some interval graph G and an ordering of its vertices so that $\chi_{\mathcal{A}}(G) \geq 3\omega(G) - 2$. We will argue by induction on $k = \omega(G)$.

If $k = 1$ then G has vertices but no edges and so we need only one color to color the vertices of G. Thus $\chi(G) = 1 = 3 * 1 - 1 = 3\omega(G) - 2$.

Assume $k \geq 2$ and for all $k' < k$ if $\mathcal{A}'$ is an online algorithm then there exists an interval graph G' with $\omega(G') = k'$ and an ordering of the vertices of G' such that if $\mathcal{A}'$ properly colors G' then it uses at least $3k' - 2$ colors. We will show that there exists a graph G with clique-size k and an ordering of $V(G)$ so that no algorithm can color G with $3k - 3$ colors.

Fix $\mathcal{A}$. By the induction hypothesis there exists a graph H with components $G_1, \ldots, G_t$, with $t = 3\binom{3k-3}{3k-5} + 1$ and $\omega(G_i) = k - 1$ such that when $\mathcal{A}$ colors H it uses at least $3k - 5$ colors on G_i, for each i. Then by the Pigeonhole Principle there exist at least 4 of the G_i on which $\mathcal{A}$ uses the same set of colors, say for $i \in \{1, 2, 3, 4\}$. We will form a new graph G from H by adding at most four new vertices so that $\mathcal{A}$ will use at least three extra colors on these vertices.

First, add two vertices x_1 and x_4 so that the interval representation of G_1 is completely contained inside I_{x_1}, G_4 is completely contained inside I_{x_4}, and x_1 and x_4 are not adjacent vertices.

Order the vertices of G so that $V(G_1) \prec V(G_2) \prec \ldots \prec V(G_t) \prec x_1 \prec x_4$ and $V(G_i)$ is ordered so that $\mathcal{A}$ uses at least $3k-5$ colors on G_i. Then since $x_1 \sim v$ for all $v \in V(G_1)$ and $x_4 \sim v$ for all $v \in V(G_4)$, we need at least one new color for x_1 and x_4.

If x_1 and x_4 are assigned different colors then add another vertex x_2 so that the interval representations of G_2 and G_3 are completely contained in I_{x_2} and $x_1 \sim x_2 \sim x_4$ (see figure below). Then since $x_2 \sim v$ for all $v \in V(G_2) \cup V(G_3)$, we will need a third color for x_2.

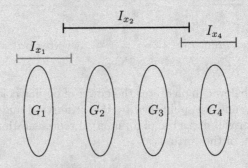

Otherwise, if x_1 and x_4 are assigned the same color, add two new vertices x_2 and x_3 so that the interval representation of G_2 is completely contained in I_{x_2}, G_3 is completely contained in I_{x_3}, $x_1 \sim x_2 \sim x_3 \sim x_4$, and $x_1 \prec x_4 \prec x_2 \prec x_3$ in the ordering of $V(G)$ (see figure below). Since $x_2 \sim x_1$ and $x_2 \sim v$ for all $v \in V(G_2)$, we will need a new color for x_2. Moreover since $x_3 \sim x_2, x_4$ and $x_3 \sim v$ for all $v \in V(G_3)$ we will need a third new color for x_3.

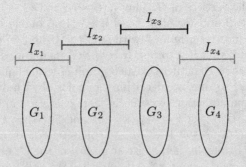

In either case, we have added at most four vertices to G', creating a graph G where $\mathcal{A}$ will need at least $3k-2$ colors.

In [58], Kierstead and Trotter also described an on-line algorithm that would use at most $3\omega(G) - 2$ colors on any interval graph. Like the example above, the proof uses induction on $\omega(G) = k$, where if $k = 1$ then the First-Fit Algorithm is used. For $k > 1$, as a new vertex v is seen in the ordering of $V(G)$, v is put into the set B provided that $\omega(G[B \cup \{v\}]) < k$ and otherwise put into a set T. If a vertex is place into the set B we color it as the algorithm would color it with respect to $G[B]$, using at most $3(k-1) - 2 = 3k - 5$ colors based on the induction hypothesis; otherwise we color v with First-Fit with respect to $G[T]$, using disjoint sets of colors for $G[B]$ and $G[T]$. The remainder of the proof is to show that at most 3 colors are used on $G[T]$ and appears in Exercise 6.23.

6.4.2 Proof of Brooks' Theorem

The proof we present of Brooks' Theorem (Theorem 6.7, restated below) is based on that of Lovász [63] and most closely resembles those in [46] and [84]. This proof is perhaps one of the more complex one we have encountered in this text. Part of the difficulty comes from the number of different scenarios we will address separately. We will first consider if the graph is not regular, then look at its connectivity. Within each case we will be creating an order of the vertices so that each vertex has at most $\Delta(G) - 1$ neighbors preceding it.

Theorem 6.7 (Brooks' Theorem) Let G be a connected graph and Δ denote the maximum degree among all vertices in G. Then $\chi(G) \leq \Delta$ as long as G is not a complete graph or an odd cycle. If G is a complete graph or an odd cycle then $\chi(G) = \Delta + 1$.

Proof: Assume G is a connected graph that is not complete nor an odd cycle. Let $k = \Delta(G)$. If $k = 0$ or $k = 1$ then G must be complete. If $k = 2$ then G is either a path or an even cycle, both of which can be colored using 2 colors. Therefore, we may assume that $k \geq 3$.

To show that we can color G using at most k colors, we will attempt to order the vertices as $v_1 \prec v_2 \prec \cdots \prec v_n$ so that each vertex has at most $k - 1$ neighbors preceding it in the list. This will ensure that if we apply First-Fit to this ordering then we will use at most k colors. We will break up the proof into cases based first on if G is regular.

Case 1: G is not regular. Then there exists some vertex with degree at most $k - 1$. Call this vertex v_n. Let N_i be the set of vertices of distance i from v_n; that is $N_0 = \{v_n\}, N_1 = N(v_n)$ and N_2 is the set of vertices whose shortest path to v_n is of length 2, and so on.

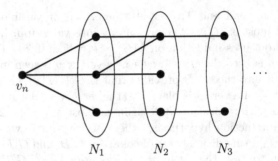

Since G is finite, there must be some value t for which N_t is the last set created. We will label the vertices in reverse order based on the N_i for which they belong; that is, v_1 up through $v_{|N_t|}$ will be from N_t, the next group of vertices from N_{t-1}, and so on until the last vertices in the list are from N_1.

$$\underbrace{v_1 \prec v_2 \prec \cdots}_{N_t} \prec \underbrace{v_i \prec v_{i+1} \prec \cdots}_{N_{t-1}} \prec \cdots \prec \underbrace{v_j \prec \ldots \prec v_{n-1}}_{N_1} \prec v_n$$

Note that every vertex (other than v_n) must have a neighbor that follows it in the list, namely the vertex along its shortest path to v_n. Thus every vertex has at most $k-1$ neighbors preceding it in the list. Thus First-Fit will use at most k colors on vertices $v_1, v_2, \ldots, v_{n-1}$ and since v_n has degree at most $k-1$, we know at least one of these colors has not been used on its neighbors and so can be used for v_n.

Case 2: G is regular and G has a cut vertex, call it x. Then every vertex has degree k and $G - x$ must have at least two components, call them $H_1, H_2, \ldots, H_t$, where $t \geq 2$. Let $G_i = H_i \cup \{x\}$, that is G_i is the graph created by putting x back into H_i along with its edges to the vertices in H_i.

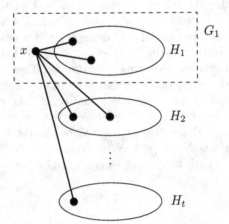

Then x must have degree at most $k-1$ in each G_i, and so we can use the method from Case 1 to properly color each G_i with at most k colors. Since we are coloring the graphs $G_1, G_2, \ldots, G_t$ independently, we can assume x gets the same color in each graph (otherwise we can permute the colors to make it so). Thus the coloring of the G_i graphs create a coloring of G using at most k colors.

Case 3: G is regular and does not contain a cut vertex. Then G must be 2-connected. Note that in any vertex ordering v_n will have k earlier neighbors. If we can ensure that two of these neighbors are given the same color then we will have a color remaining for v_n. The remainder of this case describes how to create the ordering to make this possible. Within each of the subcases below we will identify vertices $v, v_1,$ and v_2 so that $vv_1, vv_2 \in E(G)$ but $v_1v_2 \notin E(G)$ and $G - \{v_1, v_2\}$ is still connected. Let x be an arbitrarily chosen vertex of G.

Case 3a: Suppose $G - x$ is 2-connected. Since G is not complete, there must be some vertex v_2 of distance two from x. Let $x = v_1$ and choose v to be the common neighbor of x and v_2. Then $vv_1, vv_2 \in E(G)$ but $v_1v_2 \notin E(G)$. Moreover, $G - \{v_1, v_2\}$ must still be connected since $G - x = G - v_1$ is 2-connected and so v_2 cannot be a cut-vertex in $G - x$.

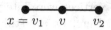

Case 3b: Suppose $G - x$ is not 2-connected. Then $G - x$ is not itself a block, and so must contain blocks $b_1, b_2, \ldots b_t$. Since the block graph $B(G - x)$ of $G - x$ is a tree (see Exercise 4.30), we know at least two blocks, say b_1 and b_t, are leaves in $B(G - x)$. This implies that both b_1 and b_t contain cut vertices of $G - x$. If either of these cut-vertices c were the only neighbor of x in the block, then $G - c$ would be disconnected. Let $v = x$ and v_1 and v_2 be a vertex from b_1 and b_t, respectively, that is not a cut-vertex of $G - x$. Then v_1 and v_2 are not adjacent and since $\deg(x) \geq 3$, we know $G - \{v_1, v_2\}$ is connected.

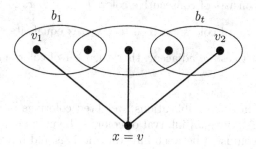

Order the vertices of G, with v_1 and v_2 identified as above and $v = v_n$. For v_{n-1}, choose a vertex that is adjacent to v that is neither v_1 nor v_2 (which we know exists since $\deg(v_n) \geq 3$). For v_{n-2} choose a vertex adjacent to either v_n or v_{n-1} that hasn't already been chosen. Note that since $G - \{v_1, v_2\}$ is connected, for each $i \in \{3, \ldots, n-1\}$ there is an available vertex $v_i \in V(G) - \{v_1, v_2, v_n, v_{n-1}, \ldots, v_{i+1}\}$ that is adjacent to at least one of $v_{i+1}, \ldots, v_n$. Continue choosing vertices in this way until all vertices have been labeled.

Now using First-Fit on this ordering of the vertices, we see that v_1 and v_2 will be given the same color 1 since these are non-adjacent vertices. Since every vertex v_i, for $3 \leq i < n$ has a neighbor after it in the order, we know it is adjacent to at most $k - 1$ vertices preceding it. Thus First-Fit will need at most k colors for G.

6.4.3 Weighted Coloring

Consider the following scenario:

> Ten families need to buy train tickets for an upcoming trip. The families vary in size but each of them needs to sit together on the train. Determine the minimum number of seats needed to accommodate the ten family trips.

This problem should sound very similar to Example 6.15 where colors were representing seats and the vertices were intervals of time indicative of when a person was on the train. Here, we are still interested in a graph coloring, but now each vertex represents a family and so has a size associated with it. In previous chapters, we used weights on the edges of a graph to indicate distance, time, or cost. For graph coloring models, weighted edges would have very little meaning. Instead, we will assigning a weight to each vertex, and finding a proper coloring will be referred to as a *weighted coloring*.

Definition 6.25 Given a weighted graph $G = (V, E, w)$, where w assigns each vertex a positive integer, a proper **weighted coloring** of G assigns each vertex a set of colors so that

(i) the set consists of consecutive colors (or numbers);

(ii) the number of colors assigned to a vertex equals its weight; and

(iii) if two vertices are adjacent, then their set of colors must be disjoint.

Note that in some publications, weighted colorings are referred to as interval colorings (since an interval of colors is being assigned to each vertex). To avoid the confusion between interval colorings and interval graphs, we use the term weighted coloring.

Before we tackle the train example, we will look at a smaller graph with a weighted coloring. We will, for the most part, use the same strategies for finding a minimum weighted coloring as we did above for unweighted coloring.

The biggest change will be to focus on locations that have large weighted cliques, which is found by adding the weights of the vertices within a complete subgraph. Thus if we have two different cliques on three vertices, one with total weight 8 and the other with total weight 10, we should initially focus on the one with the higher total weight.

Example 6.18 Find an optimal weighted coloring for the graph below where the vertices have weights as shown below.

$$w(a) = 2$$
$$w(b) = 1$$
$$w(c) = 4$$
$$w(d) = 2$$
$$w(e) = 2$$
$$w(f) = 4$$

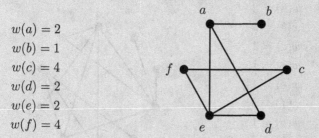

Solution: Note that $a, d,$ and e form a K_3 with total weight 8, and $c, e,$ and f form a K_3 with total weight 10. We begin by assigning weights to the vertices from the second K_3, and since e appears twice we assign it the first set of colors $\{1, 2\}$. From there we are forced to use another 4 colors on f ($\{3, 4, 5, 6\}$) and another 4 colors on c ($\{7, 8, 9, 10\}$), as shown on the left below. We can color the remaining three vertices using colors 1 through 10 as shown on the right.

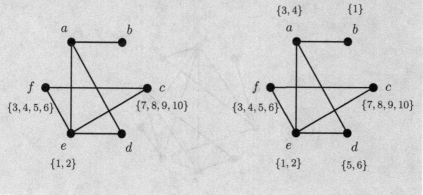

As the example above demonstrates, we focus less on a vertex degree and more on the vertex weight when we are searching for a minimal weighted coloring. This is in part due to the need for the set of colors to be consecutive. If a vertex has high weight, then it needs a larger range from which to pick the set of colors, whereas a vertex with large degree but a small weight may be able to squeeze in between the sets of colors of its neighbors.

Example 6.19 Suppose the ten families needing train tickets have the same underlying graph as that from Example 6.15 and the size of each family is noted below. Determine the minimum number of seats needed to accommodate everyone's travels.

$w(a) = 2$
$w(b) = 5$
$w(c) = 2$
$w(d) = 1$
$w(e) = 4$
$w(f) = 3$
$w(g) = 1$
$w(h) = 3$
$w(i) = 4$
$w(j) = 5$

Solution: As with the example above, we begin by looking for the largest total weight for a clique. Since the clique size is three, we want to find K_3 subgraphs with high total weight. The largest of these is with a, b, and h with total 10. Since b has the largest weight among these, we give it colors 1 through 5, assign a colors 6 and 7, and colors 8, 9, and 10 to h.

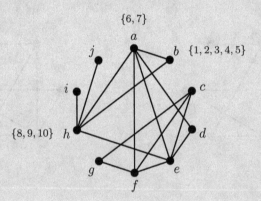

The next two cliques with a high total weight (of 9) are formed by a, e, f and c, e, f. Since a has already been assigned colors, we begin with the first clique. We can fill in the colors for e and f without introducing new colors, as shown below. Note that since e is adjacent to both a and h it could only use four consecutive colors chosen from those used on b.

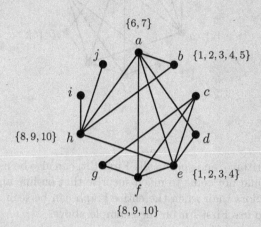

At this point we can fill in the colors for c by choosing two consecutive colors from those available (5, 6, and 7). We also fill in the colors for i and j by choosing the correct number of consecutive colors from any of 1 to 7.

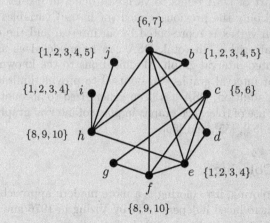

Finally, we need one color each for d and g. These can be any color not already used by one of their neighbors.

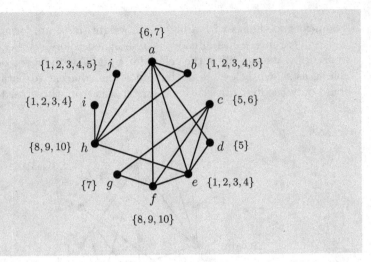

On-line algorithms, and specifically First-Fit, can also be used for weighted colorings. It should not come to much surprise that on-line algorithms generally use more colors than when the entire graph can be seen. In Exercise 6.2 you are asked to use First-Fit on the example above.

The combination of interval graphs, on-line algorithms, and weighted colorings have been extensively studied due to a very specific application, called Dynamic Storage Allocation, or DSA. The storage allocation refers to assigning variables to locations within a computer's memory, where each variable has a size associated to it. This can be thought of as the weight of a vertex. The location a variable is assigned is the set of colors a vertex is given. The dynamic part of DSA refers to variables being in use for specific intervals of time and only the previously used (or in-use) variables locations are know. Thus each vertex is represented by an interval and the coloring must use an on-line algorithm. In total, DSA can be modeled as an on-line coloring of a weighted interval graph. Modifications to the known performance of algorithms for interval graphs can be used to provide limitations on DSA performance. In addition, DSA has been generalized to account for leeway of storage, making use of tolerance graphs in place of interval graphs. For further information, see [54], [55], and [56].

6.4.4   List Coloring

Like weighted coloring, list coloring is a more modern approach to graph coloring and was introduced independently by Vizing in 1976 and Erdös, Rubin and Taylor in 1979 (see [32] and [82]). In list coloring we are concerned with assigning colors to vertices (or edges) with the added restriction that the colors must come from some predefined list. Before we dive into the theory and some surprising results, let us consider the following example.

Example 6.20 The table on the right below shows two different lists for the vertices in the graph on the left. Find a proper coloring for each version of the lists or explain why none exists.

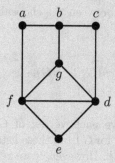

Vertex	List 1	List 2
a	$\{1,2\}$	$\{1\}$
b	$\{1,3,5\}$	$\{1,4,5\}$
c	$\{3,4\}$	$\{3,4\}$
d	$\{3,4\}$	$\{3,4\}$
e	$\{2,3\}$	$\{2,3\}$
f	$\{1,2,3\}$	$\{1,2,3\}$
g	$\{4,5\}$	$\{4,5\}$

Solution: A possible coloring from the first set of lists can be given as

$$a-1, b-4, c-3, d-4, e-3, f-2, g-5.$$

However, there is no proper coloring possible from the second set of lists. First note a must be given color 1 and so f can only be colored 2 or 3. But since those are the same colors available for e, we know one will be given color 2 and the other color 3. In either case, since d is adjacent to both f and e we know it cannot be given color 3, and so must be given color 4, forcing g to be color 5 and c to be color 3. But now b is adjacent to vertices of color 1, 3, and 5, and so cannot be given a color from its list.

The lists for the vertices need not be the same size nor contain the same items. The example above shows that finding a proper coloring can largely depend on the lists given for the vertices; the difference between the first and second set of lists above was minor (only those for a and b changed). Before we discuss some results on list coloring, we formally define what we mean by measuring a graph's ability to be list colored.

Definition 6.26 Let G be a graph where each vertex x is given a list $L(x)$ of colors. A proper *list coloring* assigns to x a color from its list $L(x)$ so that no two adjacent vertices are given the same color.

If for every collection of lists, each of size k, a proper list coloring exists then G is *k-choosable*. The minimum value for k for which G is k-choosable is called the *choosability* of G and denoted $ch(G)$.

First, we should note that choosability is not based on if there exists a singular collection of lists, each of a given size, that exhibit a proper coloring, but rather that every possible collection of lists of a given size can produce a proper coloring. This should intuitively make sense as otherwise we could find a proper coloring if each vertex has a list of size one with the lists being disjoint. However, there is a direct relationship between the choosability of a graph and its chromatic number.

Proposition 6.27 For any simple graph G, $ch(G) \geq \chi(G)$.

Proof: Let G satisfy $\chi(G) = k$ and give each vertex of G the list $\{1, 2, \ldots, k\}$. Then there is a proper coloring for G from these lists, namely the one exhibited by the fact that $\chi(G) = k$. However, if we remove the same one element from each of these lists, then G cannot be colored since otherwise $\chi(G) < k$.

This result should not be too surprising since list coloring should be more difficult than a general coloring since we are placing additional restrictions on how a color can be assigned to a vertex. However, the upper bound for the chromatic number that we discussed above (Brooks' Theorem) still holds for list coloring; the proof appears in Exercise 6.22.

Proposition 6.28 For any simple graph G, $ch(G) \leq \Delta(G) + 1$.

Graph coloring provides avenues for studying limitations and optimal strategies for resource management. As such, many additional variations of graph coloring exist that cannot be included here, such as total coloring, path coloring, and acyclic coloring.

6.5 Exercises

6.1 For each of the graphs below, complete the following.
  (a) Find the chromatic number $\chi(G)$. Include an argument why fewer colors will not suffice.
  (b) Find the chromatic index $\chi'(G)$.
  (c) Determine which graphs are perfect. Explain your answer.
  (d) Draw the line graph $L(G)$.

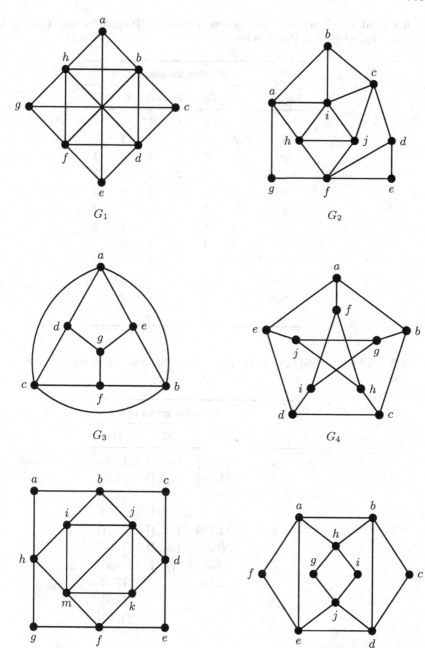

G_1

G_2

G_3

G_4

G_5

G_6

6.2 Find an optimal weighted coloring for each of the graphs from Exercise 6.1 with the weights as shown below.

Vertex	Weights for graph in 6.1					
	G_1	G_2	G_3	G_4	G_5	G_6
a	2	4	1	3	3	3
b	3	2	4	3	3	5
c	2	5	3	2	4	3
d	2	3	2	2	3	2
e	4	2	3	4	5	4
f	2	3	4	3	3	2
g	3	1	2	2	4	2
h	1	3	.	3	4	3
i	.	2	.	1	1	3
j	.	1	.	1	4	3
k	.	.	.	.	3	.
m	.	.	.	.	5	.

6.3 Find a list coloring for select graphs from Exercise 6.1 with the lists as shown below.

Vertex	Lists for graph in 6.1		
	G_1	G_3	G_5
a	{1,2,3}	{1,2,3}	{1,2}
b	{4,5,6}	{2,3,4}	{2,3}
c	{1,2}	{3,4}	{3,4}
d	{2,4}	{1,4}	{1,2}
e	{4,7,8}	{1,3,4}	{1,3,4}
f	{5,8}	{2,3,4}	{1,2,4}
g	{3}	{1,4}	{1,2}
h	{2,4,6}	.	{2,3}
i	.	.	{4,5}
j	.	.	{1,2,3}
k	.	.	{3,4}
m	.	.	{2,4}

6.4 Find a coloring of the map of the United States. Explain why four colors are necessary.

6.5 The seven committees from Exercise 5.6 need to schedule their weekly meeting. Based on the membership lists (shown below), determine the number of time slots needed so each person can make his or her committee meeting.

Committee	Members			
Benefits	Agatha	Dinah	Evan	Vlad
Computing	Evan	Nancy	Leah	Omar
Purchasing	George	Vlad	Leah	
Recruitment	Dinah	Omar	Agatha	
Refreshments	Nancy	George		
Social Media	Evan	Leah	Vlad	Omar
Travel Expenses	Agatha	Vlad	George	

6.6 Ten students in the coming semester will be taking the courses shown in the table below.

Course	Students				
Physics	Arnold	Ingrid	Fred	Bill	Jack
Mathematics	Eleanor	Arnold	Herb		
English	Arnold	David			
Geology	Carol	Bill	Fred	Herb	
Business	George	Eleanor	Carol		
Statistics	David	Ingrid	George		
Economics	Ingrid	Jack			

(a) Draw a graph modeling these conflicts.

(b) How many time periods must be allowed for these students to take the courses they want without conflicts? Include an argument why fewer time periods will not suffice.

(c) The following semester, all the students except David plan to take second courses in the same subject. David decides not to take further courses in Statistics. How many time periods will then be required?

6.7 (From [79]) A set of solar experiments is to be made at various observatories and each experiment is to be repeated for several years, as shown in the table below. Each experiment begins on a given day of the year and ends on a different given day and an observatory can perform only one experiment at a time. Determine the minimum number of observatories required to perform a given set of experiments annually.

Experiment	Start Date	End Date
A	Sept 2	Jan 3
B	Oct 15	Mar 10
C	Nov 20	Feb 17
D	Jan 23	May 30
E	Apr 4	July 28
F	Apr 30	July 28
G	June 24	Sept 30

6.8 Below is a collection of meetings that need to be assigned to conference rooms. Each group has identified when it would like to meet and how long it needs. In addition, each group has given the organizers a little leeway in the amount of time it is willing to cut short its meeting if the room is needed for another group. Model this information as a graph and determine how many conference rooms are needed.

Organization	Time	Leeway
Adam's Apples	8:30–9:30	10 minutes
Brain Teasers	9:00–11:00	45 minutes
Cookie Club	9:45–12:00	30 minutes
Disaster Readiness	10:00–1:00	45 minutes
Edison Enthusiasts	11:15–12:45	5 minutes
Fire Chiefs	12:30–3:30	30 minutes
Gary's Golfers	2:00–3:15	5 minutes
Helix Doubles	2:15–4:00	1 hour

6.9 Using the same order of the vertices for the graph from Example 6.15,
 (a) find the coloring using the First-Fit Algorithm.
 (b) find a weighted coloring using First-Fit with the weights from Example 6.19.

6.10 Use Proposition 6.9 to find an upper bound for $\chi(G_1)$ from page 276 and explain why Brook's Theorem finds a better upper bound for the chromatic number. Under what circumstances would Proposition 6.9 provide a better upper bound than Brooks' Theorem?

6.11 Show that the graph from Example 6.3 on page 284 is not chordal.

6.12 Prove that adding an edge between two vertices in a graph G can increase the chromatic number by at most 1.

6.13 Prove that $\chi(G) = \max \chi(H)$ where H is any component of G.

6.14 Prove that a graph G is bipartite if and only if $\chi(G) \leq 2$.

6.15 A graph G is *color-critical* if for all proper subgraphs H of G, $\chi(H) < \chi(G)$.
  (a) Prove that if G is color-critical with $\chi(G) = k$ then $\delta(G) \geq k - 1$.
  (b) Use this result to prove Proposition 6.12: $\chi(G) \leq 1 + \max_H \delta(H)$ for any induced subgraph H of G.

6.16 Prove that any graph with chromatic number k must have at least k vertices of degree at least $k - 1$. (Hint: try a contradiction argument)

6.17 Explain why $R(2,n) = n$.

6.18 Prove $\chi(G) \geq \dfrac{|V(G)|}{\alpha(G)}$ where $\alpha(G)$ is the independence number of G (the size of the largest independent set of vertices).

6.19 Prove that bipartite graphs are perfect.

6.20 Prove that $L(G)$ is perfect whenever G is bipartite.

6.21 Prove Proposition 6.10: If G is a graph with $l(G)$ the length of the longest path in G, then $\chi(G) \leq 1 + l(G)$.

6.22 Prove Proposition 6.28: For any simple graph G, $ch(G) \leq \Delta(G) + 1$. (Hint: use First-Fit)

6.23 Here we work through the details of the result from [58] outlined on page 323, namely that there exists an on-line algorithm $\mathcal{A}$ such that $\chi_{\mathcal{A}}(G) \leq 3\omega(G) - 2$ for all interval graphs G.

  To begin the proof we assume that $\omega(G) = k$ is fixed. Define the on-line algorithm $\mathcal{A}_k$ as follows:

  - If $k = 1$ use the First-Fit Algorithm
  - As we receive a new vertex v, put v into the set B provided that $\omega(G[B \cup \{v\}]) < k$. Otherwise put it into a set T.
  - If a vertex is placed into the set B we color it as $\mathcal{A}_{k-1}$ would color it with respect to $G[B]$. Otherwise we color v with First-Fit with respect to $G[T]$, using disjoint sets of colors for $G[B]$ and $G[T]$.

We claim that for all k and on-line interval graphs $G^{\prec}$ if $\omega(G) \leq k$ then $\chi_{\mathcal{A}_k}(G^{\prec}) \leq 3k - 2$. We argue this by induction on k.

  (a) Explain why the statement holds when When $k = 1$ (thus proving the base case of the induction argument.)
  (b) To prove the induction step, we assume $k \geq 2$. Then $\mathcal{A}_k$ uses at most $3(k-1) - 2$ colors on $G[B]$, and so it suffices to show that First-Fit uses at most 3 colors on $G[T]$. Prove the following two claims:
     (i) Claim 1: No interval of T is contained in the union of the other intervals of T. (Hint: Use a contradiction argument)

(ii) Claim 2:If $\mathcal{B}$ is a clique in an interval graph then there exists an interval $I \in \mathcal{B}$ such that every point of I is contained in at least $\frac{|\mathcal{B}|}{2}$ intervals of $\mathcal{B}$. (Hint: Argue by induction on $|B|$)

(c) To finish the proof, use the results from (b) to show $\Delta(G[T]) \leq 2$, thus proving First-Fit uses at most 3 colors on $G[T]$.

7

Planarity

Throughout this book we have not spent much time talking about the way in which we draw a graph, mainly because the drawing itself is only a visual depiction of the information stored within the edge and vertex set. In Chapter 1 we briefly discussed when two graphs are isomorphic, relating two different drawings of the same set of information. This chapter will investigate when the way in which we draw a graph can indicate underlying structural information about the graph. In particular, we will be looking at when a graph is *planar*.

> **Definition 7.1** A graph G is ***planar*** if and only if the vertices can be arranged on the page so that edges do not cross (or touch) at any point other than at a vertex.

Note that for a graph to be planar, it is only required that at least one drawing exists without edge crossings; it is not required that all possible drawings of the graph be without edge crossings. For example, below are two drawings of the graph K_4 (in the language of Section 1.3, these graphs are isomorphic). The drawing on the left is the more standard way of drawing K_4, and contains one edge crossing (ac and bd cross at a location that is not a vertex); the drawing on the right is a planar drawing of K_4 so that no edge crossings exist.

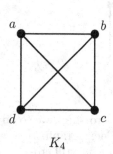

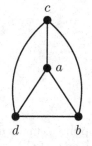

| K_4 | planar drawing of K_4 |

Often it is useful to think of taking a graph and moving around the vertices and pulling or stretching the edges so that they can be repositioned without edge crossings. We could think of obtaining the graph on the right above by rotating the entire graph on the left clockwise by 45°, moving vertex c above a, and then repositioning the edges from c to the other vertices.

Finding a planar drawing of a graph can be very tricky. In fact, simply determining if a graph is planar or not is hardly trivial. To begin, we will see an application of planarity (other than our previous interest in graph coloring) and in the following section we discuss techniques for determining planarity.

Example 7.1 Three houses are set to be built along a new city block; across the street lie access points to the three main utilities each house needs (water, electricity, and gas). Is it possible to run the lines and pipes underground without any of them crossing?

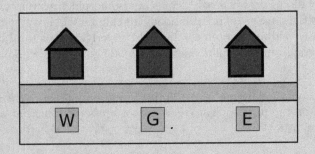

Solution: This scenario can be modeled with three vertices representing the houses ($h_1, h_2,$ and h_3) and three vertices representing the utilities ($u_1, u_2,$ and u_3). First note that if we are not concerned with edge crossings, the proper graph model is $K_{3,3}$, the complete bipartite graph with three vertices in each side of the vertex partition. The standard drawing of $K_{3,3}$, given below, clearly is not a planar drawing as there are many edge crossings; for example, h_1u_2 crosses h_2u_1.

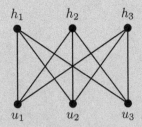

Attempting to find a drawing without edge crossings, we could stretch and move some of the edges as shown in the next graph. However, this drawing is still not planar since edges h_3u_1 and h_2u_2 still cross. In fact, no matter how you try to draw $K_{3,3}$ (try it!), there will always be at least one edge crossing. Thus the utility lines and pipes cannot be placed without any of them crossing.

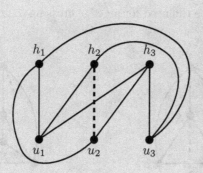

7.1   Kuratowski's Theorem

Given a specific drawing of a graph, it is easy to see if that drawing is planar or not. However, just because you cannot find a planar drawing of a graph does not mean a planar drawing does not exist. This is perhaps the most challenging part of planarity. Luckily, we have already seen one of the most important structures in showing a graph is nonplanar (namely $K_{3,3}$). The other structure we studied thoroughly in Section 2.2, namely K_5.

Example 7.2 Determine if K_5 is planar. If so, give a planar drawing; if not, explain why not.

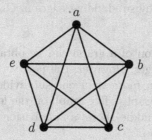

Solution: If we attempt the same procedure as we did for K_4 above, then we run into a problem. We start with the planar drawing of K_4 and add another vertex e below the edge bd (shown on the left below). We can easily add the edges from e to b, c, and d without creating crossings, but to reach a from e we would need to cross one of the edges bd, bc, or cd. Placing vertex e anywhere in the interior of any of the triangular sections of K_4 will still create the need for an edge crossing (one of which is shown on the right below).

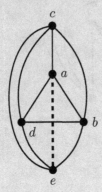 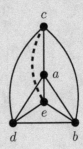

We now have two graphs that we know to be nonplanar. Moreover, having either K_5 or $K_{3,3}$ as a subgraph will guarantee that a graph is nonplanar since if a portion of a graph is nonplanar there is no way for the entire graph to be planar. What may be surprising is that these two graphs provide the basis for determining the planarity of any graph. However, it is not as simple as containing a K_5 or $K_{3,3}$ subgraph, but rather a modified version of these graphs called *subdivisions*. We introduced the notion of an edge subdivision in Section 4.3 when discussing graphs that are 2-connected.

Definition 7.2 A *subdivision* of an edge xy consists of inserting vertices so that the edge xy is replaced by a path from x to y. The subdivision of a graph G is obtained by subdividing edges in G.

Note that a subdivision of a graph can be obtained by subdividing one, two, or even all of its edges. However, the new vertices placed on the edges from G cannot appear in more than one subdivided edge. Below are three examples of subdivided graphs. The graph on the left shows a subdivision of $K_{3,3}$, the graph in the middle shows a subdivision of K_4, the graph on the right shows a subdivision of K_5.

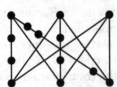 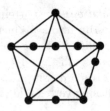

From the discussion above, we should understand why $K_{3,3}$ and K_5 subgraphs pose a problem for planarity. Adding a vertex along any of the edges (thus creating paths between the original vertices) of one of these graphs will not suddenly allow the graph to become planar. Thus containing a subdivision of $K_{3,3}$ or K_5 proves a graph is nonplanar. The Polish mathematician Kazimierz Kuratowski proved in 1930 that containing a $K_{3,3}$ or K_5 subdivision was not only enough to prove a graph was nonplanar, but more surprisingly that any nonplanar graph *must* contain a $K_{3,3}$ or K_5 subdivision.

Theorem 7.3 (Kuratowski's Theorem) A graph G is planar if and only if it does not contain a subdivision of $K_{3,3}$ or K_5.

The proof of Kuratowski's Theorem is quite technical, and will appear at the end of this section. The remainder of this section will focus on *using* Kuratowski's Theorem and other results that help determine if a given graph is planar.

In practice, it is often useful to think of moving vertices and stretching edges at the same time as looking for a $K_{3,3}$ or K_5 subdivision. Note that in order to contain a $K_{3,3}$ subdivision, a graph must have at least 6 vertices of degree 3 or greater, and in order to contain a K_5 subdivision the graph must have at least 5 vertices of degree 4 or greater. These conditions are often helpful when searching for a subdivision.

7.1.1 Euler's Formula

One major result regarding planarity that is quite useful in gaining some intuition as to the planarity of a graph was proven in 1752 by a mathematician we spent an entire section discussing, Leonhard Euler. The result was given in more geometric terms (and planarity is one area of intersection between graph theory and geometry) and uses an additional term relating to the drawing of a graph, namely a *region*.

Definition 7.4 Given a planar drawing of a graph G, a *region* is a portion of the plane completely bounded by the edges of the graph.

In practice, we can usually see the regions of a graph fairly easily, as long as we do not forget the infinite (or exterior) region. For example, the following two graphs G_1 and G_2, each have 6 vertices, but G_1 has 9 edges and 5 regions whereas G_2 has 5 edges and only one region, the infinite one.

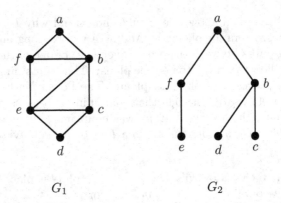

G_1 G_2

Note that every tree has exactly 1 region since no cycles exist to fully encompass a portion of the plane. As both of the graphs above are planar, they satisfy Euler's Formula below.

Theorem 7.5 (Euler's Formula) If G is a connected planar graph with n vertices, m edges, and r regions then $n - m + r = 2$.

Proof: Argue by induction on m, the number of edges in the graph. If $m = 1$ then since G is connected it is either a tree with one edge and so $n = 2$ and $r = 1$, or G is a graph with a loop, and so $n = 1$ and $r = 2$. In either case, Euler's Formula holds.

Now suppose Euler's Formula holds for all graphs with $m \geq 1$ edges and consider a graph G with $m + 1$ edges, n vertices, and r regions. We will consider the graph that is formed with the removal of an edge from G.

First, if $G' = G - e$ is not connected for any edge e in G then we know e must be a bridge of G. Thus G must be a tree, and so $n = m + 1$ and $r = 1$. Therefore $n - m + r = m + 1 - m + 1 = 2$.

Next, if $G' = G - e$ is connected for some edge e of G, then e must be a part of some cycle in G. Then there must be two different regions on the two sides of e in G, but these regions join into just one with the removal of e. Thus G' has n vertices, $r - 1$ regions, and $m - 1$ edges. By the induction hypothesis applied to G', we know $n - (m-1) + (r-1) = 2$, which simplifies to $n - m + r = 2$.

Thus by induction we know Euler's Formula holds for all connected planar graphs.

The simplicity of Euler's Formula betrays its importance. It provides incredible insight into the relationship between the two building blocks of graphs, vertices and edges, and shows their delicate balance within planar graphs.

As we investigate planar graphs, one might question if additional edges could be added to the graph while maintaining planarity (such as adding additional utility lines without creating crossings). Clearly additional edges can be added to a tree without incurring edge crossings, but what about G_1 above? We cannot add edges along the interior of the graph, but we could cut into the infinite region with some additional edges. Note: we are not considering adding multi-edges since we could theoretically add an infinite number of edges between two vertices.

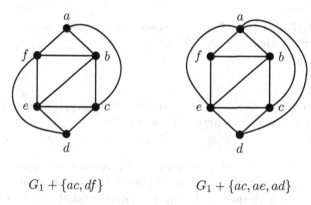

$$G_1 + \{ac, df\} \qquad\qquad G_1 + \{ac, ae, ad\}$$

Note that if we try to add any additional edges to the graph on the right that an edge-crossing would be required. This graph is called maximally planar.

Definition 7.6 A graph G is *maximally planar* if $G + e$ is nonplanar for any edge $e = xy$ for any two nonadjacent vertices $x, y \in V(G)$.

For a graph the be maximally planar, we basically need every region to be bounded by a triangle, including the infinite region. For the graph on the right above, the border for the infinite region is created by the edges between vertices a, d, and e, and so it also is triangular. This graph has 6 vertices and 12 edges.

Theorem 7.7 If G is a maximally planar simple graph with $n \geq 3$ vertices and m edges, then $m = 3n - 6$.

Proof: Assume G is maximally planar. Then every region must be bounded by a triangle, as otherwise we could add a chord to a region bounded by a longer cycle. Since every edge separates two regions, and every region is bounded by three edges, we know $r = \frac{2m}{3}$. Thus by Euler's

Formula, we have $n - m + \frac{2m}{3} = 2$, and so $3n - 3m + 2m = 6$, giving $m = 3n - 6$.

One of the implications of this theorem is that a planar graph cannot be too dense, that is there cannot be too many edges in relation to the number of vertices. The two theorems below are more useful in practice though, as they allows us to use simple counting arguments when investigating graph planarity.

Theorem 7.8 If $G = (V, E)$ is a simple planar graph with m edges and $n \geq 3$ vertices, then $m \leq 3n - 6$.

Theorem 7.9 If $G = (V, E)$ is a simple planar graph with m edges and $n \geq 3$ and no cycles of length 3, then $m \leq 2n - 4$.

These theorems are less about verifying the relationship between edges and vertices when a graph is known to be planar, but rather in determining if a graph satisfies this inequality. If a graph does not satisfy the inequality, then it must be nonplanar; however, if the graph satisfies the inequality, then it may or may not be planar and further investigations are needed to determine planarity. The proof of Theorem 7.9 appears in Exercise 7.10.

Example 7.3 Determine which of the following graphs are planar. If planar, give a drawing with no edge crossings. If nonplanar, find a $K_{3,3}$ or K_5 subdivision.

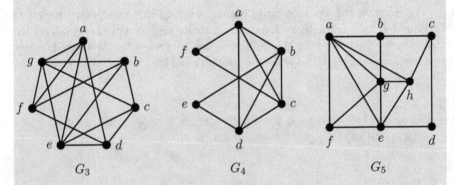

G_3 G_4 G_5

Solution: First note that all of the graphs above are drawn with edge crossings, but this does not indicate their planarity. Also, we can begin by using Euler's Theorem above to give us an indication of how likely the graph is to be planar.

For G_3, we have $|E| = 15$ and $|V| = 7$, giving us the inequality from Theorem 7.8 of $15 \leq 3 \cdot 7 - 6 = 15$. Thus the number of edges is as high as possible for a graph with 7 vertices. Although this does not guarantee the graph is nonplanar, it provides good evidence that we should search for a K_5 or $K_{3,3}$ subdivision. Also notice that every vertex has degree at least 4, so we will begin by looking for a K_5 subdivision.

To do this, we start by picking a vertex and looking at its neighbors, hoping to find as many as possible that form a complete subgraph. Beginning with a as a main vertex in K_5, we see the other main vertices would have to be either its neighbors or vertices reachable by a short path. We will start by choosing the other main vertices of the K_5 to be the neighbors of a, namely d, e, f, and g. A starting graph is shown below on the left. Next we fill in the edges between these four vertices, as shown on the right below.

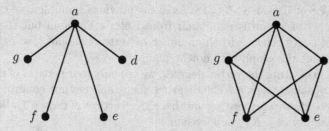

At this point, we are only missing two edges in forming K_5, namely dg and ef. We have two vertices available to use for paths between these nonadjacent vertices, and using them we find a K_5 subdivision. Thus G_3 is nonplanar.

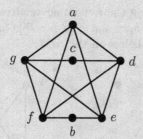

For G_4, we have $|E| = 11$ and $|V| = 6$, giving us the inequality from Theorem 7.8 of $11 \leq 3 \cdot 6 - 6 = 12$. We cannot deduce from this result that the graph is nonplanar. However, notice that two vertices have degree 2 and the remaining 4 vertices have degree 4. There cannot be a K_5 subdivision since there are not enough vertices of degree at least 4 and there cannot be a $K_{3,3}$ subdivision since there are not enough vertices of degree at least 3. Thus we can conclude that G_4 is in fact planar. A planar drawing is shown below.

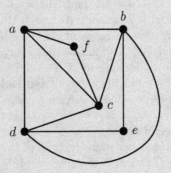

For G_5, we have $|E| = 15$ and $|V| = 8$, giving us the inequality from Theorem 7.8 of $15 \leq 3 \cdot 8 - 6 = 18$. As in the previous examples, we cannot deduce that the graph is nonplanar from Euler's Theorem but the high number of edges relative to the number of vertices should give us some suspicion that the graph may not be planar.

When inspecting the vertex degrees, we see only four vertices of degree at least 4 (namely a, e, g, h), indicating the graph cannot contain a K_5 subdivision. However, there are another three vertices of degree 3, allowing for a possibility of a $K_{3,3}$ subdivision.

Again, we start by selecting vertex a to be one of the main vertices of a possible $K_{3,3}$ subdivision. At the same time, we will look for another vertex that is adjacent to three of the neighbors of a. We see that a and g are both adjacent to b, h, e, and f. Let us begin with b, h, and e for the vertices on the other side of the $K_{3,3}$, as shown below on the left.

We now search among the remaining vertices (f, d, c) for one that is adjacent to as many of b, h, and e as possible and find c is adjacent to both b and h, as shown below on the right.

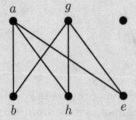 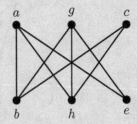

At this point we are only missing one edge to form a $K_{3,3}$ subdivision, namely ce. Luckily we can form a path from c to e using the available vertex d. Thus we have found a $K_{3,3}$ subdivision and proven that G_5 is nonplanar.

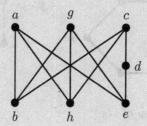

One final note of caution: subdivisions are not necessarily unique. In fact G_3 and G_5 from the example above both contain more than one subdivision, as seen in Exercises 7.3 and 7.4. Also, finding a planar drawing can be, at times nontrivial. The method described below is by no means the only option but often helps to get yourself in a good direction in finding a planar drawing.

7.1.2 Cycle-Chord Method

When a graph is drawn so the vertices are roughly arranged around a circle, it can often be easier to think about shifting their positions on the page or stretching the edges to obtain a planar drawing. But when the graph is drawn to highlight some other attribute, such as it being bipartite or showing some clumping of vertices, it can be challenging to find a planar drawing. The next few pages will detail one method for finding a planar drawing, called the *Cycle-Chord Method*. The graph G_6 below will serve as an example of how to use this method.

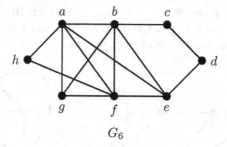

$$G_6$$

To begin, put the vertices in a circular pattern, but with some care in their arrangement. We want to find a spanning cycle (also called a hamiltonian cycle) or something approximating a spanning cycle, when placing the vertices. The edges in bold on the left represent those that are currently being placed in the planar drawing; the gray edges are ones not yet placed.

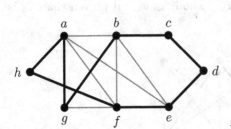

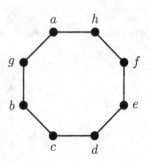

highlighted portion of G_6 planar drawing

Once we have created the spanning cycle, we attempt to place as many edges in the interior of the cycle as possible so that they do not cross. This should resemble chords of a circle. In the example G_6 below, we are making most of the interior regions of the cycle into triangulations.

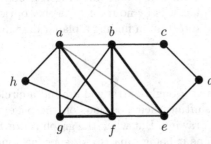

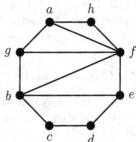

highlighted portion of G_6 planar drawing

When placing these chords, we should take care to notice any vertices that have incident edges remaining, as those will need to be placed as curves along the outside of the cycle. For example, in G_6 only two edges remain to be placed outside the spanning cycle, namely ab and ae. Having multiple edges from the same vertex left to place is often a benefit since they can be drawn in the same or opposite directions of the cycle without creating edge crossings.

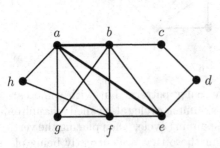

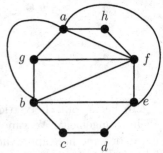

highlighted portion of G_6 planar drawing

This procedure is not perfect, and we may discover that we have made a poor choice as to which edges should be going through the interior of the cycle versus along the outside. When this happens, we simply try again using the knowledge gained from our first attempt.

Obviously this method is not written as an efficient algorithm and would be unwieldy for graphs of large size. Luckily, for the purposes of this book we will only focus on graphs of small size when finding planar drawings.

7.1.3 Proof of Kuratowski's Theorem

Now that we have some familiarity with properties of planar graphs, we return to the proof of Kuratowski's Theorem, which basically states that being non-planar is equivalent to having one of two forbidden structures: subdivisions of K_5 or $K_{3,3}$. The formal statement of the theorem is below and is written as a biconditional. Recall that biconditional statements are special in that they show both necessary and sufficient conditions for property to hold. In this case, we need only to know if a graph contains a subdivision of $K_{3,3}$, or K_5 to determine its planarity. Based on our discussion above, we should expect one direction of this theorem to be fairly easy. In our formal proof, we will use the following results, one of which was discussed in Chapter 4 when we studied connectivity but is restated below as Lemma 7.12 .

Lemma 7.10 Let H be a subgraph of G. Then G is nonplanar if H is nonplanar.

Lemma 7.11 Let G' be a subdivision of G. Then G is planar if and only if G' is planar.

Lemma 7.12 A graph G with at least 3 vertices is 2-connected if and only if for every pair of vertices x and y there exists a cycle through x and y.

As you can guess based on the inclusion of Lemma 7.12 above, we will need to make use of connectivity in our proof of Kuratowski's Theorem. In particular, we will reference a *block* of a graph, which is a maximal 2-connected subgraph (subgraph without cut-vertices).

One final note: our proof of Kuratowski's Theorem below is based on that of Dirac and Schuster [24] and is organized using the contrapositive of how we originally stated the theorem. We restate it in this form below. The general plan for the proof of the forward direction (the more difficult part) is to use the connectivity of a minimally nonplanar graph to then find subdivisions of K_5 or $K_{3,3}$.

Theorem 7.3 (Kuratowski's Theorem) A graph G is nonplanar if and only if it contains a subdivision of $K_{3,3}$ or K_5.

Proof: First suppose G contains a subdivision of $K_{3,3}$ or K_5. By our discussions above we know $K_{3,3}$ and K_5 are nonplanar. By Lemma 7.11, this means any subdivision of $K_{3,3}$ and K_5 must also be nonplanar, and so by Lemma 7.10 we know G is also nonplanar.

Conversely, suppose for a contradiction that there exists a nonplanar graph G that does not contain a subdivision of $K_{3,3}$ or K_5. Choose G to be a minimal such graph; that is, any graph with fewer vertices or edges that does not contain a subdivision of $K_{3,3}$ or K_5 must be planar. Note that since G is nonplanar then it must contain a nonplanar block B. If B is a proper subgraph of G, then G would not be minimal. Thus we know that G must itself be 2-connected.

Claim 1: Every vertex of G have degree at least 3.

Proof: Suppose for a contradiction that there exists some vertex v that does not have degree 3. We know $\deg(v) \geq 2$ since G is 2-connected, and so $\deg(v) = 2$. Let x and y be the two distinct neighbors of v. If x and y are adjacent, then $G' = G - v$ must be planar by the minimality of G. But then there must be a region for which the edge xy is part of its boundary, and the path $x\,v\,y$ can be inserted into this region, making G planar, which is a contradiction.

Thus x and y cannot be adjacent. Define $G' = G - v + xy$; that is, G is the graph obtained from G by removing vertex v and adding the edge xy. Since G' has fewer vertices than G, we know it must be planar. But then we can obtain a planar drawing of G from that of G' by subdividing the edge xy with the vertex v.

This is also a contradiction, so every vertex of G must have degree at least 3.

Claim 2: There exists an edge $e = xy$ such that $G' = G-e$ is 2-connected.

Proof: Since every vertex of G has degree at least 3, we know G cannot simply be a cycle. Moreover, since G is 2-connected we know G has an ear decomposition by Theorem 4.24. Then last path added to the decomposition must be a singular edge, as otherwise any internal vertex of the path would have degree 2 in G. Thus the graph obtained by removing this edge will remain 2-connected as it still has an ear decomposition.

Let x, y be chosen so that $G' = G - e$ is 2-connected. Then by the minimality of G we know G' must be planar. By Corollary 4.18, there exists a cycle C in G' that contains both x and y. Consider the planar drawing of G' with C chosen so that:

(i) it contains x and y,

(ii) the number of regions inside C is maximal among all planar drawings of G',

(iii) given any other cycle C' containing x and y, no planar drawing of C' has more regions inside of it than that of C.

Let $C = v_0 v_1 \cdots v_k v_{k+1} \cdots v_l v_0$, where $v_0 = x$ and $v_k = y$. Since x and y are not adjacent in G', we know $k \geq 2$.

Claim 3: There is no path P connecting two vertices in $\{v_0, v_1, \ldots v_k\}$ or in $\{v_k, v_{k+1}, \ldots v_l, v_0\}$ that lies in the exterior of C.

Proof: Suppose such a path exists, say between vertices v_i and v_j.

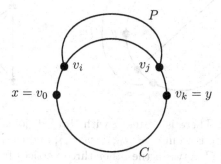

Then construct cycle C' by taking cycle C from v_0 up to v_i then the path P to v_j followed by the cycle C from v_j to v_k and then continuing back to v_0. Then the interior of C' contains all the regions of C plus the one created between P and C. This contradicts the choice of C as having the maximum number of regions in its interior.

Since G is nonplanar, we know that we cannot simply add the edge xy to the exterior of C in the planar drawing of G'. Thus there must be a path along the exterior of C that connects a vertex from the set $\{v_1, \ldots v_{k-1}\}$ to a vertex from the set $\{v_{k+1}, \ldots v_l\}$, say P is from v_i to v_j with $1 \leq i \leq k-1$ and $k+1 \leq j \leq l$.

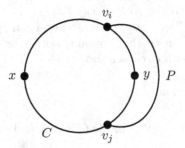

Note that no vertex of P can be adjacent to any other vertex of C as otherwise we would be in the situation we just disproved in Claim 3. Since P is placed along the exterior of C, there must be some reason why it cannot be placed in the interior of C. The four following cases are exhaustive (that is one of these must be the reason for the placement of P).

Case 1: There is a path P' between a vertex s from $\{v_0, v_1, \ldots, v_{i-1}\}$ and a vertex t from $\{v_k, \ldots, v_{j-1}\}$. Adding the edge e back in produces a $K_{3,3}$ subdivision using portions of the cycle C and paths P and P' as shown below.

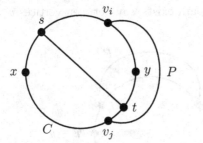 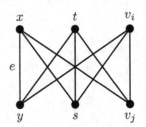

Case 2: There is vertex u with three disjoint paths to vertices of C, one of which is from $A = \{v_0, v_i, v_k, v_j\}$, and the other two vertices s and t lie along C between the other three vertices from A. One such option is shown below. Adding the edge e back in produces a $K_{3,3}$ subdivision, where the path from s to v_j uses the portion of the cycle C from s to v_i and then the path P from v_i to v_j, as shown below.

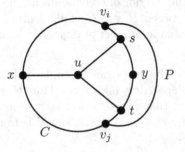 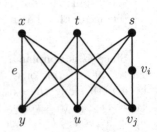

Case 3: There is a path from x to y on the interior of C with two distinct vertices s and t from which there exist disjoint paths to v_i and v_j. Adding the edge e back in produces a $K_{3,3}$ subdivision as shown below.

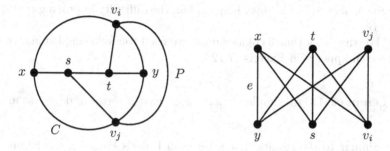

Case 4: There is a path from x to y on the interior of C with a vertex s from which there exist disjoint paths to v_i and v_j. Adding the edge e back in produces a K_5 subdivision as shown below.

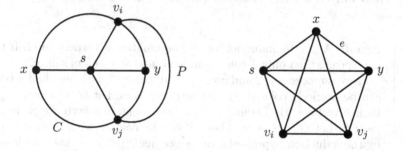

In each of the cases above we obtain a subgraph of G that is a subdivision of $K_{3,3}$ or K_5. Thus we have shown that every nonplanar graph must contain a subdivision of $K_{3,3}$ or K_5.

7.2 Graph Coloring Revisited

In Section 6.1, we discussed the famous Four Color Theorem and how it relates to graph coloring. One of the important aspects of the result was that the graphs in question are planar graphs. As we noted previously, the hunt to provide a solution to the Four Color Conjecture was a lengthy one, encompassing more than 120 years of research by some of the brightest mathematicians. Obviously we will not reproduce the proof of the Four Color Theorem here (recall the original proof was over 900 pages). Instead we will discuss the proofs that almost got us to 4 and some other results that can be deduced about planar graphs with respect to colorings. We begin with the proof that every planar graph can be colored with 6 colors and then the improvement to 5. As expected, the proof gets demonstrably harder with this minor adjustment from 6 colors to 5. This may help explain the difficulty in proving that 4 colors suffice.

For the proof that 6 colors suffice, we need the following lemma, the proof of which appears in Exercise 7.12.

Lemma 7.13 All simple planar graphs have a vertex of degree at most 5.

Similar to the results that arise from Euler's Formula, the lemma above implies that a planar graph cannot be too dense. This will prove useful when proving our two theorems about coloring planar graphs.

Theorem 7.14 Every planar graph can be colored using at most 6 colors.

Proof: Argue by induction on n, the number of vertices in G. If G has one vertex, then only 1 color is needed, and so 6 colors suffice.

Now suppose $n \geq 2$ and every planar graph G' with less than n vertices can be properly colored with at most 6 colors. Let G be a planar graph with n vertices. By Lemma 7.13, G must contain a vertex x of degree at most 5. Let $G' = G - x$. Then G' can be colored with at most 6 colors by the induction hypothesis. But since $\deg(x) \leq 5$, we know at least one of these 6 colors has not been used on any neighbor of x. Thus x can be colored with one of the 6 colors not used on its neighbors.

Therefore by induction every planar graph can be colored using at most 6 colors.

The proof that 5 colors suffice begins the same, in that we consider a vertex x of degree at most 5 and the colors assigned to each of its neighbors. As in Theorem 7.14, we will show that there is some way to guarantee a color is available for x, only this time we only have 5 colors available. We will use the technique introduced by A.B. Kempe in 1879 that was instrumental in the proof of the Four Color Theorem by Appel and Haken.[53] Note that Kempe used these ideas to attempt to prove the Four Color Theorem, though Heawood found a flaw in the argument and showed Kempe's ideas instead showed 5 colors suffice [48].

Definition 7.15 Let G be a graph in which every vertex has been colored. Then $G_{i,j}$ is the graph induced by colors i and j and a ***Kempe*** $i-j$ ***chain*** is any component of graph $G_{i,j}$.

Theorem 7.16 Every planar graph can be colored using at most 5 colors.

Proof: Argue by induction on n, the number of vertices in G. If G has fewer than 6 vertices, then at most 5 colors are needed. So we may assume G has at least $n \geq 6$ vertices and all planar graphs with less than n vertices can be colored using at most 5 colors. By Lemma 7.13, we know G has a vertex x with degree at most 5. Define $G' = G - x$. Then by the induction hypothesis, we can color G' with at most 5 colors. If the neighbors of x use at most 4 colors, then one remains for x and we can color G using at most 5 colors. If this is not the case, then $\deg(x) = 5$ and each of the neighbors of x have been given a unique color.

Let $v_1, \ldots v_5$ be the neighbors of x. Without loss of generality, we may assume that the vertices are arranged in a cyclic nature around x so that v_i has color i. Consider the graph $G_{1,3}$. This graph must contain v_1 and v_3. We will consider whether these vertices are in the same component of $G_{1,3}$. In particular if the Kempe $1-3$ chain K containing v_1 also contains v_3.

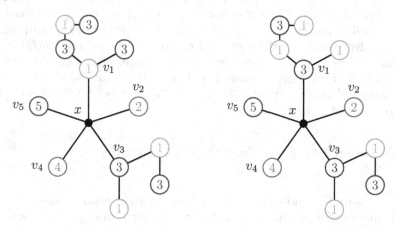

Case 1: K does not contain v_3. Then we can swap the colors along K so that vertices colored 1 are now colored 3, and vice versa. This results in a coloring of $G - x$ in which no neighbor of x has color 1, making it possible to give x color 1.

Case 2: K contains v_3. Then there must exist a path P in K from v_1 to v_3 that alternates between a vertex of color 1 and a vertex of color 3.

Together with x, we get a cycle C in which v_2 and v_4 are on separate faces. Then in $G_{2,4}$ there cannot be a Kempe $2-4$ chain that contains both v_2 and v_4 as any such path would cross C at either an edge, making G nonplanar, or a vertex, creating a vertex with more than one color. Thus we can swap the colors on the Kempe $2-4$ chain containing v_2, making v_2 have color 4 and allowing x to be given color 2.

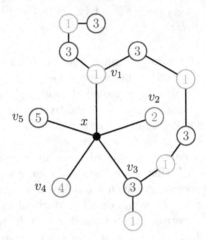

In either case, we have produced a coloring of G with 5 colors.

At the end of Chapter 6 we introduced the notion of list coloring (see Section 6.4.4). Recall that the choosability of a graph, $ch(G)$, is defined to be the smallest integer k so that any collection of lists of size k can produce a proper list coloring of the vertices of G. We also showed that $ch(G) \geq \chi(G)$, which means that the choosability of the class of planar graphs cannot be lower than 4. Carsten Thomassen proved in 1994 that the choosability of a planar graph is also at most 5 [77], though the proof below follows more closely those presented in [21] and [84].

Theorem 7.17 Every planar graph is 5-choosable.

Proof: First note that adding edges to a graph cannot reduce the choosability of a graph, so we will only consider planar graphs in which the exterior boundary is a cycle and all interior regions are triangulated. So suppose the boundary cycle is $C = v_1 v_2 \cdots v_k v_1$. Further, suppose v_1 has been given color 1, v_2 has been given color 2, all other vertices of C have lists of size at least 3, and all vertices in $G - C$ have lists of size 5. We will prove that the coloring of v_1 and v_2 can be extended to a coloring of the remaining vertices of G by inducting on $|G|$, the number of vertices in G.

Suppose $|G| = 3$. Since v_3 has a list of size 3, we know there must be a color other than 1 or 2 available for v_3, so G is list-colorable.

Now suppose $|G| \geq 4$ and assume our initial assumptions above hold for all graphs with fewer vertices. We will consider two cases for the cycle C base on the existence of a chord for C. In each of these cases we will find a subgraph of G that satisfies the induction hypothesis and explain how to extend its coloring to the entire graph G.

Case 1: C has a chord xy. Then we can form two cycles C_1 and C_2 which are contained in the cycle C together with the edge xy. Moreover, we can let edge v_1v_2 be a part of C_1, since at most one of v_1 and v_2 can equal x or y. Let G_i be the graph induced by C_i and all vertices in its interior.

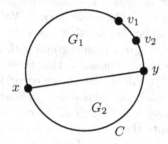

Note that G_1 has fewer vertices than G and v_1 and v_2 still have their assigned colors of 1 and 2, respectively. Thus we can apply the induction hypothesis to G_1 to obtain a proper list coloring of G_1. This fixes x and y to each have specific color, and since G_2 has fewer vertices than G, applying the induction hypothesis to G_2 with x and y playing the role of v_1 and v_2, we get a proper list coloring of G_2. Combining these two colorings produces a proper list coloring of G.

Case 2: C does not have any chords. Let $v_1, u_1, u_2, \ldots, u_m, v_{k-1}$ be the neighbors of v_k in their cyclic order around v_k. Then by how C was defined, each of the u_i lie in the interior of C and since each of the regions in the interior of C are triangulated we know there exists a path from v_1 to v_{k-1} using the u_i vertices, that is $P = v_1 u_1 \cdots u_m v_{k-1}$. Let C' be the cycle formed by removing v_k from C and adding in the path P.

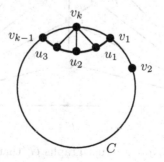

Since v_k has a list of size 3, we know at least two of these s, t are not 1, namely the color of v_1. Remove s, t from the list of each u_i vertex. Then v_1 and v_2 are on C', with colors 1 and 2, respectively, and all other vertices of C' have lists of size at least 3. Then by the induction hypothesis we can properly list color $G - \{v_k\}$. Since s and t are not used for v_1 and any u_i vertex, at most one of these two colors can be used on v_{k-1}, leaving the other available for v_k. Thus we have obtained a proper list coloring for G.

Note that the theorem above cannot be strengthened, that is there exist planar graphs that are not 4-choosable. Finding one such example is nontrivial; in fact, it was an open problem in graph theory for almost 15 years, and first proven by Margit Voigt in 1993 [83] (though his proof appeared just one year before Thomassen published the result above). Voigt's construction relies on making complex triangular figures that are embedded into larger complex triangular figures, resulting in a planar graph with 238 vertices, each of which has a list of size 4 specifically chosen so that there is no way to find a proper coloring of the graph. We will outline the construction of the graph below and the proof that this graph is not 4-choosable appears in the Exercises.

To begin, we start with the graph G_v below, where all vertices are given the lists $\{1, 2, 3, 4\}$. Note that some of the triangles are marked with $*$ and some vertices are larger, what we will call marked.

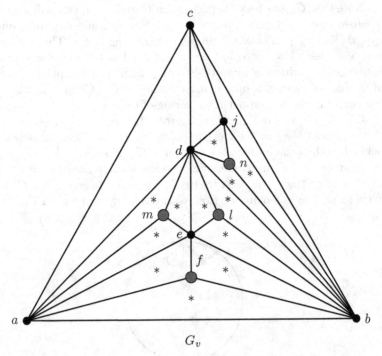

G_v

We will show how to construct graphs G' that will fit into each of these triangles so that the marked vertex of G' will coincide with the marked vertex of G_v. This will create the graph G_p.

Next, let H be the graph shown below, with f as the marked vertex. Note that vertex p also has a list $\{1, 2, 3, 4\}$. Inside of the region created by afp we will insert a copy of graph H_1, and inside the region created by bfp we will insert a copy of graph H_2.

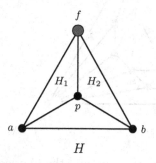

H

Below are the graphs H_1 and H_2, where the vertices have lists as shown on their right. Both H_1 and H_2 have a marked vertex f which corresponds to the vertex f in graph H above.

Note that H_1 and H_2 are the same underlying graph, but with different lists assigned to their vertices. These graphs are embedded into the graph H, creating a graph G' that is then placed into each of the $*$ triangles of G_v, where the marked vertex f from H corresponds to the marked vertex from G_v.

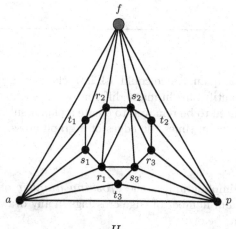

H_1

Vertex	List
r_1	$\{2, 3, 5, 6\}$
r_2	$\{1, 3, 5, 6\}$
r_3	$\{4, 3, 5, 6\}$
s_1	$\{2, 3, 5, 6\}$
s_2	$\{1, 3, 5, 6\}$
s_3	$\{4, 3, 5, 6\}$
t_1	$\{1, 3, 5, 6\}$
t_2	$\{1, 3, 4, 5\}$
t_3	$\{2, 3, 4, 5\}$

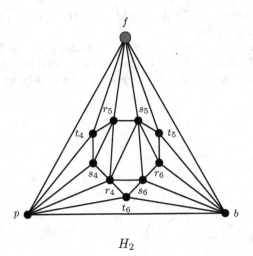

Vertex	List
r_4	$\{3, 4, 5, 6\}$
r_5	$\{1, 3, 5, 6\}$
r_6	$\{2, 3, 5, 6\}$
s_4	$\{3, 4, 5, 6\}$
s_5	$\{1, 3, 5, 6\}$
s_6	$\{2, 3, 5, 6\}$
t_4	$\{1, 3, 4, 5\}$
t_5	$\{1, 2, 3, 5\}$
t_6	$\{2, 3, 4, 5\}$

$$H_2$$

The resulting graph G_p then contains the 10 original vertices and 12 copies of the graph G', each of which contains an additional 19 vertices, resulting in a graph with 238 total vertices. The proof that this graph is not 4-choosable is surprisingly not too difficult (see Exercise 7.22). A smaller example containing only 63 vertices was published by the Iranian mathematician Maryam Mirzakhani in 1996 [66] and appears in Exercise 7.23 (Mirzakhani was the first woman awarded the prestigious Fields Medal prize and died of breast cancer at the age of 40).

7.3 Edge-Crossing

Once a graph is known to be nonplanar, one could ask how close the graph is to being planar. You could quantify this in many different ways, such as how many vertices or edges would need to be removed to create a planar subgraph. Here we consider instead how many times the edges of a graph cross, called the *crossing number*.

Definition 7.18 For any simple graph G the **crossing number** of G, denoted $cr(G)$, is the minimum number of edge crossings in any drawing of G satisfying the conditions below:

(i) no edge crosses another more than once, and

(ii) at most two edges cross at a given point.

Note that we need to be a bit precise when defining the crossing number, since we could theoretically draw edges as complex curves requiring a large number of crossings. For example the graph G_7 on the left below violates (i) since the edge ef crosses bc more than once, whereas the graph G_8 on the right below violates (ii) since edges ae, bf, cg, and dh all cross at a point (not a vertex) in the center of the graph.

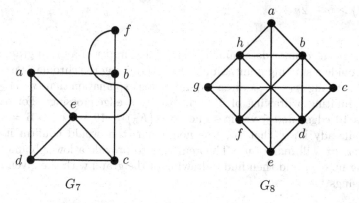

$$G_7 \qquad\qquad G_8$$

It should be clear that $cr(G) = 0$ if and only if G is planar. But what about K_5 and $K_{3,3}$, the two graphs that are instrumental in determining planarity?

Example 7.4 Determine the crossing numbers for K_5 and $K_{3,3}$.

Solution: Since we know K_5 and $K_{3,3}$ are not planar, $cr(K_5), cr(K_{3,3}) \geq 1$. A drawing of each of these adhering to the criteria above is shown below, proving they each have crossing number 1.

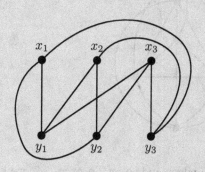 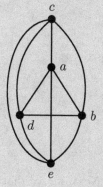

Recall that Theorem 7.7 showed that the number of edges in a maximally planar graph on n vertices must equal $3n - 6$ and Theorem 7.9 showed the maximum number of edges is $2n - 4$ for a graph without any 3-cycles. These imply a nonplanar graph must not have too many edges in relation to the number of vertices and provides motivation for the following result for the crossing number of a graph.

Theorem 7.19 Let G be a simple graph with m edges and n vertices. Then $cr(G) \geq m - 3n + 6$. Moreover, if G is bipartite then $cr(G) \geq m - 2n + 4$.

This result cannot determine the crossing number of a given graph, but it does provide a starting point. Similar to determining the chromatic number of a graph, stating $cr(G) = k$ often requires some explanation that $cr(G) \geq k$ and then exhibiting a drawing of G with exactly k edge-crossings. For example, K_5 has 10 edges and 5 vertices and so $cr(K_5) \geq 10 - 3 * 5 + 6 = 1$. If we didn't already know that K_5 was nonplanar, this would confirm it for us. However, we will mainly use Theorem 7.19 to provide a lower bound for the crossing number and then find a drawing of the graph with the given number of crossings.

Example 7.5 Find the crossing number for K_6 and $K_{4,4}$.

Solution: First note that K_6 has 15 edges. Then by Theorem 7.7, we know $cr(K_6) \geq 15 - 3 * 6 + 6 = 3$. Below is a drawing of K_6 with 3 edge crossings, and so we know $cr(K_6) = 3$.

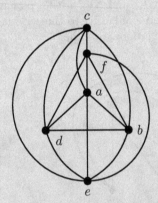

Next, we see that $K_{4,4}$ has 16 edges and so by Theorem 7.7, we know $cr(K_{4,4}) \geq 16 - 2 * 8 + 4 = 4$. Since the drawing below of $K_{4,4}$ has 4 edge crossings, we know $cr(K_{4,4}) = 4$.

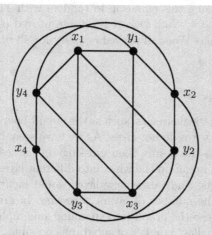

Unfortunately, there is no clear formula for the crossing number of a graph, even for complete graphs! The two results below give upper bounds for K_n and $K_{m,n}$ in terms of the floor function (see Appendix B).

Theorem 7.20

$$cr(K_n) \leq \frac{1}{4} \left\lfloor \frac{n}{2} \right\rfloor \left\lfloor \frac{n-1}{2} \right\rfloor \left\lfloor \frac{n-2}{2} \right\rfloor \left\lfloor \frac{n-3}{2} \right\rfloor$$

In 1972 Richard Guy proved this bound and in fact conjectured that equality holds for all n [44]. He proved equality for $n \leq 10$ and in 2007 Shengjun Pan and Bruce Richter showed equality holds for $n \leq 12$ [69].

Note that for K_6, the inequality above gives $cr \leq 3$ and from Theorem 7.19 we have $cr(K_6) \geq 3$. Thus we know the crossing number without exhibiting a specific drawing of K_6. Unfortunately these bounds quickly diverge, as even moving up just one size to K_7 we get $6 \leq cr(K_7) \leq 9$.

The Polish mathematician Kazimierz Zarankiewicz proved a similar upper bound for bipartite graphs (shown below) in 1954, though he conjectured this quantity should also be the exact formula for the crossing number of complete bipartite graphs [87].

Theorem 7.21

$$cr(K_{m,n}) \leq \left\lfloor \frac{m}{2} \right\rfloor \left\lfloor \frac{m-1}{2} \right\rfloor \left\lfloor \frac{n}{2} \right\rfloor \left\lfloor \frac{n-1}{2} \right\rfloor$$

Similar to K_6, using this equation in tandem with Theorem 7.19, we can determine $cr(K_{4,4}) = 4$. Note Daniel Kleitman proved equality for $K_{5,n}$ and $K_{6,n}$, whereas Douglas Woodall proved equality for all $m, n \leq 7$ as well as for $K_{7,9}$ [59][86].

7.3.1 Thickness

When designing electrical circuits, such as the circuit board within a computer, it is important that wires do not cross. As we have seen, if there are too many connections between the points then crossings are necessary on a plane. To combat this, we can break up the wires into different layers, each one of which contains no crossings. Since each new layer would incur additional cost, we want to minimize the number of layers necessary. In graph theoretic terms, we want to decompose the graph into spanning subgraphs, each of which are planar, using the smallest number of subgraphs possible. This minimum value is called the *thickness* of a graph.

> **Definition 7.22** Let $T = \{H_1, H_2, \ldots, H_t\}$ be a set of spanning subgraphs of G so that each H_i is planar and every edge of G appears in exactly one graph from T. The **thickness** of a graph G, denoted $\theta(G)$, is the minimum size of T among all possible such collections.

Clearly $\theta(G) = 1$ if and only if G is planar, since T would contain only G itself. Below is a decomposition of K_6 into two planar spanning subgraphs and since we know that K_6 is not planar we have shown $\theta(K_6) = 2$.

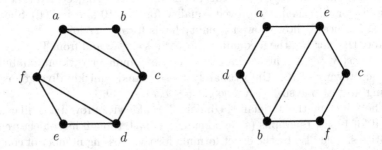

Exercise 7.17 looks at the interplay between $cr = 2$ and θ. The following two corollaries use the results on the number of edges in planar graphs from Theorems 7.7 and 7.9 from Section 7.1.1.

Corollary 7.23 Let G be a connected simple graph with n vertices and m edges. Then

$$\theta(G) \geq \left\lceil \frac{m}{3n-6} \right\rceil$$

Corollary 7.24 Let G be a connected simple bipartite graph with n vertices and m edges. Then

$$\theta(G) \geq \left\lceil \frac{m}{2n-4} \right\rceil$$

While a general formula is not known for the thickness of a graph, the theorem below does establish the thickness for a complete graph (and so could serve as an upper bound for any graph on n vertices). This result is based on the work from [1],[3],[4], and [81].

Theorem 7.25

$$\theta(K_n) = \begin{cases} \left\lfloor \dfrac{n+7}{6} \right\rfloor & n \neq 9, 10 \\ 3 & n = 9, 10 \end{cases}$$

7.4   Exercises

7.1 Draw a subdivision of K_3 where one edge has two vertices inserted, another edge has one vertex inserted, and the last edge is not subdivided. What is another name for this graph? What can you conclude about C_n, the cycle on n vertices?

7.2 Determine if the following graphs are subdivisions of $K_{3,3}$ and explain your answer.

(a)

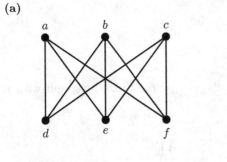

(b)

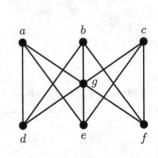

(c)

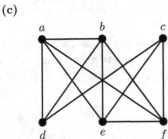

(d)

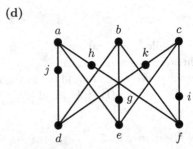

7.3 Using the vertices b, c, e, f, g as the main vertices, find a K_5 subdivision for G_3 from Example 7.3.

7.4 Find a different $K_{3,3}$ subdivision for G_5 from Example 7.3.

7.5 For each of the graphs below, determine if it is planar or nonplanar. If planar, give a drawing with no edge crossings. If nonplanar, find a $K_{3,3}$ or K_5 subdivision.

(a)

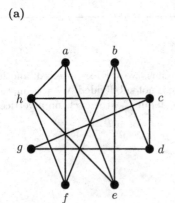

(b)

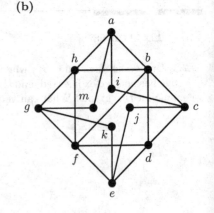

(c)   **(d)**

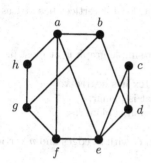

(e)

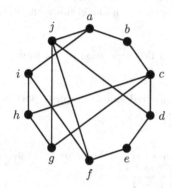

7.6 Determine the crossing number for K_7 and $K_{3,4}$ and give a drawing with that number of edge-crossings.

7.7 Explain why any maximally planar graph is chordal. Use this to prove that any maximally planar graph on more than three vertices has clique-size 4.

7.8 Prove the following statement or give a counterexample: Every graph with $\chi(G) \leq 4$ must be planar.

7.9 Use Theorems 7.8 and 7.9 to give formal proofs that neither $K_{3,3}$ nor K_5 are planar.

7.10 Prove Theorem 7.9: Every planar simple graph with no cycles of length 3 satisfies $m \leq 2n - 4$.

7.11 Prove the following corollary to Euler's Formula: If G is a planar graph with n vertices, m edges, and r regions then $n - m + r = 1 + c(G)$ where $c(G)$ denotes the number of components of G.

7.12 Prove Lemma 7.13 that every planar simple graph contains a vertex of degree at most 5.

7.13 Prove that every planar simple graph with at least 4 vertices has at least 4 vertices with degree at most 5.

7.14 Prove that the average degree of any planar simple graph is at most 6.

7.15 Recall that a graph is k-regular if every vertex has degree k.
(a) Find a 3-regular simple bipartite graph that is planar.
(b) Prove that no 4-regular simple bipartite graph is planar.

7.16 Prove Theorem 7.19 that every simple graph G with m edges and n vertices satisfies $cr(G) \geq m - 3n + 6$.

7.17 (a) Prove that if graph G has $cr(G) = 1$ then $\theta(G) = 2$.
(b) Show that the converse is not true; that is, find a graph with $\theta(G) = 2$ and $cr(G) > 1$.

7.18 Prove that $\theta(K_8) = 2$ by finding a decomposition of K_8 into two planar subgraphs.

7.19 Prove Corollary 7.23 that any connected simple graph G with n vertices and m edges satisfies $\theta(G) \geq \left\lceil \dfrac{m}{3n - 6} \right\rceil$.

7.20 Prove Corollary 7.24 that any connected simple bipartite graph G with n vertices and m edges satisfies $\theta(G) \geq \left\lceil \dfrac{m}{2n - 4} \right\rceil$.

7.21 Prove that a 4-regular graph satisfies $\theta(G) \leq 2$.

7.22 In order to prove the graph G_p constructed on 362 is not 4-choosable, we must show some vertex cannot be colored using a color from its list. Use the steps below to create the proof.
(a) Use the K_4 created from vertices a, b, c, d to prove one of the marked vertices f, l, m, n is colored 1.
(b) Without loss of generality, assume f is given color 1. Explain why one of the triangles afe, efb, and afb must be colored with colors $1, 2, 3$.
(c) Without loss of generality, assume a and b are colored using colors 2 and 3. In either case p must have color 4. If a has color 2, explain why at least one of T_1, T_2, T_3 cannot be colored using their lists.
(d) If b has color 2, explain why at least one of T_4, T_5, T_6 cannot be colored using their lists.

7.23 To better understand the graph on 63 vertices found by Miriam Mirzakhani that is not 4-choosable, we begin with a smaller example then move to her larger graph. For each of the vertices in these graphs, we are labeling them with a set of colors $\bar{i}$ which consists of a set of colors chosen from a larger set S with color i removed (so if $S = \{1, 2, 3, 4, 5\}$ then $\bar{1} = \{2, 3, 4, 5\}$, $\bar{2} = \{1, 3, 4, 5\}$, etc.).

(a) Prove that the graph below does not have a list coloring based on the lists given for each vertex, where $S = \{1, 2, 3, 4\}$.

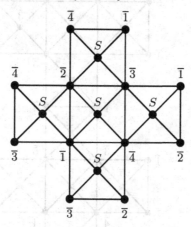

(b) Let G be the graph on the next page, with vertices given lists from a larger set $S = \{1, 2, 3, 4, 5\}$. Define G' to be the graph formed by adding one additional vertex with list $\bar{1}$ that is adjacent to all of the vertices on the outer edge of the graph. Prove G' does not have a list coloring [66].

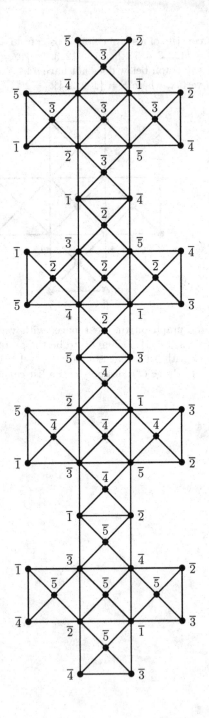

Appendix

A graph is essentially a mathematical structure built upon the relationship between objects within a set. At various times throughout this text we used set theory, function properties, and matrix operations. The sections to follow review some of these items, as well as provide the pseudocode for various algorithms appearing in the book.

A   Set Theory

While graphs are defined in terms of a set of vertices and edges, and these two sets uniquely determine the graph, this section will focus more generally on the basics of set theory.

> **Definition A.1** A *set* is a collection of objects. We denote sets with capital letters $(A, B, C, \ldots)$ and the objects within a set A are often called *elements*, denoted $x \in A$. If x is not an element of A we write $x \notin A$.

Suppose A is the set of digits, that is the set of integers from 0 to 9. We could write this as

$$A = \{0, 1, 2, 3, 4, 5, 6, 7, 8, 9\}$$

which is called *roster notation* since we are explicitly listing the elements in the set. We could also write A in *set-builder notation* as follows:

$$A = \{x \in \mathbb{Z} : 0 \leq x \leq 9\}$$

Here we are giving a rule for when an element belongs in a set. The advantage of set-builder notation is when there are a large number of elements in the set and listing them would be excessively time consuming; however, this only works when there is a clear way to describe the elements in the set. For small sets, roster notation is preferred.

There are numerous operations one can perform on a set. We will be concerned with a handful of these, mainly unions, intersections, complements, and subsets. To do this, we must first define our domain for a given problem, which is called the *universal set*.

Definition A.2 Given a universal set U and sets A and B within U we define the following:

- The **union** of A and B, denoted $A \cup B$, is the set of elements contained in A or B. Thus $A \cup B = \{x \in U : x \in A \text{ or } x \in B\}$.

- The **intersection** of A and B, denoted $A \cap B$, is the set of elements contained in A and B. Thus $A \cap B = \{x \in U : x \in A \text{ and } x \in B\}$.

- The **complement** of A, denoted $\overline{A}$, is the set of elements in the universal set that are not in A. Thus $\overline{A} = \{x \in U : x \notin A\}$.

- The **difference** of two sets, denoted $A - B$, is the set of elements contained in A but not B. Thus $A - B = \{x \in U : x \in A \text{ and } x \notin B\}$.

- We say A is a **subset** of B if and only if every element of A is also an element of B. Thus $A \subseteq B$ if and only if for all $x \in A$ we have $x \in B$.

We use the notation $\emptyset$ for the set with no elements.

The example that follows provides some practice with the definitions above.

Example A.1 Let $U = \{x \in \mathbb{Z} : 1 \leq x \leq 12\}$ and

$$A = \{1, 2, 3, 5, 6, 7\}$$
$$B = \{1, 8, 9, 12\}$$
$$C = \{1, 2, 5, 9, 10\}$$

Find each of the following:

(a) $A \cap B$ (d) $A \cap B \cap C$ (g) $\overline{A} \cap B$

(b) $A \cup C$ (e) $(A \cap B) - C$ (h) $\overline{A} \cap B \cap C$

(c) $A - C$ (f) $\overline{A}$

Solution:

(a) $A \cap B = \{1\}$ as 1 is the only element in both A and B.

(b) $A \cup C = \{1, 2, 3, 5, 6, 7, 9, 10\}$. Note that we do not write $1, 2$, or 5 twice since we only care if an element is in A or C, not how many times it appears in the sets.

(c) $A - C = \{3, 6, 7\}$ since these are the three elements of A that are not in C.

(d) $A \cap B \cap C = \{1\}$ since this is the only element in all of A, B and C.

(e) $(A \cap B) - C = \emptyset$ since there are no elements in both A and B that are not also in C.

(f) $\overline{A} = \{4, 8, 9, 10, 11, 12\}$ since these are the integers between 1 and 12 that are not in A.

(g) $\overline{A} \cap B = \{8, 9, 12\}$ since these are the elements in both $\overline{A}$ and B.

(h) $\overline{A} \cap B \cap C = \{9\}$ since 9 is the only element in both B and C but not A.

Visually we can display the relationship between sets in terms of a Venn diagram. Note that these are only useful when we are considering at most 3 sets within the same universal set. The sets described in the example above could be displayed as follows:

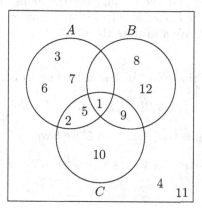

An entire course could be spent studying set theory. We will conclude this section with two important results on the union of sets. The first shows how the complement of a union is equivalent to the intersection of the complements, whereas the second relates the size of a union to the sizes of the sets that make up a union.

Theorem A.3 (De Morgan's Laws) For all sets A and B,

$$\overline{A \cup B} = \overline{A} \cap \overline{B} \text{ and } \overline{A \cap B} = \overline{A} \cup \overline{B}$$

In lieu of the proof (which can be found in [31]), we will show by example how De Morgan's Laws holds.

Example A.2 Using the sets from Example A.1, find $\overline{A \cup B}, \overline{A \cap B}, \overline{A} \cap \overline{B}$ and $\overline{A} \cup \overline{B}$, and verify the relationships from De Morgan's Laws.

Solution: First note that $A \cap B = \{1\}$ and $A \cup B = \{1, 2, 3, 5, 6, 7, 8, 9, 12\}$. Thus $\overline{A \cup B}$ is all the elements not in $A \cup B$ and $\overline{A \cap B}$ contains all the elements not in $A \cap B$, namely

$$\overline{A \cup B} = \{4, 10, 11\} \text{ and } \overline{A \cap B} = \{2, 3, 4, 5, 6, 7, 8, 9, 10, 11, 12\}.$$

Also, $\overline{A} = \{4, 8, 9, 10, 11, 12\}$ and $\overline{B} = \{2, 3, 4, 5, 6, 7, 10, 11\}$. Thus

$$\overline{A} \cap \overline{B} = \{4, 10, 11\} \text{ and } \overline{A} \cup \overline{B} = \{2, 3, 4, 5, 6, 7, 8, 9, 10, 11, 12\}.$$

Thus we have $\overline{A \cup B} = \overline{A} \cap \overline{B}$ and $\overline{A \cap B} = \overline{A} \cup \overline{B}$.

Finally, when counting the size of a union, we cannot simply add the sizes of the sets that encompass the union, since the items in the intersection will be double counted. The Principle of Inclusion-Exclusion shows how to account for this double count. We also state the three set version below.

Theorem A.4 (Principle of Inclusion-Exclusion) For all sets A and B,

$$|A \cup B| = |A| + |B| - |A \cap B|$$

Moreover, given three sets A, B, and C we have

$$|A \cup B \cup C| = |A| + |B| + |C| - (|A \cap B| + |A \cap C| + |B \cap C|) + |A \cap B \cap C|$$

Although we stated the Principle of Inclusion-Exclusion in terms of at most three sets, the result can be generalized to any number of sets.

Example A.3 Using the sets from Example A.1, verify the formulas from the Principle of Inclusion-Exclusion.

Solution: We have $|A| = 6, |B| = 4$, and $|C| = 5$. Moreover, $|A \cap B| = 1, |A \cap C| = 3, |B \cap C| = 2$ and $|A \cap B \cap C| = 1$. Thus by the Principle of Inclusion-Exclusion we have

$$|A \cup B| = 6 + 4 - 1 = 9$$
$$|A \cup B \cup C| = 6 + 4 + 5 - (1 + 3 + 2) + 1 = 10$$

Note that these quantities can be verified directly by examining the Venn diagram above.

B Functions

Functions are one of the building blocks of mathematics. While there are many ways to study and use functions, this section will review a few concepts commonly used in graph theory: one-to-one, onto, and bijections.

As graphs are themselves discrete objects, the functions discussed here will be defined on finite sets of discrete objects. The function g below takes the objects from set X to set Y as described by the arrows.

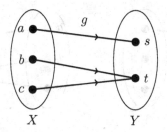

Definition B.1 A *function* f is a relation that assigns each input x exactly one output, denoted $f(x)$. The *domain* X is the set of all inputs and the *range* Y is the set of possible outputs, written as $f : X \to Y$.

In calculus or other areas of mathematics concerned with continuity, we often see functions written in terms of a specific rule, such as $f : \mathbb{R} \to \mathbb{R}$ given by $f(x) = x^2$. The example above cannot be written as such, and so we would explicitly list how the function behaves:

$$g(a) = s$$
$$g(b) = t$$
$$g(c) = t$$

This is not to say all discrete functions must be written this way – some can be given an explicit rule, such as $f : \mathbb{N} \to \{0,1\}$ where $f(x) = 0$ if x is even and 1 otherwise. Of greater importance for us is some specific properties of functions.

Definition B.2 Let $f : X \to Y$ be a function. Then f is

- *onto* if and only if for every $y \in Y$ there exists some $x \in X$ such that $f(x) = y$;

- *one-to-one* if and only if whenever $f(x_1) = f(x_2)$ implies $x_1 = x_2$.

If f is both one-to-one and onto, we say f is a *bijection*.

In practice we can view onto as ensuring every element of the range is reached by the function. The function g above is onto since all elements of Y are hit by an arrow. A function is one-to-one if two inputs never have the same output, and so g above is not one-to-one since both b and c are sent to the element t. Below are some visuals with the various options for a function to be one-to-one, onto, both or neither.

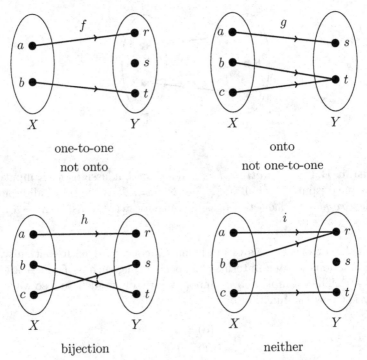

In practice we can view onto as ensuring every element of the range is

Function properties appear at various times in this book. Bijections appear in Section 1.3 since proving two graphs are isomorphic requires a clear correspondence between the vertices and edges. Network flow (see Section 4.4) looks at the interplay between two functions applied to the edges of a graph. In addition, finding a matching in a graph (see Chapter 5) and coloring the vertices of a graph (see Chapter 6) both overlay a function on the graph that must meet certain requirements (prefect matchings as bijections, for example). Two additional functions, called the *floor* and *ceiling*, are used in Chapter 7.

Definition B.3 The *floor* of a real number x, denoted $\lfloor x \rfloor$, is the unique integer n such that $n \leq x < n+1$.

The *ceiling* of a real number x, denoted $\lceil x \rceil$, is the unique integer n such that $n - 1 < x \leq n$.

Intuitively, we can think of the floor function as rounding down and the ceiling function as rounding up.

Example B.1 Calculate the floor and ceiling of the given quantities listed below.

(a) 3.14 (b) 12 (c) -1.16

Solution:

(a) $\lfloor 3.14 \rfloor = 3$ and $\lceil 3.14 \rceil = 4$ since $3 < 3.14 < 4$.

(b) $\lfloor 12 \rfloor = 12$ and $\lceil 12 \rceil = 12$. Note for any integer n, $\lfloor n \rfloor = \lceil n \rceil = n$.

(c) $\lfloor -1.16 \rfloor = -2$ and $\lceil -1.16 \rceil = -1$ since $-2 < -1.16 < -1$.

C Matrix Operations

Matrices are useful in graph theory when wanting to encode a graph for use within a computer program. Although we could do this in terms of the vertex-set and edge-set, matrices lend themselves more naturally to many operations on graphs. Section 1.4 described how to form the adjacency matrix of a graph and one use of this was seen in Section 2.3.2. This section will talk through the very basics of matrix operations; for more information on matrices and the large amount of theory or applications surrounding them, consult one of the numerous linear algebra texts.

> **Definition C.1** A *matrix* is an array of objects, most commonly num-
> bers, and denoted using capital letters $(A, B, C, \ldots)$. Entries in a matrix A
> are denoted by a_{ij} where i indicates the row and j the column. A matrix
> with n rows and m columns is said to have **dimension** $n \times m$.

Below are four matrices. In the matrix A, entry $a_{21} = 4$ since 4 is in the
second row and first column and entry $a_{13} = 5$ since 5 is in the first row and
third column. Similarly, $b_{21} = 2$ and $b_{13} = 3$.

$$A = \begin{bmatrix} 0 & 1 & 5 \\ 4 & 1 & 0 \\ 2 & 0 & 1 \end{bmatrix} \qquad C = \begin{bmatrix} 3 & 1 & 4 \\ 1 & 5 & 9 \end{bmatrix}$$

$$B = \begin{bmatrix} 0 & 1 & 3 \\ 2 & 1 & 0 \end{bmatrix} \qquad D = \begin{bmatrix} 0 & 1 & 2 & 5 \\ 2 & 0 & 7 & 4 \\ 9 & 2 & 5 & 0 \end{bmatrix}$$

The matrix A above has dimension 3×3 since it has 3 rows and 3 columns.
The matrices B and C both have dimension 2×3, and D has dimension 3×4.

When a matrix has the same number of rows as columns we call it *square*,
so A is the only square matrix listed. For many graph applications we use the
adjacency matrix which assigns a row and column to each vertex, making it
a square matrix.

We can perform a variety of operations on matrices, though here we will
only discuss addition, subtraction, and multiplication. Matrix dimension is
important here since operations such as addition and multiplication only work
under the correct dimensions of the matrices.

> **Definition C.2** Given two matrices A and B of the same dimension, we
> can define their sum $A + B$ as the sum of their respective entries; that is
>
> $$A + B_{ij} = a_{ij} + b_{ij} \text{ over all values } i \text{ and } j$$
>
> Similarly, the difference $A - B$ is defined as the difference of their
> respective entries:
>
> $$(A - B)_{ij} = a_{ij} - b_{ij} \text{ over all values } i \text{ and } j$$

Example C.1 Given the matrices $A, B,$ and C above, find the sum or difference noted below or explain why it doesn't exist.

(a) $B + C$ (c) $A + B$

(b) $B - C$ (d) $A + A$

Solution:

(a) Below are the calculations needed to find the sum. Future calculations will omit this level of detail.

$$B + C = \begin{bmatrix} 0 & 1 & 3 \\ 2 & 1 & 0 \end{bmatrix} + \begin{bmatrix} 3 & 1 & 4 \\ 1 & 5 & 9 \end{bmatrix}$$

$$= \begin{bmatrix} 0+3 & 1+1 & 3+4 \\ 2+1 & 1+5 & 0+9 \end{bmatrix}$$

$$= \begin{bmatrix} 3 & 2 & 7 \\ 3 & 6 & 9 \end{bmatrix}$$

(b) $B - C = \begin{bmatrix} -3 & 0 & -1 \\ 1 & -4 & -9 \end{bmatrix}$

(c) $A + B$ does not exist because they have different dimensions.

(d) $A + A$ can also be written as $2A$, which is found by multiplying each entry of A by 2. This is called scalar multiplication of a matrix.

$$2A = \begin{bmatrix} 0 & 2 & 10 \\ 8 & 2 & 0 \\ 4 & 0 & 2 \end{bmatrix}$$

Unlike matrix addition and subtraction, matrix multiplication is not done componentwise. Instead, use the correct row and column to find a given entry in a matrix. For example, if we want to find the product $B \times A$, we would naturally start by computing the entry in the first row and column. This would be done by taking the first row of B and multiplying it by the first column of A, as shown below:

$$\begin{bmatrix} 0 & 1 & 3 \end{bmatrix} \cdot \begin{bmatrix} 0 \\ 4 \\ 2 \end{bmatrix} = 0 * 0 + 1 * 4 + 3 * 2 = 10$$

In order for these vectors to be multiplied, they must have the same number of entries, and so the number of columns of B must match the number of rows in A.

Definition C.3 Given a matrix A of dimension $n \times m$ and a matrix B of dimension $m \times r$, we can define their product AB to be the matrix of dimension $n \times r$ where entry $(AB)_{ij}$ is defined as:

$$a_{i1}b_{1j} + a_{i2}b_{2j} + \cdots + a_{im}b_{mj}$$

for all $1 \leq i \leq n$ and all $1 \leq j \leq r$.

We denote AA as A^2. For any $n \geq 1$, we let A^n to be the matrix obtained by multiplying A by itself n times.

Example C.2 Given the matrices A, B, and D above, find the product noted below or explain why it doesn't exist.

(a) BA (d) A^2

(b) AB (e) A^3

(c) BD (f) B^2

Solution:

(a) First note that $B \times A$ will be a 2×3 matrix. We have already calculated the entry in the first row and first column to be $(ba)_{11} = 10$. To find the entry in the first row and second column, $(ba)_{12}$, we need to multiply the first row of B by the second column of A, as shown below:

$$\begin{bmatrix} 0 & 1 & 3 \end{bmatrix} \cdot \begin{bmatrix} 1 \\ 1 \\ 0 \end{bmatrix} = 0*1 + 1*1 + 3*0 = 1$$

Now to find the entry in the second row and first column, $(ba)_{21}$, we need to multiply the second row of B by the first column of A, as shown below:

$$\begin{bmatrix} 2 & 1 & 0 \end{bmatrix} \cdot \begin{bmatrix} 0 \\ 4 \\ 2 \end{bmatrix} = 2*0 + 1*4 + 0*2 = 4$$

At this point we have the following entries in $B \times A$:

$$\begin{bmatrix} 10 & 1 & \cdot \\ 4 & \cdot & \cdot \end{bmatrix}$$

Using a similar procedure we can fill in the remaining entries to get

$$B \times A = \begin{bmatrix} 10 & 1 & 3 \\ 4 & 3 & 10 \end{bmatrix}$$

(b) This matrix does not exist since the dimensions do not correctly match (A has 3 columns but B has only 2 rows).

(c) $B \times D$ is a 2×4 matrix as shown below:

$$B \times D = \begin{bmatrix} 29 & 6 & 22 & 4 \\ 2 & 2 & 11 & 14 \end{bmatrix}$$

(d) A^2 is a 3×3 matrix as shown below:

$$A^2 = \begin{bmatrix} 14 & 1 & 5 \\ 4 & 5 & 20 \\ 2 & 2 & 11 \end{bmatrix}$$

(e) A^3 can be found by multiplying A^2 above by A:

$$A^3 = \begin{bmatrix} 14 & 15 & 75 \\ 60 & 9 & 40 \\ 30 & 4 & 21 \end{bmatrix}$$

(f) B^2 does not exist since only a square matrix can be multiplied to itself.

As can be seen above, matrix multiplication is not commutative, that is we cannot change the order (AB versus BA) and expect the same outcome.

D   Algorithm Efficiency

In Section 2.2 we spent time discussing efficiency of algorithms, in particular the challenges of using Brute Force to solve a Traveling Salesman problem. At other times throughout this book, algorithm efficiency and performance were mentioned to explain why a specific problem was difficult to solve. This section will elaborate on what algorithm efficiency means and how mathematicians determine what makes one algorithm more efficient than another, though from a nontechnical approach.

The algorithms we have studied throughout this book have been written so that *you*, the reader, could perform the necessary computations to find a solution. However, all of these algorithms can be written in such a way as to have a computer perform the calculations and provide the answer. When doing so, the information from a graph needs to be encoded in such a way as to allow the algorithm to pull the requisite information needed for the computations. For example, the Traveling Salesman Problem would need as an input the name of the n cities and the weights of the $\frac{n(n-1)}{2}$ edges in the complete graph K_n. It is possible for a given problem to have more than one way to encode the required information, so when evaluating the performance of an algorithm we will assume a standard method of encoding the input has been set and that this is done as efficiently as possible.

In addition, algorithm performance is generally evaluated based on a worst case scenario. Although an algorithm may quickly solve one instance of a problem, it is still possible to be quite slow for solving another instance. We will be concerned with how poorly an algorithm can perform over all possible instances and this evaluation is often given in terms of the running time. While the running time is dependent on the encoding scheme and computing power, changing either of these parameters does not change the complexity of the algorithm; that is, as we will see later in this section, improvements in computing power will not drastically affect the overall running time for an inefficient algorithm with large inputs.

When discussing the Brute Force Algorithm for the Traveling Salesman Problem in Section 2.2.1, a rough estimate for the number of required calculations was determined to be $\frac{(n+1)!}{2n}$. The estimate is given in terms of n, which is the number of cities represented in the complete graph K_n. Essentially, algorithm performance is a function $f(n)$ whose input is the size of the graph and whose output is the number of required calculations.

Definition D.1 The *performance function* $f(n)$ for a graph theory algorithm is a function whose input is n, the number of vertices in the graph, and whose output is the number of required calculations to complete the algorithm.

Using this language, the performance function for Brute Force is $f(n) = \frac{(n+1)!}{2n}$.

Consider for a moment two algorithms, one whose performance function is $n^2 + 3n + 1$ and another whose performance function is n^2. When n is small, these would give slightly different outputs for running time, but as n gets larger the addition of $3n + 1$ matters much less than the common n^2 component. In essence, we consider these two algorithms to have the same complexity due to the common highest power of their performance functions. More formally, these functions would be of the same *order*.

> **Definition D.2** Given a function f, it has **order at most** g, denoted $f(n)$ is $O(g(n))$, if there exists a constant c and a nonnegative integer a so that $|f(n)| \le c|g(n)|$ for all $x \ge a$.

Using this notation $f(n) = n^2 + 3n + 1$ is $O(n^2)$ since for $n \ge 1$ we have $n^2 + 3n + 1 \le n^2 + 3n^2 + n^2 = 5n^2$ (so our constants from the definition above are $c = 5$ and $a = 1$).

Algorithms with performance functions of order at most n^m, for some integer m, are called **polynomial-time algorithms**. These are, in some sense, considered "good" algorithms as they run fairly fast even as the size of the input grows. The table below gives an analysis of how various values of m impact the running time as the size of the input grows (values are rounded slightly). Note that these were calculated in the same manner as those for the Brute Force Algorithm on page 86, where we are using the best known supercomputer at the time of publication.

n	n^2	n^5	n^{10}
10	3×10^{-15} seconds	3×10^{-12} seconds	3×10^{-8} seconds
50	7.5×10^{-14} seconds	9.5×10^{-9} seconds	0.3 seconds
100	3×10^{-13} seconds	3×10^{-7} seconds	50.5 minutes
500	7.5×10^{-12} seconds	9.5×10^{-4} seconds	938 years
1000	3×10^{-11} seconds	0.03 seconds	$960,903$ years

Notice that when the polynomial has degree 5, an algorithm with input size 1000 can still be completed in less than a second! Also, even when the polynomial has degree 10, run-time only becomes infeasible after 100 inputs. In fact, only after the input size is 250 does the run-time take longer than a year and at 200 the run-time is around 1 month.

If polynomial-time algorithms are considered "good" algorithms, what constitutes a "bad" algorithm? One such type is called an **exponential-time algorithm**, where the performance functions are of order at most k^n for some constant k. For example $f(n) = 2^n + 5n + 1$ is of order at most 2^n since for all $n \ge 0$ we have $2^n + 5n + 1 \le 2^n + 5 \cdot 2^n + 2^n = 7 \cdot 2^n$ (so our constants from the definition above are $c = 7$ and $a = 0$). An algorithm with this performance function is considered an inefficient algorithm since small increases in the input size result in large increases in the running time, as shown in the following table.

The calculations we computed for Brute Force, gave a performance function in terms of a factorial, which is another type of inefficient algorithm called a *factorial-time algorithm* since it is of order at most $n!$. Factorial-time algorithms are in fact one of the worst in terms of their performance function. Note that there exists a more complex algorithm for solving a Traveling Salesman Problem (the Held-Karp algorithm) that has complexity $n^2 2^n$, but this still puts it in the category of inefficient algorithms since it is within the class of exponential-time algorithms.

n	2^n	3^n	$n!$
10	3×10^{-14} seconds	1.8×10^{-12} seconds	1×10^{-10} seconds
50	0.3 seconds	252 days	2.9×10^{40} years
100	1.2 million years	4.9×10^{23} years	8.9×10^{133} years
500	3×10^{126} years	3.4×10^{214} years	1.1×10^{1110} years
1000	1×10^{277} years	1.2×10^{453} years	3.8×10^{2543} years

Notice how smaller inputs for these types of algorithms quickly produce infeasible run-times. For example, even with input size 50 all of the polynomial-time algorithms in the table above had run times under a second; however, only the 2^n algorithm is reasonable. To understand the scale of the entries above, scientists believe the age of the earth to be 4.54 billion years, which is 4.54×10^9 years. Examining these algorithms with an input size of 100 shows the widening gap between polynomial-time and exponential-time or factorial-time algorithms.

All of the calculations above are based on a fixed computing power. One might ask if major improvements in computers would drastically change the feasibility of these algorithms. For example, if we are able to increase the number of calculations per second by a factor of 1000, how much faster would a n^2 algorithm run versus a 2^n algorithm? A few categories of algorithms are shown below with various factors of increase. These are all evaluated at $n = 100$.

Factor increase	n^{10}	2^n	$n!$
1	50.5 minutes	1.2 million years	8.9×10^{133} years
100	30.3 seconds	$12{,}180$ years	8.9×10^{131} years
1000	3.03 seconds	$1{,}218$ years	8.9×10^{130} years
1000000	0.003 seconds	1.2 years	8.9×10^{127} years

As you can see, polynomial-time algorithms gain more of a savings from these increase factors. Is it surprising that the exponential-time and factorial-time algorithms do not gain as much of a savings? In fact, even with a computer that is a million times faster than the current best supercomputer, a factorial-time algorithm remains infeasible with only an input size of 100.

The algorithm complexities described above were chosen due to their prevalence in graph theory problems (for example, Dijkstra's Algorithm, from Section 2.3.1 is of order n^2). However, other time complexities exist and can be similarly analyzed. A *linear-time algorithm* has order n and a *logarithmic-time algorithm* has order $\log n$. Algorithms of these orders are even more efficient than polynomial-time algorithms and fall into the class of "good" algorithms. Additionally, some graph algorithm complexities are given in terms of both the number of vertices and the number of edges. For example, Kruskal's Algorithm is of order $m \log n$, where m is the number of edges and n is the number of vertices. However, since the number of edges in a simple graph is no more than the number of edges in a complete graph ($\frac{n(n-1)}{2}$) any performance function using m, the number of edges, as an input can replace m with n^2 to give a rough estimate only in terms of the number of vertices.

When mathematicians and computer scientists describe algorithm complexity, they are often concerned with the overall class to which the algorithm belongs. Complexity classes are used to describe which algorithms are "good" and which are "bad." The two most commonly referenced classes are P and NP.

> **Definition D.3** Problems that can be solved by a deterministic sequential machine using at worst a polynomial-time algorithm belong to *class P*.
>
> Problems for which there is no known polynomial-time solution algorithm but for which a proposed solution can be verified in polynomial-time belong to *class NP*.

The definition above uses the term *deterministic sequential machine* which, roughly speaking, is the equivalent of the modern computer; given a set of steps, such as those in an algorithm, a deterministic sequential machine can solve a problem and produce an output. Polynomial-time, linear-time, and logarithmic-time algorithms fall into class P, whereas exponential-time and factorial-time algorithms fall into class NP. For a more technical discussion of complexity theory, see [73].

Many of the problems discussed in this book are from class P, for example finding an Eulerian circuit (Fleury's Algorithm), a minimum spanning tree (Kruskal's or Prim's Algorithm), a shortest path (Dijkstra's Algorithm), or a maximum matching (Augmenting Path Algorithm). However, other problems are known to be in NP, such as the Traveling Salesman Problem and finding an optimal coloring. In fact, the Traveling Salesman Problem is a classic example of an NP problem that belongs to a subclass of NP, called

N P-Complete. NP-Complete problems are considered to have equivalent complexities since if any one of the problems can be solved in polynomial-time, then all others can be solved in polynomial-time.

A major question in complexity theory is whether $P = NP$ or $P \neq NP$. Another way of phrasing this is if any problem that can be verified in polynomial-time can also be solved in polynomial-time. In fact, this problem is of such importance to the fields of mathematics and computer science that it was named one of the Millennium Problems by the Clay Mathematics Institute (CMI) in 2000. CMI chose seven problems believed to be of great importance to mathematics in the new millennium, and where a prize of one million dollars would be awarded for a published and verifiable solution to any of these problems. The details behind P vs. NP get quite technical, especially the deeper into the research you dive, but a final note is warranted. If $P = NP$, then our current method of Internet encryption (which is based on integer factorization, a known NP problem) would be essentially useless.

E   Pseudocode

Throughout this book, we have discussed not just the theory of graph theory, but also the applications and various algorithms used to find solutions to topics of interest. This Appendix provides pseudocode to some of the algorithms mentioned in this book, most of which are written in the same format as from [52].

Fleury's Algorithm
From Section 2.1.4 (see page 55)

procedure FLEURY(G)
    if the graph is not eulerian or semi-eulerian **then**
        return None
    else
        tour $\leftarrow \emptyset$
        if the graph has no edges **then**
            return tour
        if the graph is semi-eulerian **then**
            $u \leftarrow$ either of the two odd vertices
        else
            $u \leftarrow$ any vertex
        while there are edges in the graph **do**
            found $\leftarrow$ False
            for v adjacent to u **do**
                Remove edge (u, v) from A
                if A is connected **then**
                    found $\leftarrow$ True
                else
                    Reinsert edge (u, v) into A
            if not found **then**
                Remove edge (u, v) from A
                Remove vertex u from A
            Append edge (u, v) to tour
            $u \leftarrow v$
    return tour

Hierholzer's Algorithm
From Section 2.1.4 (see page 59)

procedure HIERHOLZER(G)
    if the graph is not eulerian or semi-eulerian **then**
        return none
    if graph is semi-eulerian **then**
        $u \leftarrow$ either of the two odd vertices
    else
        $u \leftarrow$ any vertex
    subtour $\leftarrow \emptyset$
    tour $\leftarrow u$
    repeat
        $u \leftarrow$ vertex in tour with unvisited edge
        subtour $\leftarrow u$
        $v = u$
        while $w \neq u$ **do**
            $(v, w) \leftarrow$ take unvisited edge leaving v
            subtour $\leftarrow$ subtour $\cup \{w\}$
            $v = w$
        Integrate subtour in tour
    until tour is eulerian circuit or trail

Dijkstra's Algorithm
From Section 2.3.1 (see page 102)

procedure DIJKSTRA($G, w, s; d$)
    $d(s) \leftarrow 0, T \leftarrow V$
    for $v \in V \setminus \{s\}$ **do**
        $d(v) \leftarrow \infty;$
    while $T \neq \emptyset$ **do**
        find some $u \in T$ such that $d(u)$ is minimal
        $T \leftarrow T \setminus \{u\}$
        for $v \in T \cap A_u$ **do**
            $d(v) \leftarrow \min\{d(v), d(u) + w(uv)\}$

Kruskal's Algorithm

From Section 3.1.1 (see page 126)

---

   procedure KRUSKAL(G, v)
      $F \leftarrow E$;
      $A \leftarrow \varnothing$;
      while $|A| < n - 1$ **do**
         find $e \in F$ such that $v(e)$ is minimum;
         $F \leftarrow F - \{e\}$;
         if $G(A \cup \{e\})$ acyclic **then**
            $A \leftarrow A \cup \{e\}$;
      -- $G(A)$ is a minimum spanning tree

Prim's Algorithm

From Section 3.1.1 (see page 131)

---

   procedure PRIM($G, w; T$)
      $g(1) \leftarrow 0$;
      $S \leftarrow \varnothing$;
      $T \leftarrow \varnothing$;
      for $i \leftarrow 2$ **to** n **do**
         $g(i) \leftarrow \infty$;
      while $S \neq V$ **do**
         choose $i \in V \setminus S$ such that $g(i)$ is minimal;
         $S \leftarrow S \cup \{i\}$;
         if $i \neq 1$ **then**
            $T \leftarrow T \cup \{e(i)\}$;
         for $j \in A_i \cap (V \setminus S)$ **do**
            if $g(j) > w(ij)$ **then**
               $g(j) \leftarrow w(ij)$;
               $e(j) \leftarrow ij$;

Depth First Search
From Section 3.3.1 (see page 147)

> **procedure** DFS($G, s; nr, p$)
>> **for** $v \in V$ **do**
>>> $nr(v) \leftarrow 0$;
>>> $p(v) \leftarrow 0$;
>>
>> **for** $e \in E$ **do**
>>> $u(e) \leftarrow$ false;
>>
>> $i \leftarrow 1$;
>> $v \leftarrow s$;
>> $nr(s) \leftarrow 1$;
>> **repeat**
>>> **while** there exists $w \in A_v$ with $u(vw) =$ false **do**
>>>> choose some $w \in A_v$ with $u(vw) =$ false;
>>>> $u(vw) \leftarrow$ true;
>>>> **if** $nr(w) = 0$ **then**
>>>>> $p(w) \leftarrow v$;
>>>>> $i \leftarrow i + 1$;
>>>>> $nr(w) \leftarrow i$;
>>>>> $v \leftarrow w$;
>>>
>>> $v \leftarrow p(v)$;
>> **until** $v = s$ **and** $u(sw) =$ true for all $w \in A_s$;

Breadth First Search
From Section 3.3.2 (see page 150)

> **procedure** BFS($G, s; d$)
>> $Q \leftarrow \varnothing$;
>> $d(s) \leftarrow 0$;
>> append s to Q;
>> **while** $Q \neq \varnothing$ **do**
>>> remove the first vertex v from Q;
>>> **for** $w \in A_v$ **do**
>>>> **if** $d(w)$ is undefined **then**
>>>>> $d(w) \leftarrow d(v) + 1$;
>>>>> append w to Q;

Edmonds-Karp Algorithm
From Section 4.4 (see page 191)

procedure EDMONDS-KARP(G, c, s, t)
    for $e \in E$ **do**
        $f(e) \leftarrow 0$
    Label s with $(-, \infty)$;
    for $v \in V$ **do**
        $u(v) \leftarrow false$
        $d(v) \leftarrow \infty$
    repeat
        among all vertices with $u(V) = false$, let v be the vertex which was labelled first
        for $e \in \{e \in E : e^- = v\}$ **do**
            if $w = e^+$ is not labelled **and** $f(e) > 0$ **then**
                $d(w) \leftarrow \min\{c(e) - f(e), d(v)\}$
                label w with $(v, +, d(w))$
        for $e \in \{e \in E : e^+ = v\}$ **do**
            if $w = e^-$ is not labelled **and** $f(e) > 0$ **then**
                $d(w) \leftarrow \min\{f(e), d(v)\}$
                label w with $(v, -, d(w))$
        $u(v) \leftarrow true$;
        if t is labelled **then**
            let d be the last component of the label of t;
            $w \leftarrow t$;
            while $w \neq s$ **do**
                find the first component v of the label of w;
                if the second component of the label of w is $+$ **then**
                    set $f(e) \leftarrow f(e) + d$ for $e = vw$;
                else
                    set $f(e) \leftarrow f(e) - d$ for $e = wv$
                $w \leftarrow v$
            delete all labels except for the label of s;
            for $v \in V$ **do**
                 $d(v) \leftarrow \infty$; $u(v) \leftarrow false$
    until $u(v) = true$ for all vertices v which are labelled;
    S is set of vertices that are labelled and put $T \leftarrow V - S$.

Hungarian Algorithm
From Section 5.1.1 (see page 224)

```
procedure HUNGARIAN(n, w; mate)
    for v ∈ V do
        mate(v) ← 0;
    for i ← 1 to n do
        u_i ← MAX{w_ij : j ← 1, ..., n};
        v_i ← 0;
    nrex ← n;
    while nrex ≠ 0 do
        for i ← 1 to n do
            m(i) ← false;
            p(i) ← 0;
            δ_i ← ∞;
        aug ← false;
        Q ← {i ∈ S : mate(i) ← 0};
        repeat
            remove an arbitary vertex i from Q;
            m(i) ← true;
            j ← 1;
            while aug = false and j ≤ n do
                if mate(i) ≠ j' then
                    if u_i + v_j − w_ij < δ_j then
                        δ_j ← u_i + v_j − w_ij;
                        p(j) ← i;
                        if δ_j = 0 then
                            if mate(j') = 0 then
                                AUGMENT(mate, p, j'; mate);
                                aug ← true;
                                nrex ← nrex − 1;
                            else
                                Q ← Q ∪ mate(j');
                j ← j + 1;
```

Hungarian Algorithm (continued)

$$\begin{aligned}
&\textbf{if } aug = \text{false} \textbf{ and } Q = \varnothing \textbf{ then}\\
&\quad J \leftarrow \{i \in S : m(i) \leftarrow \text{true}\};\\
&\quad K \leftarrow \{j' \in T : \delta_j \leftarrow 0\};\\
&\quad \delta \leftarrow \text{MIN}\{\delta_j : j' \in T \setminus K\};\\
&\quad \textbf{for } i \in J \textbf{ do}\\
&\qquad u_i \leftarrow u_i - \delta;\\
&\quad \textbf{for } j' \in K \textbf{ do}\\
&\qquad v_j \leftarrow v_j + \delta;\\
&\quad \textbf{for } j' \in T \setminus K \textbf{ do}\\
&\qquad \delta_j \leftarrow \delta_j - \delta;\\
&\quad X \leftarrow \{j' \in T \setminus K : \delta_j \leftarrow 0\};\\
&\quad \textbf{if } mate(j') \neq 0 \text{ for all } j' \in X \textbf{ then}\\
&\qquad \textbf{for } j' \in X \textbf{ do}\\
&\qquad\quad Q \leftarrow Q \cup \{mate(j')\};\\
&\quad \textbf{else}\\
&\qquad \text{choose } j' \in X \text{ with } mate(j') = 0;\\
&\qquad \text{AUGMENT}(mate, p, j'; mate);\\
&\qquad aug \leftarrow \text{true};\\
&\qquad nrex \leftarrow nrex - 1;\\
&\textbf{until } aug = \text{true};
\end{aligned}$$

Augment (for Hungarian Algorithm)

$$\begin{aligned}
&\textbf{procedure } \text{AUGMENT}(mate, p, j'; mate)\\
&\quad \textbf{repeat}\\
&\qquad i \leftarrow p(j)\\
&\qquad mate(j') \leftarrow i\\
&\qquad next \leftarrow mate(i)\\
&\qquad mate(i) \leftarrow j'\\
&\qquad \textbf{if } next \neq 0 \textbf{ then}\\
&\qquad\quad j' \leftarrow next\\
&\quad \textbf{until } next = 0
\end{aligned}$$

Edmond's Blossom Algorithm
From Section 5.2.1 (see page 244)

procedure EDMONDS(G; $mate$, $nrex$)
    $nrex \leftarrow n$
    for $i = 1$ **to** n **do**
        $mate(i) \leftarrow 0$
    for $k = 1$ **to** $n - 1$ **do**
        if $mate(k) = 0$ **and** there exists $j \in A_k$ with $mate(j = 0)$ **then**
            choose $j \in A_k$ with $mate(j) = 0$;
            $mate(j) \leftarrow k$; $mate(k) \leftarrow j$; $nrex \leftarrow nrex - 2$
    $r \leftarrow 0$
    while $nrex \geq 2$ **and** $r \leq n - 2$ **do**
        $r \leftarrow r + 1$
        if $mate(r) = 0$ **then**
            $Q \leftarrow \emptyset$; $aug \leftarrow false$; $m \leftarrow 0$
            for $v \in V$ **do**
                $p(v) \leftarrow 0$; $d(v) \leftarrow -1$; $a(v) \leftarrow true$
                $CA(v) \leftarrow A_v$
            $d(r) \leftarrow 0$; append r to Q
            while $aug = false$ **and** $Q \neq \emptyset$ **do**
                remove the first vertex x of Q
                if $a(x) = true$ **then**
                    $cont \leftarrow false$
                    for $y \in CA(x)$ **do**
                        $u(y) \leftarrow false$
                repeat
                  choose $y \in CA(x)$ with $u(y) = false$; $u(y) \leftarrow true$
                  if $a(y) = true$ **then**
                    if $d(y)$ even **then**
                      $m \leftarrow m + 1$
                      BLOSSOM$(x, y; B, w)$
                      CONTRACT(B, m, w)
                else
                  if $d(y) = -1$ **then**
                    if $mate(y) = 0$ **then**
                      AUGMENT(x, y)
                  else
                    $z \leftarrow mate(y)$
                    $p(y) \leftarrow x$; $d(y) \leftarrow d(x) + 1$
                    $p(z) \leftarrow y$; $d(z) \leftarrow d(y) + 1$
                    insert z with priority $d(z)$ into Q
                until $u(y) = true$ for all $y \in CA(v)$ **or** $aug = true$
                or $cont = true$

Blossom (from Edmond's Blossom Algorithm)

---

procedure BLOSSOM$(x, y; B, w)$
    $P \leftarrow \{x\}$;
    $P' \leftarrow \{y\}$;
    $u \leftarrow x$;
    $v \leftarrow y$;
    repeat
        $P \leftarrow P \cup \{p(u)\}$;
        $u \leftarrow p(u)$;
    until $p(u) = r$;
    repeat
        $P' \leftarrow P' \cup \{p(v)\}$;
        $v \leftarrow p(v)$;
    until $v = r$;
    $S \leftarrow P \cap P'$;
    let w be the element of S for which $d(w) \geq d(z)$ for all $z \in S$;
    $B \leftarrow ((P \cup P') \setminus S) \cup \{w\}$;

Contract (from Edmond's Blossom Algorithm)

---

procedure CONTRACT(B, m, w)
    $b \leftarrow n + m$
    $a(b) \leftarrow true$
    $p(b) \leftarrow p(w)$
    $d(b) \leftarrow d(w)$
    $mate(b) \leftarrow mate(w)$
    insert b into Q with priority $d(b)$
    $CA(b) \leftarrow \cup_{z \in B} CA(z)$
    for $z \in CA(b)$ **do**
        $Ca(z) \leftarrow CA(z) \cup \{b\}$
    for $z \in B$ **do**
        $a(z) \leftarrow false$
    for $z \in CA(b)$ **do**
        if $a(z) = true$ **and** $p(z) \in B$ **then**
            $d(z) \leftarrow d(b) + 1$
            $p(z) \leftarrow b$
            $d(mate(z)) \leftarrow d(z) + 1$
    $cont \leftarrow true$

Augment (from Edmond's Blossom Algorithm)

---

procedure AUGMENT(x, y)
    $P \leftarrow \{y, x\}$;
    $v \leftarrow x$;
    while $p(v) \neq 0$ **do**
        $P \leftarrow P \cup \{p(v)\}$;
        $v \leftarrow p(v)$;
    while there exists $b \in P$ with $b > n$ **do**
        choose the largest $b \in P$ with $b > n$
        $B \leftarrow B(b - n)$; $w \leftarrow w(b - n)$; $z \leftarrow mate(w)$
        let q be the neighbor of b on P different from z
        choose some $q' \in B \cap CA(q)$
        determine the alternating path B' of even length in B from w to q'
        replace b by w in P
        insert B' into P between w and q
    $u \leftarrow y$; $v \leftarrow x$
    while $v \neq r$ **do**
        $z \leftarrow mate(v)$; $mate(v) \leftarrow u$; $mate(u) \leftarrow v$
        $u \leftarrow z$; let v be the successor of z on P
    $mate(v) \leftarrow u$; $mate(u) \leftarrow v$
    $nrex \leftarrow nrex - 2$; $aug \leftarrow true$

Gale-Shapley Algorithm
From Section 5.3 (see page 252)

---

procedure GALE-SHAPLEY$(G = (M \cup W, E))$
    Initialize all $m \in M$ and $w \in W$ to *free*
    while *free* man m who still has a woman w to propose to **do**
        w = first woman on m's list to whom m has not yet proposed
        if w is free **then**
            (m, w) become engaged
        else
            some pair (m', w) already exists
            if w prefers m to m' **then**
                m' becomes free
                (m, w) become engaged
            else
                (m', w) remain engaged

First-Fit Coloring Algorithm
From Section 6.4.1 (see page 310)

procedure $FF(G,\ V = \{v_1, \ldots, v_n\})$
    for $i = 1$ to n **do**
      $L(i) = \{1, \ldots, i\}$
      for $i = 1$ to n **do**
        Set $c(i) =$ the first color in $L(i)$
        for j with $i < j$ and $v_i v_j \in E(G)$ **do**
          $L(j) := L(j) - c(i)$
    return $c(1), \ldots c(n)$

Selected Hints and Solutions

Chapter 1: Graph Models, Terminology, and Proofs

1.1 (b) No. Loop at c and multi-edges between d and e.
(c) $\deg(a) = 2, \deg(b) = 2, \deg(c) = 3, \deg(d) = 2, \deg(e) = 3, \deg(f) = 0$
(d) ab, bc
(e) $N(a) = \{b, e\}$

1.3 (b) Yes;   **(c)** Yes.

1.7 (a) Isomorphic $(a \to s, b \to y, c \to z, d \to w, e \to u, f \to t, g \to x, h \to v)$
(b) Not isomorphic (look at vertex degrees)

1.8 score sequences: $0, 1, 2, 3$; $0, 2, 2, 2$; $1, 1, 2, 2$; $1, 1, 1, 3$.

1.9 (a), (b), (d), (e), (f), and (h) are tournaments.

1.11 (a), (d), and (f) are strong; (b), (e), and (h) are not strong.

1.13 The matrix can only have 1 and 0 entries, with 0 entries on the diagonal.

1.15 Let G be a graph with m edges. Then by the Handshaking Lemma, we know $\sum_{v \in V(G)} \deg(v) = 2m$. Since m is an integer, we know this sum is even.

1.16 Use the Handshaking Lemma.

1.17 $2mn$

1.18 The maximum number of edges in a bipartite graph occurs in a complete bipartite graph $K_{m,n}$ and with 15 vertices this would be when $m = 8$ and $n = 7$. But this graph would have $8 * 7 = 56$ edges.

1.20 (b) C_4 **(d)** Use Corollary 1.22

Chapter 2: Graph Routes

2.1 (b) yes **(c)** no
(d) $\deg(a) = 3$, $\deg(b) = 3$, $\deg(c) = 3$, $\deg(d) = 2$, $\deg(e) = 4$
(e) ab, be, cb
(f) b, e
(g) Answers may vary: walk: $abae$ trail: $aeab$ path:$aebc$
(h) Answers may vary: closed walk ebe circuit: $eaebcde$ cycle: $ebcde$
(i) semi-eulerian since there are two odd vertices.

2.2 eulerian circuits: (a), (b), (d); hamiltonian cycles: (c), (e)

2.3 (a) (i) $\deg(a) = 4$, $\deg(b) = 4$, $\deg(c) = 4$, $\deg(d) = 2$, $\deg(e) = 6$, $\deg(f) = 3$, $\deg(g) = 4$, $\deg(h) = 3$; (ii) semi-eulerian; (iii) one solution: $fabcaegbefghcdeh$
(c) (i) $\deg(a) = 2$, $\deg(b) = 2$, $\deg(c) = 4$, $\deg(d) = 2$, $\deg(e) = 2$, $\deg(f) = 3$, $\deg(g) = 2$, $\deg(h) = 3$; (ii) neither
(e) (i) $\deg(a) = 4$, $\deg(b) = 4$, $\deg(c) = 2$, $\deg(d) = 2$, $\deg(e) = 4$, $\deg(f) = 2$; (ii) eulerian; (iii) one solution: $abfebcadea$

2.4 (a) $abcdefbecfa$
(c) $jbcdefghabihmfkdjimmkjm$

2.5 hamiltonian: (b), (c), and (d); non-hamiltonian: (a), (e), and (f)

2.6 (a) (i) $adbcea$, weight $= 21$; $bdaceb \rightarrow acebda$, weight $= 19$; $cedabc \rightarrow abceda$, weight $= 18$; $dabced \rightarrow abceda$, weight $= 18$; $ecbade \rightarrow abceda$, weight $= 18$; (ii) $adbcea$, weight $= 21$; (iii) $abceda$, weight $= 18$
(c) (i) $acbfdea$, weight $= 23$; $bcdeafb$, weight $= 27$; $cbfdeac$, weight $= 23$; $debcafd$, weight $= 27$; $edcbfae$, weight $= 27$; $fbcdeaf$, weight $= 27$ (answers may vary in the case of ties); (ii) $aedcbfa$, weight $= 27$; (iii) $acbfdea$, weight $= 23$
(e) (i) $jnkmopj$, weight $= 1548$; $kmjnpok$, weight $= 1442$; $mknjopm$, weight $= 1483$; $njmkopn$, weight $= 1442$; $opnjmko$, weight $= 1351$; $pomknjp$, weight $= 1548$ (ii) $jmkopnj$, weight $= 1442$; (iii) $jnkmpoj$, weight $=1483$

2.9 A graph is eulerian if and only if it is connected and all vertices are even. A graph is semi-eulerian if and only if it is connected and exactly two vertices are odd. A graph cannot have all even vertices while still having two odd vertices. Thus a graph cannot be both eulerian and semi-eulerian.

2.15 (a) P_k (b) Start with a path P_{k-4} and attach enough edges to each vertex to give it degree at least k. At the other endpoint of these edges attach a copy of K_{k-1}.

2.23 $m = n$

Chapter 3: Trees

3.1 (d) and (e) are both trees; (a), (f), and (g) are all not trees; (b) and (c) may or may not be trees.

3.3 Weights of minimum spanning trees: (a) 11 (b) 65 (c) 21 (d) 28 (e) 1018.

3.4 Weight: 17

3.5 (a) $(1, 2, 6, 2, 1)$ (b) $(7, 7, 7, 7, 7, 2)$

3.7 (a) 0; 2; 3; 4 (b) 4
(c) none; r; a; b; f
(d) a, b; d, e, f; g, h; i; j, k
(e) none; r; r, b; r, a, c; r, b, f, i
(f) $a, b, c, d, e, f, g, h, i, j, k$; c, g, h; none; none; none
(g) b; d, e; g; none

3.9 Total weight is 1608.

3.12 If a graph is very sparse (the number of edges is close to $n - 1$), then Reverse Delete would only need a few steps to obtain a tree. However, if a graph is very dense (so the number of edges is much greater than $n - 1$), then using Reverse Delete would require many more steps than Kruskal's to obtain a minimum spanning tree.

3.13 (a) spanning forest of the disconnected graph
(b) spanning tree of the component containing the root vertex

3.14 Insert a step 0 that chooses the required edge. Continue with the algorithms as before. The resulting tree might be minimum but cannot be guaranteed.

3.15 (a) Each component is either an isolated vertex or a tree with at least two vertices.
(b) Each component of a forest is a tree and so contains one fewer edge then the number of vertices in the component.

3.16 Show for all vertices in $T - v$ there exists a path between them.

3.17 Use the Handshaking Lemma.

3.18 Use Corollary 3.7.

3.19 Show two neighbors of the non-leaf x are in different components of $T - x$.

3.21 Remove a vertex x of degree k and consider what the components of $T - x$ must be.

3.22 (a) T has an odd number of vertices, not all of which can be of odd degree.

3.28 Use the hydrogen-depleted graph. For pentane there are 3 possible trees on 5 vertices and for hexane there are 5 possible trees on 6 vertices (with maximum degree 4).

Chapter 4: Connectivity and Flow

4.3 (a) blocks: $\{a,c\}, \{b,c\}, \{c,d\}, \{d,e,f,g\}$
(b) blocks: $\{a,b,e\}, \{c,e\}, \{d,e\}, \{e,f\}, \{f,g,h,i\}$

4.5 (a) $\kappa = 1, \kappa' = 1$ (c) $\kappa = 3, \kappa' = 3$

4.6 (a) Flow $= 16$, $P = \{s\}, \overline{P} = \{a,b,c,d,e,f,g,h,t\}$
(b) Flow $= 20$, $P = \{s,a,b,c,d\}, \overline{P} = \{e,f,g,h,i,t\}$

4.7 $K_{1,5}$; $K_{1,k}$

4.9 $\kappa = \kappa' = 1$

4.12 $K_{1,8}$

4.13 For S to be a cut-set of G, there must be at least two vertices that are disconnected in $G - S$.

4.15 (a) Consider an endpoint of the bridge.
(b) Not true. Use a bow-tie graph.

4.16 Consider if the graph is connected or disconnected. If connected, use an eulerian circuit.

4.17 If e lies on every $x - y$ path then removing it will disconnect x and y. Use a contradiction argument for the converse (Assume e is a bridge but for all vertices x and y there exists an $x - y$ path not containing e. Prove this means e is not a bridge)

4.20 Use Whitney's Theorem.

4.23 Since G is k-connected, we know that there exist k internally disjoint paths between any two vertices of G. Subdividing an edge along any of these paths does not effect the number of internally disjoint paths.

Chapter 5: Matching and Factors

5.1 (a) one solution: $a \cdots b - c \cdots i - k \cdots h - d$; not augmenting since a is saturated by M
(b) one solution: $d - h \cdots k - j$
(c) not perfect, not maximum, not maximal; add edge de to M to get a maximum matching

5.3 Answers may vary.
(a) ag, ci, dh **(b)** aj, bg, ch, di, ef
(c) ah, bg, cf, ej **(d)** ab, cd, ef, gh, ij
(e) ab, cf, de, gh **(f)** af, bg, ch, di, ej

5.4 (a) (a), (b), (d), and (f) are bipartite.

5.6 Answers may vary. Benefits $\leftrightarrow$ Agatha, Computing $\leftrightarrow$ Leah, Purchasing $\leftrightarrow$ George, Recruitment $\leftrightarrow$ Dinah, Refreshments $\leftrightarrow$ Nancy, Social Media $\leftrightarrow$ Evan, Travel Expenses $\leftrightarrow$ Vlad

5.7 Answers may vary. Adam $\leftrightarrow$ Statistics, Chris $\leftrightarrow$ Calculus, Dave $\leftrightarrow$ Real Analysis, Maggie $\leftrightarrow$ Abstract Algebra, Roland $\leftrightarrow$ Geometry, Hannah $\leftrightarrow$ Topology

5.8 (a) Make two vertices for each professor.

5.10 (a) Rich $\leftrightarrow$ Alice, Stefan $\leftrightarrow$ Dahlia, Tom $\leftrightarrow$ Beth, Victor $\leftrightarrow$ Cindy
(b) Alice $\leftrightarrow$ Rich, Beth $\leftrightarrow$ Stefan, Cindy $\leftrightarrow$ Victor, Dahlia $\leftrightarrow$ Tom

5.16 $C_n, P_n, K_n : \lfloor \frac{n}{2} \rfloor$; $K_{m,n} = \min\{m, n\}$

5.17 A hamiltonian graph has a spanning cycle, which is a 2-factor. If a graph is disconnected then it cannot be hamiltonian, even if it has a 2-factor. Also, the graph obtained by joining two copies of K_3 by an edge will be connected, have a 2-factor, but no hamiltonian cycle.

5.24 Argue by induction on k. If $k = 1$ then the graph is itself a collection of independent edges, and so is a 1-factor. Suppose now that $k \geq 2$ and for all regular bipartite graphs with degree less than k has a 1-factorization. Let G be a k-regular bipartite graph. Then by Corollary 5.5 we know G has a perfect matching M. Remove M to form a new graph G'. Then G' is still bipartite with each vertex having degree $k - 1$. Thus by the induction hypothesis G' has a 1-factorization. Together with M we have a 1-factorization of G.

Chapter 6: Graph Coloring

6.1 (a) $\chi(G_1) = 4$, $\chi(G_2) = 4$, $\chi(G_3) = 4$, $\chi(G_4) = 3$, $\chi(G_5) = 3$, $\chi(G_6) = 3$.
(b) $\chi'(G_1) = 5$, $\chi'(G_2) = 5$, $\chi'(G_3) = 4$, $\chi'(G_4) = 4$, $\chi'(G_5) = 5$, $\chi'(G_6) = 4$.
(c) perfect: G_1, G_3; not perfect: G_2, G_4, G_5, G_6

6.2 Answers may vary. Optimal number of colors is listed.
G_1: 8 colors. $a = \{5,6\}\, b = \{1,2,3\}\, c = \{7,8\}\, d = \{5,6\}\, e = \{1,2,3,4\}\, f = \{7,8\}\, g = \{1,2,3\}\, h = \{4\}$
G_2: 10 colors. $a = \{3,4,5,6\}\, b = \{8,9\}\, c = \{3,4,5,6,7\}\, d = \{8,9,10\}\, e = \{1,2\}\, f = \{3,4,5\}\, g = \{1\}\, h = \{7,8,9\}\, i = \{1,2\}\, j = \{10\}$
G_3: 11 colors. $a = \{8\}\, b = \{1,2,3,4\}\, c = \{9,10,11\}\, d = \{3,4\}\, e = \{5,6,7\}\, f = \{5,6,7,8\}\, g = \{1,2\}$

6.3 Answers may vary.
G_1: $a - 1, b - 4, c - 1, d - 2, e - 7, f - 5, g - 3, h - 6$
G_3: $a - 2, b - 3, c - 4, d - 1, e - 1, f - 2, g - 4$
G_5: $a - 1, b - 2, c - 3, d - 2, e - 1, f - 4, g - 2, h - 3, i - 4, j - 1, k - 2, m - 2$

6.4 Look for a wheel W_5

6.5 $\chi = 4$

6.10 Proposition 6.9 is better when there is a single vertex of large degree

6.11 Cycle $d\,e\,h\,f\,j\,n$ has no chord.

6.12 If the endpoints of the edge are already given different colors then the coloring remains proper when adding the edge. Otherwise at most one more color would be needed for one of the endpoints.

6.14 If G is bipartite then we can use a color for each partite set. If $\chi \leq 2$ we can split the vertices into sets X and Y based on the color given.

6.15 (a) Let $x \in V(G)$. Then $G - x$ can be colored in $k - 1$ colors. Suppose $\deg_G(x) < k - 1$. Then there is some color not used on any of the neighbors of x in $G - x$ and that color can be used for x in G, contradicting that $\chi(G) = k$. Thus $\delta(G) \geq k - 1$.

6.19 Bipartite graphs do not have odd cycles, and so cannot contain an induced odd cycle. Thus by the Strong Perfect Graph Theorem we know bipartite graphs are perfect.

-OR- We know $\chi(G) = 2 = \omega(G)$ for all bipartite graphs. Since any induced subgraph of a bipartite graph remains bipartite, we know that G is perfect.

6.22 Let each vertex have list size $\Delta(G) + 1$ and order the vertices as $v_1, v_2, \ldots, v_n$. Then applying First-Fit to this order, we know that each vertex has at most $\Delta(G)$ neighbors preceding it and so must have at least one color left in its list. Thus G can be colored with lists of size $\Delta(G) + 1$.

Chapter 7: Planarity

7.1 C_6; Every C_n (for $n \geq 3$) is a subdivision of K_3.

7.2 (a) and **(d)** are subdivisions; **(b)** is not since vertex g is on more than one path between vertices; **(c)** is not since extra edges were added to the $K_{3,3}$ graph.

7.5 (b), (c), (d) are planar; (a), (e) are nonplanar

7.6 $cr(K_7) = 9$; $cr(K_{3,4}) = 2$

7.7 Maximally planar means all regions are enclosed by a triangle, so no induced cycles of length more than three. So a graph with more than three vertices will always contain a K_4.

7.8 The statement is false.

7.12 Use the Handshaking Lemma and Theorem 7.8.

7.14 If G is maximally planar then number of edges is at most $3n - 6$, and so the degree sum is at most $6n - 12$. Thus the average degree is at most $6 - \frac{12}{n}$.

7.15 (a) $K_{4,4}$ with matching removed
(b) Use the Handshaking Lemma (do not claim it contains a $K_{3,3}$ subgraph)

7.17 (b) K_7

7.19 Use Theorem 7.7.

7.20 Use Theorem 7.9.

Bibliography

[1] V B Alekseev and V S Gončakov. "The thickness of an arbitrary complete graph." In: *Mathematics of the USSR-Sbornik* 30.2 (1976), pp. 187–202.

[2] Alexandru T. Balaban. "Applications of Graph Theory in Chemistry." In: *J. Chem. Inf. Comput. Sci.* 25 (1985), pp. 334–343.

[3] Lowell W. Beineke. "The decomposition of complete graphs into planar subgraphs." In: *Graph Theory and Theoretical Physics*. Academic Press, London, 1967, pp. 139–153.

[4] Lowell W. Beineke and Frank Harary. "The thickness of the complete graph." In: *Canadian J. Math.* 17 (1965), pp. 850–859. ISSN: 0008-414X.

[5] Claude Berge. "Farbung von Graphen, deren samtliche bzw. deren ungerade Kreise starr sind." In: *Wissenschaftliche Zeitschrift* (1961).

[6] Norman L. Biggs, E. Keith Lloyd, and Robin J. Wilson. *Graph Theory 1736-1936*. Oxford: Clarendon Press, 1976. ISBN: 0-19-853901-0.

[7] J.A. Bondy and V. Chvátal. "A method in graph theory." In: *Discrete Mathematics* 15 (1976), pp. 111–136.

[8] J.A. Bondy and U.S.R. Murty. *Graph Theory*. New York: Springer, 2008. ISBN: 978-1-84628-969-9.

[9] C. W. Borchardt. "Über eine Interpolationsformel für eine Art Symmetrischer Functionen und über Deren Anwendung." In: *Math. Abh. der Akademie der Wissenschaften zu Berlin* (1860), pp. 1–20.

[10] Rowland Leonard Brooks. "On colouring the nodes of a network." In: *Mathematical Proceedings of the Cambridge Philosophical Society*. Vol. 37. 2. Cambridge University Press. 1941, pp. 194–197.

[11] Arthur Cayley. "A theorem on trees." In: *Quart. J. Math.* 23 (1889), pp. 376–378.

[12] G. Chartrand, L. Lesniak, and P. Zhang. *Graphs & Digraphs, Fifth Edition*. Chapman & Hall book. ISBN: 9781439826270.

[13] Gary Chartrand. *Introductory Graph Theory*. New York: Dover, 1984. ISBN: 978-0486247755.

[14] Maria Chudnovsky et al. "The strong perfect graph theorem." In: *Annals of mathematics* (2006), pp. 51–229.

[15]   V. Chvátal. "On Hamilton's ideals." In: *J. Combinatorial Theory Ser. B* 12 (1972), pp. 163–168.

[16]   William Cook. *The Traveling Salesman Problem.* 2015. URL: http://www.math.uwaterloo.ca/tsp/index.html (visited on 06/08/2015).

[17]   William J. Cook. *In Pursuit of the Traveling Salesman.* Princeton, NJ: Princeton University Press, 2012. ISBN: 978-0-691-15270-7.

[18]   Béla Csaba et al. "Proof of the 1-factorization and Hamilton decomposition conjectures." In: *Mem. Amer. Math. Soc.* 244.1154 (2016), pp. v+164. ISSN: 0065-9266.

[19]   G. Dantzig, R. Fulkerson, and S. Johnson. "Solution of a large-scale traveling-salesman problem." In: *J. Operations Res. Soc. Amer.* 2 (1954), pp. 393–410. ISSN: 0160-5682.

[20]   Keith Devlin. *The Unfinished Game: Pascal, Fermat and the Seventeenth-Century Letter that Made the World Mondern.* New York, NY: Basic Books, 2008. ISBN: 978-0465018963.

[21]   Reinhard Diestel. *Graph Theory.* 3rd ed. New York: Springer, 2005. ISBN: 978-3-540-26182-7.

[22]   Edsger W. Dijkstra. "A note on two problems in connexion with graphs." In: *Numerishe Mathematik* 1 (1959), pp. 269–271.

[23]   Edsger. W. Dijkstra. "Reflections on [[22]]." Circulated privately. 1982. URL: http://www.cs.utexas.edu/users/EWD/ewd08xx/EWD841a.PDF.

[24]   G. A. Dirac and S. Schuster. "A theorem of Kuratowski." In: *Nederl. Akad. Wetensch. Proc. Ser. A. Indagationes Math.* 16 (1954), pp. 343–348.

[25]   G.A. Dirac. "Some theorems on abstract graphs." In: *Proc. Lond. Math. Soc.* 2 (1952), pp. 69–81.

[26]   Gregory D. Dreifus et al. "Path Optimization Along Lattices in Additive Manufacturing Using the Chinese Postman Problem." In: *3D Printing and Additive Manufacturing* 4.2 (June 2017). ISSN: 2329-7662.

[27]   David Easley and Jon Kleinberg. *Networks, Crowds and Markets.* New Tork, NY: Cambridge University Press, 2010. ISBN: 978-0-521-19533.

[28]   Jack Edmonds. "Paths, trees, and flowers." In: *Canadian Journal of mathematics* 17 (1965), pp. 449–467.

[29]   Jack Edmonds and Ellis L. Johnson. "Matching, Euler tours and the Chinese postman." In: *Mathematical Programming* 5 (1973), pp. 88–124.

[30]   Jack Edmonds and Richard M. Karp. "Theoretical Improvements in Algorithmic Efficiency for Network Flow Problems." In: 19.2 (1972). ISSN: 0004-5411.

[31] Susanna Epp. *Discrete mathematics with applications*. 5th ed. Boston, MA: Cengage Learning, 2020. ISBN: 978-1337694193.

[32] Paul Erdos, Arthur L Rubin, and Herbert Taylor. "Choosability in graphs." In: *Proc. West Coast Conf. on Combinatorics, Graph Theory and Computing, Congressus Numerantium*. Vol. 26. 1979, pp. 125–157.

[33] Leonhard Euler. "Solutio problematis ad geometriam situs pertinentis." In: *Commentarii Academiae Scientiarum Imperialis Petropolitanae* 8 (1736), pp. 128–140.

[34] Mark Fernandez. *High Performance Computing*. 2011. URL: http://en. community.dell.com/techcenter/high-performance-computing/w/ wiki/2329 (visited on 06/26/2015).

[35] Fleury. "Deux problèmes de géométrie de situation." In: *Journal de mathématiques élémentaires* 2 (1883), pp. 257–261.

[36] L. R. Ford Jr. and D. R. Fulkerson. "Maximal flow through a network." In: *Canadian J. Math.* 8 (1956), pp. 399–404. ISSN: 0008-414X.

[37] J.C. Fournier. *Graphs Theory and Applications: With Exercises and Problems*. ISTE. ISBN: 9781118623091.

[38] D. Gale and L.S. Shapley. "College Admissions and the Stability of Marriage." In: *The American Mathematical Monthly* 69 (1962), pp. 9–15.

[39] Martin Charles Golumbic and Ann N. Trenk. *Tolerance Graphs*. New Tork, NY: Cambridge University Press, 2004. ISBN: 0-521-82758-2.

[40] R.L. Graham and Pavol Hell. "On the history of the minimum spanning tree problem." In: *Annals of the History of Computing* 7 (1985), pp. 43–57.

[41] J.L. Gross, J. Yellen, and M. Anderson. *Graph Theory and Its Applications*. Textbooks in Mathematics. CRC Press, 2018. ISBN: 9780429757099.

[42] Mèï-gu Guan. "Graphical method of the odd-and-even point." In: *Sci. Sinica* 12 (1963), pp. 281–287. ISSN: 0582-236x.

[43] Dan Gusfield and Robert W. Irving. *The Stable Marriage Problem*. Cambridge, MA: The MIT Press, 2012. ISBN: 0-262-07118-5.

[44] Richard K. Guy. "Crossing numbers of graphs." In: *Graph theory and applications (Proc. Conf., Western Michigan Univ., Kalamazoo, Mich., 1972; dedicated to the memory of J. W. T. Youngs)*. 1972, pp. 111–124.

[45] P. Hall. "On Representatives of Subsets." In: *Journal of the London Mathematical Society* 10.1 (1935), pp. 26–30.

[46] John M. Harris, Jeffry L. Hirst, and Michael J. Mossinghoff. *Combinatorics and Graph Theory*. 2nd ed. New York: Springer, 2008. ISBN: 978-0-387-797710-6.

[47]   N. Hartsfield and G. Ringel. *Pearls in Graph Theory: A Comprehensive Introduction.* Dover Books on Mathematics. ISBN: 9780486315522.

[48]   JP Heawood. "On the four-color map problem." In: *Quart. J. Pure Math* 29 (1898), pp. 270–285.

[49]   Carl Hierholzer. "Über die möglichekeit, einen linienzug ohne wiederholung und ohne unterbrechung zu umfahren." In: *Mathematische Annalen* 6 (1873), pp. 30–32.

[50]   George Hutchinson. "Evaluation of Polymer Sequence Fragment Data Using Graph Theory." In: *Bulletin of Mathematical Biophysics* 31 (1969), 541–562.

[51]   Intel. *Processors - Intel Microprocessor Export Compliance Marks.* 2014. URL: http://www.intel.com/support/processors/sb/CS-032813.htm (visited on 07/10/2015).

[52]   Dieter Jungnickel. *Graphs, Networks and Algorithms.* 4th ed. New York: Springer, 2013. ISBN: 978-3-642-32277-8.

[53]   Alfred B Kempe. "On the geographical problem of the four colours." In: *American journal of mathematics* 2.3 (1879), pp. 193–200.

[54]   H.A. Kierstead. "A polynomial time approximation algorithm for dynamic storage allocation." In: *Discrete Appl. Math.* 88 (1991), pp. 231–237.

[55]   H.A. Kierstead and Karin R. Saoub. "First-Fit coloring of bounded tolerance graphs." In: *Discrete Appl. Math.* 159 (2011), pp. 605–611.

[56]   H.A. Kierstead and Karin R. Saoub. "Generalized Dynamic Storage Allocation." In: *Discrete Mathematics & Theoretical Computer Science* 16 (2014), pp. 253–262.

[57]   H.A. Kierstead, D. Smith, and W. Trotter. "First-Fit coloring of interval graphs has performance ratio at least 5." In: *European J. of Combin.* 51 (2016), pp. 236–254.

[58]   H.A. Kierstead and W. Trotter. "An extremal problem in recursive combinatorics." In: *Congr. Numer.* 33 (1981), pp. 143–153.

[59]   Daniel J. Kleitman. "The crossing number of $K_{5,n}$." In: *J. Combinatorial Theory* 9 (1970), pp. 315–323. ISSN: 0021-9800.

[60]   Joseph B Kruskal. "On the shortest spanning subtree of a graph and the traveling salesman problem." In: *Proceedings of the American Mathematical society* 7.1 (1956), pp. 48–50.

[61]   Harold Kuhn. "The Hungarian Method for the assignment problem." In: *Naval Research Logistics Quarterly* 2 (1955), pp. 83–97.

[62]   László Lovász. "Normal hypergraphs and the perfect graph conjecture." In: *Discrete Mathematics* 2.3 (1972), pp. 253–267.

[63]   László Lovász. "Three short proofs in graph theory." In: *Journal of Combinatorial Theory, Series B* 19.3 (1975), pp. 269–271.

[64]   Roger Mallion. "A contemporary Eulerian walk over the bridges of Kaliningrad." In: *BSHM Bulletin* 23 (2008), pp. 24–36.

[65]   Karl Menger. "Zur allgemeinen kurventheorie." In: *Fundamenta Mathematicae* 10.1 (1927), pp. 96–115.

[66]   Maryam Mirzakhani. "A small non-4-choosable planar graph." In: *Bull. Inst. Combin. Appl.* 17 (1996), pp. 15–18. ISSN: 1183-1278.

[67]   J. Mycielski. "Sur le coloriage des graphs." In: *Colloq. Math.* 3 (1955), pp. 161–162. ISSN: 0010-1354.

[68]   Oystein Ore. "A note on hamiltonian circuits." In: *American Mathematical Monthly* 67 (1960), p. 55.

[69]   Shengjun Pan and R. Bruce Richter. "The crossing number of K_{11} is 100." In: *J. Graph Theory* 56.2 (2007), pp. 128–134. ISSN: 0364-9024.

[70]   H Prüfer. "Neuer bewis eines Satzes uber Permutationnen." In: *Arch. Math. Phys.* 27 (1918), pp. 742–744.

[71]   F.P. Ramsey. "On a problem in formal logic." In: *Proc. London Math. Soc.* 30 (1929), pp. 264–286.

[72]   D. Salsburg. *The Lady Tasting Tea: How Statistics Revolutionized Science in the Twentieth Century.* Henry Holt and Company, 2002. ISBN: 9781466801783.

[73]   Michael Sipser. *Introduction to the Theory of Computation.* 3rd ed. Boston, MA: Cengage Learning, 2013. ISBN: 978-1-133-18779-0.

[74]   *Small Ramsey Numbers.* 2017. URL: http : / / www . combinatorics . org / ojs / index . php / eljc / article / view / DS1 / pdf (visited on 03/11/2017).

[75]   *The Merriam Webster Dictionary.* Springfield, MA: Merriam Webster, 2005. ISBN: 978-0-8777-9636-7.

[76]   *The Sveriges Riksbank Prize in Economic Science in Memory of Alfred Nobel.* 2013. URL: http://www.nrmp.org/wp-content/uploads/2013/08/The-Sveriges-Riksbank-Prize-in-Economic-Sciences-in-Memory-of-Alfred-Nobel1.pdf (visited on 11/05/2015).

[77]   Carsten Thomassen. "Every planar graph is 5-choosable." In: *J. Combin. Theory Ser. B* 62.1 (1994), pp. 180–181. ISSN: 0095-8956.

[78]   Top500.org. *High Performance Computing.* 2014. URL: http://top500.org/ (visited on 06/26/2015).

[79]   Alan Tucker. *Applied Combinatorics.* 6th ed. Hoboken, NJ: Wiley, 2012. ISBN: 978-0-470-45838-9.

[80]   William T Tutte. "The factorization of linear graphs." In: *Journal of the London Mathematical Society* 1.2 (1947), pp. 107–111.

[81]   John Michael Vasak. *The thickness of the complete graph*. Thesis
       (Ph.D.)–University of Illinois at Urbana-Champaign. ProQuest LLC,
       Ann Arbor, MI, 1976, p. 70.

[82]   Vadim G Vizing. "Coloring the vertices of a graph in prescribed colors."
       In: *Diskret. Analiz* 29.3 (1976), p. 10.

[83]   Margit Voigt. "List colourings of planar graphs." In: *Discrete Math.*
       120.1-3 (1993), pp. 215–219. ISSN: 0012-365X.

[84]   Douglas B. West. *Introduction to Graph Theory*. 2nd ed. Upper Saddle
       River, NJ: Prentice Hall, 2001. ISBN: 0-13-014400-2.

[85]   Robin Wilson. *Four Colors Suffice*. Princeton, NJ: Princeton University
       Press, 2002. ISBN: 0-691-11533-8.

[86]   D. R. Woodall. "Cyclic-order graphs and Zarankiewicz's crossing-
       number conjecture." In: *J. Graph Theory* 17.6 (1993), pp. 657–671. ISSN:
       0364-9024.

[87]   K. Zarankiewicz. "On a problem of P. Turan concerning graphs." In:
       Fund. Math. 41 (1954), pp. 137–145. ISSN: 0016-2736.

Image Credits

Most of the figures that appear in this book were created electronically by the author. For the other images that were used, either with explicit permission or via public domain use, credit is given here, organized by order of appearance.

- The map of Königsberg on page 46 is a public domain image, file *Image-Koenigsberg,_Map_by_Merian-Erben_1652.jpg*.

- The map on page 316 is a public domain image, courtesy of author Jkan997 of Wikimedia Commons, file *Amtrak_Cascades.svg*, released under the Creative Commons Attribution-ShareAlike 3.0 Unreported License.

Index

Printed in the United States
by Baker & Taylor Publisher Services

Printed in the United States
By Bookmasters